# Decomposability of Tensors

# Decomposability of Tensors

Special Issue Editor
**Luca Chiantini**

MDPI • Basel • Beijing • Wuhan • Barcelona • Belgrade

*Special Issue Editor*
Luca Chiantini
Università degli Studi di Siena
Italy

*Editorial Office*
MDPI
St. Alban-Anlage 66
4052 Basel, Switzerland

This is a reprint of articles from the Special Issue published online in the open access journal *Mathematics* (ISSN 2227-7390) in 2018 (available at: https://www.mdpi.com/journal/mathematics/special_issues/Mathematics_Tensor_Decomposition).

For citation purposes, cite each article independently as indicated on the article page online and as indicated below:

LastName, A.A.; LastName, B.B.; LastName, C.C. Article Title. *Journal Name* **Year**, *Article Number, Page Range.*

**ISBN 978-3-03897-590-8 (Pbk)**
**ISBN 978-3-03897-591-5 (PDF)**

© 2019 by the authors. Articles in this book are Open Access and distributed under the Creative Commons Attribution (CC BY) license, which allows users to download, copy and build upon published articles, as long as the author and publisher are properly credited, which ensures maximum dissemination and a wider impact of our publications.

The book as a whole is distributed by MDPI under the terms and conditions of the Creative Commons license CC BY-NC-ND.

# Contents

About the Special Issue Editor . . . . . . . . . . . . . . . . . . . . . . . . . . . . . . . . . . . vii

Preface to "Decomposability of Tensors" . . . . . . . . . . . . . . . . . . . . . . . . . . . . . ix

**Yang Qi**
A Very Brief Introduction to Nonnegative Tensors from the Geometric Viewpoint
Reprinted from: *Mathematics* **2018**, *6*, 230, doi:10.3390/math6110230 . . . . . . . . . . . . . . . . . 1

**Edoardo Ballico**
Set Evincing the Ranks with Respect to an Embedded Variety (Symmetric Tensor Rank and Tensor Rank
Reprinted from: *Mathematics* **2018**, *6*, 140, doi:10.3390/math6080140 . . . . . . . . . . . . . . . . . 20

**Alex Casarotti, Alex Massarenti and Massimiliano Mella**
On Comon's and Strassen's Conjectures
Reprinted from: *Mathematics* **2018**, *6*, 217, doi:10.3390/math6110217 . . . . . . . . . . . . . . . . . 29

**Alessandro De Paris**
Seeking for the Maximum Symmetric Rank
Reprinted from: *Mathematics* **2018**, *6*, 247, doi:10.3390/math6110247 . . . . . . . . . . . . . . . . . 42

**Alessandra Bernardi, Enrico Carlini, Maria Virginia Catalisano, Alessandro Gimigliano and Alessandro Oneto**
The Hitchhiker Guide to: Secant Varieties and Tensor Decomposition
Reprinted from: *Mathematics* **2018**, *6*, 314, doi:10.3390/math6120314 . . . . . . . . . . . . . . . . . 63

# About the Special Issue Editor

**Luca Chiantini** was born in 1957 and received his degree in Mathematics in 1979. He is Full Professor of Geometry at the University of Siena, Department of Information Engineering and Mathematical Sciences. He has published more than 90 research papers in Algebraic Geometry and Commutative Algebra, and he has edited books and conference proceedings on these topics. He has been a member of the scientific committee of several international meetings, including the joint meeting of the Unione Matematica Italiana and the real Sociedad Matematica Espanola. He participated in the editorial board of several Italian mathematical journals. Actually, he is a member of the editorial board of the Bolletino dell'Unione Matematica Italiana, and he is a member of the panel of the Scuola Matematica Interuniversitaria. His recent research interests focus on the geometry of special projective varieties, like Veronese and Segre varieties, and their applications in tensor analysis and algebraic statistics.

# Preface to "Decomposability of Tensors"

The Special Issue "tensor decomposition" is devoted to collecting papers on a subject that is rapidly developing in recent years, with (unexpected) connections between different areas of mathematics. Though tensor analysis is a topic that, for a long time, has been considered a chapter of multilinear algebra with a view towards numerical analysis, it turned out recently that methods of projective geometry, often arising from a classical point of view, have a strong connection with the theory of tensors and can produce advances that are also valuable in applicative domains. In particular, the study of tensor rank, i.e., the complexity of a tensor, was invigorated by the introduction of techniques based on the background of projective geometry. If one considers elementary tensors as a tensor product of vectors, then the computation of the rank corresponds to finding the minimum k, such that a tensor belongs to the k-secant variety of a product variety. In projective terms, the main notions of tensor analysis can be defined modulo scalar products, so most problems can be translated in terms of points in projective spaces, and product varieties become Segre embeddings of products of projective spaces (or Veronese embeddings of projective spaces, in the symmetric setting). In this circle of ideas, a decomposition of a tensor T corresponds to a set Z of projective points, such that T belongs to the linear span of Z. Thus, the geometry of secant varieties to Segre and Veronese varieties provides basic tools for understanding the decomposition of a given tensor. Notions like the uniqueness and minimality of a given decomposition found a natural formulation in terms of projective geometry. Also, special decompositions can be described in terms of proprieties of secant spaces. Altogether, the initial features of these new perspectives were described in a series of books and papers. Yet, as the theory develops quickly, it is useful to make frequent reports on the status of the art.

This book collects papers that contain both surveys on the actual main achievements on some classical problems on the decomposition of tensors, like the best-known bounds on the rank of symmetric tensors, together with results on special decompositions and extensions to the generalized study of decompositions with respect to any subvarieties.

We hope that the content of this book will provide a helpful collection of geometric perspectives on tensor analysis and tensor decomposition, which are necessary both to create a solid starting point for future developments and to establish a background of geometric methods for people who arrived to work in the subject coming from different points of view.

Luca Chiantini
*Special Issue Editor*

*Review*

# A Very Brief Introduction to Nonnegative Tensors from the Geometric Viewpoint

## Yang Qi

Department of Mathematics, University of Chicago, 5734 S. University Avenue, Chicago, IL 60637, USA; yangqi@math.uchicago.edu

Received: 27 September 2018; Accepted: 24 October 2018; Published: 30 October 2018

**Abstract:** This note is a short survey of nonnegative tensors, primarily from the geometric point of view. In addition to basic definitions, we discuss properties of and questions about nonnegative tensors, which may be of interest to geometers.

**Keywords:** nonnegative tensors; low-rank approximations; uniqueness and identifiability; spectral theory; EM algorithm; semialgebraic geometry

## 1. Introduction

Tensors are ubiquitous in mathematics and sciences. In the study of complex and real tensors, algebraic geometry has demonstrated its power [1,2]. On the other hand, tensor computations also help people understand classical algebraic varieties, such as the secant varieties of Segre varieties and Veronese varieties, and raise interesting and challenging questions in algebraic geometry [3,4]. Traditionally, geometers tend to study tensors in a coordinate-free way. However, in applications, practitioners must work with coordinates. Among those tensors widely used in practice, a large number of them are nonnegative tensors, i.e., tensors with nonnegative entries. In this case, most powerful geometric tools developed for complex tensors can not be applied directly due to the fact that the Euclidean closure of tensors with rank no greater than a fixed integer is no longer a variety, but a semialgebraic set. This forces us to investigate the semialgebraic geometry of nonnegative tensors. In this note, we will review some important properties of nonnegative tensors obtained by studying the semialgebraic geometry, and propose several open problems which are pivotal in understanding nonnegative tensors and also may be interesting to geometers.

## 2. Definitions

Nonnegative tensors arise naturally in many areas, such as hyperspectral imaging, statistics, spectroscopy, computer vision, phylogenetics, and so on. See [5–8] and the references therein. Before further investigations, let us recall basic definitions of tensors.

**Definition 1.** *Let $V_1, \ldots, V_d$ be vector spaces over a field $\mathbb{K}$. The tensor product $\mathbb{V} = V_1 \otimes \cdots \otimes V_d$ is the free linear space spanned by $V_1 \times \cdots \times V_d$ quotient by the equivalence relation:*

$$(v_1, \ldots, \alpha v_i + \beta v'_i, \ldots, v_d) \sim \alpha(v_1, \ldots, v_i, \ldots, v_d) + \beta(v_1, \ldots, v'_i, \ldots, v_d) \tag{1}$$

*for every $v_i, v'_i \in V_i$, $\alpha_i, \beta_i \in \mathbb{K}$, and $i = 1, \ldots, d$. An element of $V_1 \otimes \cdots \otimes V_d$ is called a tensor.*

Equivalently, $V_1 \otimes \cdots \otimes V_d$ is the vector space of multilinear functions:

$$V_1^* \times \cdots \times V_d^* \to \mathbb{K},$$

where $V_i^*$ is the dual space of $V_i$ for $i = 1, \ldots, d$. A representative of the equivalence class of $(v_1, \ldots, v_d)$ is called a *decomposable tensor* and denoted by $v_1 \otimes \cdots \otimes v_d$.

The *rank* of a given tensor $T \in V_1 \otimes \cdots \otimes V_d$ is the minimum integer $r$ such that $T$ is a sum of $r$ decomposable tensors, i.e.,

$$T = \sum_{i=1}^{r} v_{1,i} \otimes \cdots \otimes v_{d,i}, \tag{2}$$

where $v_{j,i} \in V_j$ for $j = 1, \ldots, d$ and $i = 1, \ldots, r$. Such a decomposition is called a *rank decomposition* (or canonical polyadic decomposition or CP decomposition).

Now we focus on the case $\mathbb{K} = \mathbb{R}$, and for each $V_i$ we fix a basis, which enables us to work with coordinates. Let $\mathbb{R}_+$ be the semiring of nonnegative real numbers. A nonnegative tensor in $V_1 \otimes \cdots \otimes V_d$ is a tensor whose coordinates are nonnegative. Let $V_i^+$ denote the set of nonnegative vectors in $V_i$ for each $i = 1, \ldots, d$, and $\mathbb{V}^+$ denote the set of nonnegative tensors in $\mathbb{V}$.

**Definition 2.** *For $T \in \mathbb{V}^+$, the nonnegative rank of $T$ is the minimum integer $r$ so that there exist nonnegative vectors $v_{i,j} \in V_i^+$ for $i = 1, \ldots, d$ and $j = 1, \ldots, r$ making Equation (2) holds.*

It is clear that $\mathrm{rank}_+(T) \geq \mathrm{rank}(T)$ for every $T \in \mathbb{V}^+$. Besides, there exists some $T \in \mathbb{V}^+$ such that $\mathrm{rank}_+(T) > \mathrm{rank}(T)$. For example, let

$$T = e_1 \otimes e_1 \otimes e_1 + e_1 \otimes e_2 \otimes e_2 + e_2 \otimes e_2 \otimes e_1 + e_2 \otimes e_1 \otimes e_2 \in \mathbb{R}^2 \otimes \mathbb{R}^2 \otimes \mathbb{R}^2, \tag{3}$$

where $e_1 = [1,0]^\top$ and $e_2 = [0,1]^\top$, then $\mathrm{rank}_+(T) = 4 > 2 = \mathrm{rank}_{\mathbb{R}}(T)$.

## 3. Applications

One reason that nonnegative tensors are popular is due to the statistical interpretation behind—a Bayesian network [9–11]. More precisely, assume a joint distribution of several random variables $x_i$ is given by:

$$p(x_1, \ldots, x_d) = \int \prod_{i=1}^{d} p(x_i \mid \theta) \, d\mu_\theta \tag{4}$$

where $\theta$ is a latent variable. When $x_1, \ldots, x_d$ and $\theta$ are discrete, (4) becomes:

$$t_{i_1, \ldots, i_d} = \sum_{p=1}^{r} \lambda_r u_{i_1, p} \cdots u_{i_d, p}, \tag{5}$$

i.e., a nonnegative rank decomposition [5,12]. Such a model, for instance, has been applied in clustering [13].

As another application, nonnegative tensors have shown their powers in image processing. Usually hyperspectral images are processed as nonnegative matrices $M \in \mathbb{R}_+^{n \times m}$, where $n$ is the number of pixels and $m$ denotes the number of spectral bands. By the sensor developments, it is possible to collect time series of hyperspectral data, which can be understood as nonnegative tensors, namely $A \in \mathbb{R}_+^{n \times m \times d}$, where $d$ is the dimensionality of the time or multiangle ways [14]. A nonnegative rank decomposition of $A$ gives a blind spectral unmixing of hyperspectral data.

Recently, tensor methods have also used in isogeometric analysis (IGA) [15–17]. A Galerkin-based approach of IGA studies tensor-product B-splines. To obtain Galerkin matrices effectively, low rank approximations of integral kernels are employed [16]. In many cases, the constructed mass tensor is a positive tensor.

## 4. Algorithms

Due to the broad and important applications, nonnegative tensor decomposition (NTD) and nonnegative matrix factorization (NMF) have received vast research on their algorithms. Perhaps the most popular algorithm for NMF is the multiplicative updating rule [18], and since then, numerous more efficient algorithms have been proposed, such as the algorithms using the alternating nonnegative least squares [19–21], the algorithms using the hierarchical alternating least squares [22], the algorithms using deflation [23], and many more algorithms based on other methods, for example [24,25], etc. The main ideas of some algorithms have been naturally generalized to decompose nonnegative tensors, for example [23,26–28]. Since the main purpose of this note is to introduce the geometric properties of nonnegative tensors, we invite those readers who are interested in algorithms to read comprehensive surveys on algorithms of NTD, for example [29–31].

## 5. Nonnegative Rank Decompositions

It is known that when $d > 2$, rank decompositions (2) are often unique over $\mathbb{C}$ and $\mathbb{R}$, which is very important in applications. For nonnegative tensors, it is also an important issue to investigate the identifiability property. Before studying the identifiability of nonnegative tensors, let us recall fundamental definitions and known results of complex and real tensors.

For any tuple of positive integers $(n_1, \ldots, n_d)$, there is a unique integer $r_g(n_1, \ldots, n_d)$, which only depends on $n_1, \ldots, n_d$ such that the set of complex rank-$r_g(n_1, \ldots, n_d)$ tensors in $\mathbb{C}^{n_1} \otimes \cdots \otimes \mathbb{C}^{n_d}$ contains a Zariski open subset of $\mathbb{C}^{n_1} \otimes \cdots \otimes \mathbb{C}^{n_d}$. In fact, $r_g(n_1, \ldots, n_d)$ is the minimum integer $r$ such that the $r$th secant variety of the Segre variety $\mathrm{Seg}(\mathbb{P}^{n_1-1} \times \cdots \times \mathbb{P}^{n_d-1})$ is the ambient space $\mathbb{P}^{n_1 \cdots n_d - 1}$. $r_g(n_1, \ldots, n_d)$ is called the *generic rank* of $\mathbb{C}^{n_1} \otimes \cdots \otimes \mathbb{C}^{n_d}$. It is not always the case that the generic rank $r_g(n_1, \ldots, n_d)$ equals $\lceil \frac{n_1 \cdots n_d}{n_1 + \cdots + n_d - d + 1} \rceil$, which leads us to the following definition.

**Definition 3.** *If the $\mathbb{K}$-dimension of the set of rank-$r$ tensors in $\mathbb{K}^{n_1} \otimes \cdots \otimes \mathbb{K}^{n_d}$ is strictly less than $\min\{r(n_1 + \cdots + n_d - d + 1), n_1 \cdots n_d\}$, then $\mathbb{K}^{n_1} \otimes \cdots \otimes \mathbb{K}^{n_d}$ is called $r$-defective.*

When $\mathbb{K}$ is algebraically closed, $\mathbb{K}^{n_1} \otimes \cdots \otimes \mathbb{K}^{n_d}$ is not $r$-defective, which implies a general rank-$r$ tensor $T$ has finitely many rank decompositions. If we require further that $T$ has a unique decomposition, we will arrive at the following definition.

**Definition 4.** *If a general rank-$r$ tensor in $\mathbb{K}^{n_1} \otimes \cdots \otimes \mathbb{K}^{n_d}$ has a unique rank-$r$ decomposition over $\mathbb{K}$, then $\mathbb{K}^{n_1} \otimes \cdots \otimes \mathbb{K}^{n_d}$ is called $r$-identifiable.*

There has been a large amount of research on defectivity [32–34] and identifiability [1,35–41]. Here, we highlight three notable results.

**Theorem 1** (Kruskal)**.** *Let $V_1, \ldots, V_d$ be finite dimensional vector spaces over a field $\mathbb{K}$ [35], and*

$$T = \sum_{i=1}^{r} v_{1,i} \otimes \cdots \otimes v_{d,i} \in V_1 \otimes \cdots \otimes V_d. \tag{6}$$

*If $\kappa_1 + \cdots + \kappa_d \geq 2r + d - 1$, then $\mathrm{rank}(T) = r$ and $T$ has a unique rank-$r$ decomposition (6), where $\kappa_i$ is the maximum integer such that every subset of $\{v_{i,1}, \ldots, v_{i,r}\}$ with $\kappa_i$ elements is linearly independent for $i = 1, \ldots, d$.*

**Theorem 2** (Bocci–Chiantini–Ottaviani [1])**.** *Assume $n_1 \leq \cdots \leq n_d$. Then $\mathbb{C}^{n_1} \otimes \cdots \otimes \mathbb{C}^{n_d}$ is $r$-identifiable when:*

$$r \leq \frac{\prod_{j=1}^{d} n_j - (n_1 + n_2 + n_3 - 2) \prod_{j=3}^{d} n_j}{1 + \sum_{j=1}^{d}(n_j - 1)}.$$

**Theorem 3** (Chiantini–Ottaviani–Vannieuwenhoven). $\mathbb{C}^{n_1} \otimes \cdots \otimes \mathbb{C}^{n_d}$ is $r$-identifiable when [40]:

$$r < \left\lceil \frac{\prod_{j=1}^{d} n_j}{1 + \sum_{j=1}^{d}(n_j - 1)} \right\rceil$$

and $\prod_{j=1}^{d} n_j \leq 15000$, except the following cases:

| $(n_1,\ldots,n_d)$ | $r$ | Type |
|---|---|---|
| $(4,4,3)$ | 5 | defective |
| $(4,4,4)$ | 6 | sporadic |
| $(6,6,3)$ | 8 | sporadic |
| $(n,n,2,2)$ | $2n-1$ | defective |
| $(2,2,2,2,2)$ | 5 | sporadic |
| $n_1 > \prod_{i=2}^{d} n_i - \sum_{i=2}^{d}(n_i - 1)$ | $r \geq \prod_{i=2}^{d} n_i - \sum_{i=2}^{d}(n_i - 1)$ | unbalanced |

Note that Theorems 2 and 3 focus on complex tensors; however, with the help of the following lemma, we are able to extend these results to real tensors.

**Lemma 1.** *If $\mathbb{C}^{n_1} \otimes \cdots \otimes \mathbb{C}^{n_d}$ is $r$-identifiable, then $\mathbb{R}^{n_1} \otimes \cdots \otimes \mathbb{R}^{n_d}$ is $r$-identifiable when $r < r_g(n_1, \ldots, n_d)$ [42].*

As an example, we have the following corollary.

**Corollary 1.** $\mathbb{R}^{n_1 \times \cdots \times n_d}$ *is $r$-identifiable if:*

$$r < \left\lceil \frac{\prod_{i=1}^{d} n_i}{1 + \sum_{i=1}^{d}(n_i - 1)} \right\rceil,$$

$\prod_{i=1}^{d} n_i \leq 15000$, and $(n_1, \ldots, n_d, r)$ is not one of the following cases:

| $(n_1,\ldots,n_d)$ | $r$ |
|---|---|
| $(4,4,3)$ | 5 |
| $(4,4,4)$ | 6 |
| $(6,6,3)$ | 8 |
| $(n,n,2,2)$ | $2n-1$ |
| $(2,2,2,2,2)$ | 5 |
| $n_1 > \prod_{i=2}^{d} n_i - \sum_{i=2}^{d}(n_i - 1)$ | $r \geq \prod_{i=2}^{d} n_i - \sum_{i=2}^{d}(n_i - 1)$ |

In fact, for the above exceptional cases, we can derive more information from Theorem 3.

**Corollary 2.**

- $\mathbb{R}^{4 \times 4 \times 3}$ is 5-defective.
- For any $n \geq 2$, $\mathbb{R}^{n \times n \times 2 \times 2}$ is $(2n-1)$-defective.
- For $n_1 \geq \cdots \geq n_d \geq 2$, $\mathbb{R}^{n_1 \times \cdots \times n_d}$ is $r$-defective if

$$n_1 > \prod_{i=2}^{d} n_i - \sum_{i=2}^{d}(n_i - 1) \quad \text{and} \quad r \geq \prod_{i=2}^{d} n_i - \sum_{i=2}^{d}(n_i - 1).$$

Recall that for a symmetric tensor $T \in S^d(V)$ over $\mathbb{K}$, the *symmetric rank* of $T$ is the minimum integer $r$ such that

$$T = \sum_{i=1}^{r} \lambda_i v_i^{\otimes d},$$

where $\lambda_i \in \mathbb{K}, v_i \in V$ for $i = 1, \ldots, r$. For symmetric tensors, we can also have the definitions of generic rank, $r$-defectivity, and $r$-identifiability. More precisely, the generic symmetric rank, $r_g(n;d)$, is defined to be the minimum integer $r$ such that the $r$th secant variety of the Veronese variety $\nu_d(\mathbb{P}^{n-1})$ fills in the ambient space over $\mathbb{C}$. If the $\mathbb{K}$-dimension of the set of symmetric rank-$r$ tensors in $S^d(\mathbb{K}^n)$ is strictly less than $\min\{rn, \binom{n+d-1}{d}\}$, then $S^d(\mathbb{K}^n)$ is called $r$-defective. If a general symmetric rank-$r$ tensor has a unique decomposition over $\mathbb{K}$, $S^d(\mathbb{K}^n)$ is called $r$-identifiable. The defectivity problem has been completely solved in [43].

**Theorem 4** (Alexander–Hirschowitz). *The generic rank [43]:*

$$r_g(n;d) = \left\lceil \frac{\binom{n+d-1}{d}}{n} \right\rceil$$

*except the following case:*

- When $d = 2$, $r_g(n;d) = n$.
- When $(d,n) = (3,5), (4,3), (4,4), (4,5)$, $r_g(n;d) = \lceil \frac{\binom{n+d-1}{d}}{n} \rceil + 1$.

The identifiability problem of $S^d(\mathbb{C}^n)$ has been addressed in [2,44–46], and the complete solution was given in [2].

**Theorem 5** (Chiantini–Ottaviani–Vannieuwenhoven). *$S^d(\mathbb{C}^{n+1})$ is $r$-identifiable when [2]*

$$r < \left\lceil \frac{\binom{n+d}{d}}{n+1} \right\rceil$$

*and $d \geq 3$, except that $(d,n,r) \in \{(6,2,9), (4,3,8), (3,5,9)\}$ where a general complex symmetric rank-$r$ tensor has exactly two symmetric rank decompositions.*

Similar to Lemma 1, we have the following lemma for real symmetric tensors.

**Lemma 2.** *Let $r < r_g(n;d)$. If $S^d(\mathbb{C}^n)$ is $r$-identifiable, then $S^d(\mathbb{R}^n)$ is $r$-identifiable [42].*

As an example, we have

**Corollary 3.** *$S^d(\mathbb{R}^{n+1})$ is $r$-identifiable when:*

$$r < \left\lceil \frac{\binom{n+d}{d}}{n+1} \right\rceil$$

*and if $(d,n,r) \notin \{(6,2,9), (4,3,8), (3,5,9)\}$.*

Now, we are in a position to study the relations among complex, real, and nonnegative ranks. Given real vector spaces $V_1, \ldots, V_d$ of dimensions $n_1, \ldots, n_d$, respectively, let $\mathbb{V} := V_1 \otimes \cdots \otimes V_d$ and $\mathbb{V}_\mathbb{C}$ be the complexification of $\mathbb{V}$. For any positive integer $r$, let

$$D_r^+ = \{X \in \mathbb{V}^+ \mid \mathrm{rank}_+(X) \leq r\}$$

denote the set of nonnegative tensors with nonnegative ranks not greater than $r$.

**Theorem 6.** *Let $r < r_g(n_1, \ldots, n_d)$. For a general $T \in D_r^+$, its real rank and complex rank are also $r$. If $\mathbb{V}_\mathbb{C}$ is $r$-identifiable, then $T$ has a unique rank-$r$ decomposition, which is nonnegative [42].*

For nonnegative tensors, when $r \geq r_g(n_1, \ldots, n_d)$, the set of nonnegative rank-$r$ tensors may contain a nonempty open subset of $\mathbb{V}^+$ under the Euclidean topology. If so, $r$ is called a *nonnegative* typical rank. A Similar phenomenon happens in the real case which motivates the definition of *real typical rank*, i.e., $r$ is a real typical rank if the set of real rank-$r$ tensors contains a nonempty open subset of $\mathbb{V}$. We will illustrate the difference between nonnegative ranks and real ranks by the following example, where $\mathbb{R}_+^{n \times n \times n}$ denotes the set of nonnegative tensors in $\mathbb{R}^{n \times n \times n}$.

**Proposition 1.** *[42]*

- The nonnegative typical ranks of $\mathbb{R}_+^{2 \times 2 \times 2}$ are $2, 3, 4$.
- The nonnegative typical ranks of $\mathbb{R}_+^{3 \times 3 \times 3}$ are all integers $m$ satisfying:

$$5 \leq m \leq 9.$$

- When $n \geq 4$, the nonnegative typical ranks of $\mathbb{R}_+^{n \times n \times n}$ consist of all integers $m$ satisfying:

$$\left\lceil \frac{n^3}{3n-2} \right\rceil \leq m \leq n^2.$$

Theorem 6 and Proposition 1 reveal that for a general nonnegative rank-$r$ tensor $T$, the true difference among its complex, real, and nonnegative ranks appears when $r \geq r_g(n_1, \ldots, n_d)$, namely when $r < r_g(n_1, \ldots, n_d)$, the complex and real ranks of $T$ are also $r$, but when $r > r_g(n_1, \ldots, n_d)$, $D_r^+$ contains a nonempty open subset $\mathcal{U}$ of $\mathbb{V}^+$ such that for each $T \in \mathcal{U}$,

$$r_g(n_1, \ldots, n_d) = \mathrm{rank}_{\mathbb{C}}(T) \leq \mathrm{rank}_{\mathbb{R}}(T) < \mathrm{rank}_+(T) = r.$$

More concretely, let $T$ be the tensor defined in (3). Then there exists a nonempty open neighborhood $\mathcal{U}$ of $T$ in $\mathbb{R}_+^{2 \times 2 \times 2}$ such that for any $A \in \mathcal{U}$,

$$\mathrm{rank}_{\mathbb{C}}(A) = 2 < 4 = \mathrm{rank}_+(A).$$

## 6. Low Rank Approximations

Given $T \in \mathbb{V}^+$ with $r \leq \mathrm{rank}_+(T)$, let:

$$\delta(T) = \inf_{X \in D_r^+} \|T - X\|,$$

where $\|\cdot\|$ is the *Hilbert–Schmidt norm*.

It is known that the set $D_r = \{X \in V_1 \otimes \cdots \otimes V_d \mid \mathrm{rank}(X) \leq r\}$ is not closed under the Euclidean topology over $\mathbb{R}$ or $\mathbb{C}$ when $r > 1$ [47]. However, for nonnegative tensors, we can show:

**Proposition 2.** $D_r^+$ *is a closed semialgebraic set under the Euclidean topology* [48].

Since $D_r^+$ is closed, for any $T \notin D_r^+$, there is always some $T_0 \in D_r^+$ such that $\|T - T_0\| = \delta(T)$, i.e., the optimization problem:

$$\min_{\mathrm{rank}_+(X) \leq r} \|T - X\| \tag{7}$$

makes sense. Furthermore, we can have the following result.

**Proposition 3.** *A general $T \in \mathbb{V}^+$ has a unique best nonnegative low-rank approximation* [48].

Before studying nonnegative rank approximations, let us recall the following useful lemma.

**Lemma 3.** *For $T \in V$ over $\mathbb{R}$, assume $\mathrm{rank}(T) > r$ and $\lambda \sum_{l=1}^{r} T_l$ is a best rank-r approximation, where $T_l = v_{1,l} \otimes \cdots \otimes v_{d,l}$ and $\left\| \sum_{l=1}^{r} T_l \right\| = 1$. Then:*

$$\langle T, v_{1,j} \otimes \cdots \otimes \widehat{v_{i,j}} \otimes \cdots \otimes v_{d,j} \rangle = \lambda \left\langle \sum_{l=1}^{r} T_l, v_{1,j} \otimes \cdots \otimes \widehat{v_{i,j}} \otimes \cdots \otimes v_{d,j} \right\rangle, \qquad (8)$$

*where $i = 1, \ldots, d$, and $j = 1, \ldots, r$, where $\lambda = \langle T, \sum_{l=1}^{r} T_l \rangle$, and $\langle, \rangle$ denotes tensor contraction.*

The *support* of a vector $u \in V$ is defined to be:

$$\mathrm{supp}(u) := \{ j = \{1, \ldots, \dim_{\mathbb{R}} V\} \mid \text{the } j\text{th coordinate of } u \text{ is nonzero} \}.$$

Then for a nonnegative tensor $T$, Lemma 3 becomes

**Lemma 4.** *Let $T \in V^+$ with $\mathrm{rank}_+(T) > r$ and $Y = \sum_{j=1}^{s} v_{1,j} \otimes \cdots \otimes v_{d,j}$ be a solution of (7). Then:*

$$\langle T, v_{1,j} \otimes \cdots \otimes w_{i,j} \otimes \cdots \otimes v_{d,j} \rangle \leq \left\langle Y, v_{1,j} \otimes \cdots \otimes w_{i,j} \otimes \cdots \otimes v_{d,j} \right\rangle \qquad (9)$$

*where $w_{i,j} \in V_i^+$, $i = 1, \ldots, d$, and $j = 1, \ldots, s$. For every pair $(i, j)$, define:*

$$\widetilde{V}_{i,j} := \{ v \in V_i : \mathrm{supp}(v) \subseteq \mathrm{supp}(v_{i,j}) \}.$$

*Then:*

$$\langle T, v_{1,j} \otimes \cdots \otimes w_{i,j} \otimes \cdots \otimes v_{d,j} \rangle = \left\langle Y, v_{1,j} \otimes \cdots \otimes w_{i,j} \otimes \cdots \otimes v_{d,j} \right\rangle \qquad (10)$$

*for $w_{i,j} \in \widetilde{V}_{i,j}$.*

Lemma 4 guarantees us the following result.

**Proposition 4.** *Let $T \in V^+$ with $\mathrm{rank}_+(T) > r$ and $X$ be a solution of (7). Then $\mathrm{rank}_+(X) = r$ [48].*

Proposition 4 shows that it is indeed appropriate to call a solution of (7) a best nonnegative rank-r approximation.

By Proposition 3, we know a general nonnegative tensor has a unique best nonnegative rank-r approximation. However, it is still unclear if this best approximation has a unique nonnegative rank-r decomposition. Below is an example where we have the uniqueness. On the other hand, the general case is not known yet.

**Proposition 5.** *Let $r = 2$ or $3$ and let $n_1, \ldots, n_d \geq 3$. Then for a general $T \in \mathbb{R}_+^{n_1 \times \cdots \times n_d}$, its unique best nonnegative rank-r approximation has a unique nonnegative rank-r decomposition [42].*

**Question 1.** *Assume $V$ is r-identifiable. Given a general $T \in V^+$, is it true that the unique best nonnegative rank-r approximation of $T$ has a unique nonnegative decomposition?*

## 7. Spectral Theory

In this section, we start with nonnegative rank-one approximations, which lead us to the spectral theory of nonnegative tensors.

**Proposition 6.** *Given $T \in V^+$, let $u_1 \otimes \cdots \otimes u_d \in V$ be a best real rank-one approximation of $T$. Then $u_1, \ldots, u_d$ can be chosen in the form $u_1 \in V_1^+, \ldots, u_d \in V_d^+$.*

By Proposition 6, for a nonnegative tensor $T$, we will not distinguish a best real rank-one approximation and a best nonnegative rank-one approximation. By Lemma 3, a best real rank-one approximation of a real tensor is a solution of (8), which motivates us the following definition.

**Definition 5.** *Let $V_1, \ldots, V_d$ be vector spaces over $\mathbb{K}$ of dimensions $n_1, \ldots, n_d$. For $T \in V_1 \otimes \cdots \otimes V_d$, we call $(\lambda, u_1, \ldots, u_d) \in \mathbb{K} \times V_1 \times \cdots \times V_d$ a normalized singular pair of $T$ if:*

$$\begin{cases} \langle T, u_1 \otimes \cdots \otimes \widehat{u}_i \otimes \cdots \otimes u_d \rangle = \lambda u_i, \\ \langle u_i, u_i \rangle = 1, \end{cases} \quad (11)$$

*where $i = 1, \ldots, d$. Then, $\lambda$ is called a normalized singular value and $(u_1, \ldots, u_d)$ is called a normalized singular vector tuple. When $\mathbb{K} = \mathbb{R}$, $\lambda \geq 0$, and $u_i \in V_i^+$, we call $(\lambda, u_1, \ldots, u_d)$ a nonnegative normalized singular pair of $T$.*

Similar definitions have been proposed by several authors; for example, the following projective variant was introduced in [49].

**Definition 6.** *Given vector spaces $W_1, \ldots, W_d$ over $\mathbb{K}$ of dimensions $n_1, \ldots, n_d$, for $T \in W_1 \otimes \cdots \otimes W_d$, $([v_1], \ldots, [v_d]) \in \mathbb{P}W_1 \times \cdots \times \mathbb{P}W_d$ is called a projective singular vector tuple if [49]:*

$$\langle T, v_1 \otimes \cdots \otimes \widehat{v}_i \otimes \cdots \otimes v_d \rangle = \lambda_i v_i \quad (12)$$

*for some $\lambda_i \in \mathbb{K}$, where $1 \leq i \leq d$.*

The number of projective singular vector tuples of a general complex tensor was calculated in [49], which is the Euclidean Distance (ED) degree of $\mathbb{P}W_1 \times \cdots \times \mathbb{P}W_d$ by [50].

**Theorem 7.** *Let $T$ be a generic tensor in $W_1 \otimes \cdots \otimes W_d$ over $\mathbb{C}$. Then $T$ has exactly $c(n_1, \ldots, n_d)$ simple projective singular vector tuples corresponding to nonzero singular values, where $c(n_1, \ldots, n_d)$ is the coefficient of the monomial $\prod_{i=1}^d t_i^{n_i-1}$ in the polynomial [49]:*

$$\prod_{i=1}^d \frac{\widehat{t}_i^{n_i} - t_i^{n_i}}{\widehat{t}_i - t_i}, \text{ where } \widehat{t}_i = \sum_{j \neq i} t_j, i = 1, \ldots, d.$$

Over $\mathbb{R}$, there are several nonempty open subsets $\mathcal{U}_1, \ldots, \mathcal{U}_k$ of $V_1 \otimes \cdots \otimes V_d$ such that the number of projective singular vector tuples is constant on each $\mathcal{U}_i$, denoted by $m_i$, for $i = 1, \ldots, k$, but $m_i \neq m_j$ if $i \neq j$. One way to describe the number of projective singular vector tuples by using a single number is to impose certain probability distribution on $V_1 \otimes \cdots \otimes V_d$ and compute the expected number of projective singular vector tuples of $T$ when $T$ is randomly drawn under the given distribution.

**Theorem 8.** *Let $T \in \mathbb{R}^{n_1 \times \cdots \times n_d}$ be a random tensor drawn under the Gaussian distribution. Then the expected number of projective singular vector tuples of $T$ is given by [51]:*

$$\frac{(2\pi)^{d/2}}{2^{n/2}} \frac{1}{\prod_{i=1}^d \Gamma(\frac{n_i}{2})} \int_W |\det \mathcal{C}| \, d\mu_W,$$

where $n = \sum_i n_i$, $\Gamma$ is Euler's gamma function,

$$C = \begin{bmatrix} \lambda I_{n_1-1} & A_{1,2} & \cdots & A_{1,d} \\ A_{1,2}^\top & \lambda I_{n_2-1} & \cdots & A_{2,d} \\ \vdots & \vdots & \ddots & \vdots \\ A_{1,d}^\top & A_{2,d}^\top & \cdots & \lambda I_{n_d-1} \end{bmatrix},$$

and W is the vector space formed by $\lambda$ and $A_{i,j}$ with $i < j$.

Coming back to nonnegative tensors, we may have more information about their singular pairs than real tensors. Before studying singular pairs of nonnegative tensors, let us recall the following well-known Perron–Frobenius Theorem. See for example [52] for more details.

**Theorem 9.** *Given a nonnegative square matrix M,*

- *its spectral radius $r(M)$ is an eigenvalue.*
- *there is some nonnegative vector $v \neq 0$ such that $Mv = r(M)v$.*
- *$r(M) > 0$ if M is irreducible.*
- *there is some positive vector $u > 0$ such that $Mu = r(M)u$ if M is irreducible.*
- *if M is irreducible, then $\lambda$ is an eigenvalue of M with a nonnegative eigenvector if and only if $\lambda = r(M)$.*
- *$r(M)$ is simple if M is irreducible.*
- *every eigenvalue $\lambda$ satisfies $|\lambda| \leq r(M)$ if M is irreducible.*

The next three results, namely Lemmas 5–7, give an analogue of the tensorial Perron–Frobenius Theorem [53–56] for nonnegative normalized singular pairs, which will help us learn more about best rank-one approximations.

**Lemma 5** (Existence). *Any nonnegative tensor has (at least) a nonnegative normalized singular pair.*

**Definition 7.** *A tensor is called positive if all its entries are positive.*

**Lemma 6** (Positivity). *A positive tensor has a positive normalized singular pair.*

Recall that the *spectral norm* for a tensor, which is NP-hard to compute or approximate [57], is defined as follows.

**Definition 8.** *For $T \in V_1 \otimes \cdots \otimes V_d$ over $\mathbb{R}$, let $\|T\|_\sigma := \max\{|\langle T, u_1 \otimes \cdots \otimes u_d \rangle| : \|u_1\| = \cdots = \|u_d\| = 1\}$ be the spectral norm of T.*

**Lemma 7** (Generic Uniqueness). *Let T be a general real tensor. Then T has a unique normalized singular pair $(\lambda, u_1, \ldots, u_d)$ such that $\lambda = \|T\|_\sigma$.*

Lemma 7 motivates the following open question.

**Question 2.** *Can we give a sufficient condition such that any nonnegative tensor satisfying this condition has a unique normalized singular pair with $\lambda = \|T\|_\sigma$ and this condition can be satisfied by a general nonnegative tensor?*

For matrices over $\mathbb{R}$ or $\mathbb{C}$, by the Eckart–Young Theorem, best low-rank approximations can be obtained from successive best rank-one approximations. However, for tensors, this 'deflation procedure' does not work [48,58,59].

Besides singular vector tuples, eigenvalues and eigenvectors of a tensor can be also defined. Unlike matrices, there are several ways to define eigenvalues and eigenvectors of a tensor [60]. In this note, we will use the following one which was firstly introduced in [55,61].

**Definition 9.** *For $T \in V^{\otimes d}$ over $\mathbb{K}$, if:*

$$\langle T, u^{\otimes(d-1)} \rangle = \lambda u,$$

*then $\lambda \in \mathbb{K}$ is called an eigenvalue of $T$, and $u \in V$ is called an eigenvector. The pair $(\lambda, u)$ is called an eigenpair. Two eigenpairs $(\lambda, u)$ and $(\mu, v)$ of $T$ is said to be equivalent if $t^{d-2}\lambda = \mu$ and $tu = v$ for some $t \in \mathbb{K}$.*

When $T$ is a real or complex symmetric tensor, a best rank-one approximation of $T$ can be always chosen to be symmetric [62,63], and thus is an eigenvector of $T$.

**Theorem 10.** *Let $T \in V^{\otimes(d+1)}$ be a real random tensor under the Gaussian distribution. Then the expected number of equivalence classes of eigenpairs of $T$ is given by [64]:*

$$\frac{2^{n-1}\sqrt{d}^n \Gamma(n-\frac{1}{2})}{\sqrt{\pi}(d+1)^{n-\frac{1}{2}}\Gamma(n)}\left[2(n-1)\,_2F_1\left(1, n-\frac{1}{2}; \frac{3}{2}; \frac{d-1}{d+1}\right) + \,_2F_1\left(1, n-\frac{1}{2}; \frac{n+1}{2}; \frac{1}{d+1}\right)\right],$$

*where $_2F_1$ is the Gaussian hypergeometric function.*

The number of equivalence classes of eigenpairs of a generic complex symmetric tensor has been calculated in [65]. See [66] for another proof. This number is the ED degree of the Veronese variety [50].

**Theorem 11.** *Let $V$ be an n-dimensional complex vector space. Let $T \in S^d(V)$ be a symmetric tensor whose equivalence classes of eigenpairs are finitely many. Then, $T$ has [65]:*

$$\frac{(d-1)^n - 1}{d - 2}$$

*equivalence classes of eigenpairs, counted with multiplicities.*

For the real case, again, we may impose a probability distribution on $S^d(V)$ and compute the expected number of equivalence classes of eigenvalues of a random symmetric tensor.

**Theorem 12.** *Let $V$ be an n-dimensional real vector space of dimension n and $T \in S^d(V)$ be drawn under the Gaussian distribution. Then, the expected number of equivalence classes of eigenpairs of $T$ is [51]:*

$$\frac{1}{2^{(n^2+3n-2)/4}\prod_{j=1}^n \Gamma(j/2)} \int_{\mu_2 \leq \cdots \leq \mu_n} \int_{-\infty}^{+\infty} \left(\prod_{j=2}^n |\sqrt{d}\lambda - \sqrt{d-1}\mu_j|\right)$$

$$\left(\prod_{i<j}(\mu_i - \mu_j)\right) e^{-\lambda^2/2 - \sum_{j=2}^n \mu_j^2/4} \, d\lambda \, d\mu_2 \cdots d\mu_d.$$

A closed formula of the above integral was given in [67]. Other variants of eigenvalues and eigenvectors can be found in [60,61]. For our purpose, we will focus on the following definition introduced in [48].

**Definition 10.** *For $T \in S^d(V)$ over $\mathbb{K}$, $(\lambda, v) \in \mathbb{K} \times V$ is called a normalized eigenpair of $T$ if the following equations hold:*

**Cases 1.**
$$\langle T, v^{\otimes(d-1)} \rangle = \lambda v, \langle v, v \rangle = 1.$$

In particular, $\lambda$ is called a normalized eigenvalue. We say two normalized eigenpairs $(\alpha, u)$ and $(\beta, v)$ of $T$ are equivalent if:
$$(\alpha, u) = (\beta, v)$$

or if
$$(-1)^{d-2}\alpha = \beta \text{ and } u = -v.$$

In the following, we would like to investigate sufficient conditions to ensure a tensor to have a unique rank-one approximation. First we recall the definition of the *multipolynomial resultant* [68,69]. For any given $n+1$ homogeneous polynomials $F_0, \ldots, F_n \in \mathbb{C}[x_0, \ldots, x_n]$ with positive total degrees $d_0, \ldots, d_n$, let $F_i = \sum_{|\alpha|=d_i} c_{i,\alpha} x_0^{\alpha_0} \cdots x_n^{\alpha_n}$, where $\alpha = (\alpha_0, \ldots, \alpha_n)$ and $|\alpha| = \alpha_0 + \cdots + \alpha_n$. Associate every pair $(i, \alpha)$ with a variable $u_{i,\alpha}$. For a polynomial $P$ in the variables $u_{i,\alpha}$, denote by $P(F_0, \ldots, F_n)$ the result obtained by letting $u_{i,\alpha} = c_{i,\alpha}$. Then we have the following classical result [68,69].

**Theorem 13.** *There is a unique polynomial, denoted by* Res, *in $u_{i,\alpha}$'s with integer coefficients, where $i = 0, \ldots, n$, and $|\alpha| \in \{d_0, \ldots, d_n\}$, that has the following properties:*

- $F_0 = \cdots = F_n = 0$ *has a nonzero solution over* $\mathbb{C}$ *if and only if* Res $(F_0, \ldots, F_n) = 0$.
- Res $(x_0^{d_0}, \ldots, x_n^{d_n}) = 1$.
- Res *is irreducible over* $\mathbb{C}$.

**Definition 11.** Res $(F_0, \ldots, F_n) \in \mathbb{C}$ *is called the resultant of $F_0, \ldots, F_n$.*

**Definition 12.** *For a symmetric tensor $T$, the resultant $\psi_T(\lambda)$ of the following polynomials is called the characteristic polynomial of $T$ [70].*

- For $T \in S^{2d-1}(V)$,
$$\langle T, v^{\otimes(d-1)} \rangle - \lambda x^{d-2} v = 0 \quad \text{and} \quad x^2 - \langle v, v \rangle = 0.$$

- For $T \in S^{2d}(V)$,
$$\langle T, v^{\otimes(2d-1)} \rangle - \lambda \langle v, v \rangle^{d-1} v = 0.$$

Note the resultant $\psi_T(\lambda)$ is a (univariate) polynomial in $\lambda$. We call the resultant of $\psi_T(\lambda)$ and its derivative $\psi_T'(\lambda)$, denoted by $D_{\text{eig}}(T)$, the *eigen discriminant*.

**Proposition 7.** *Let $V$ be a real vector space, and $\rho = \|T\|_\sigma$ [48]. Define*
$$H_\rho := \{T \in S^d(V) \mid \rho \text{ is not a simple eigenvalue of } T\}.$$

*Then, $H_\rho$ is a real hypersurface in $S^d(V)$.*

Let $W = V \otimes_\mathbb{R} \mathbb{C}$ be the complexification of $V$. Then we have:

**Theorem 14.** $D_{\text{eig}}(T) = 0$ *is a defining equation of the complex hypersurface [48]*
$$H_{\text{disc}} := \{T \in S^d(W) \mid T \text{ has a non-simple normalized eigenvalue}\}.$$

In fact $H_\rho$ consists of some components of the real points of $H_{\text{disc}}$. In the sense of [50], Theorem 14 shows the ED discriminant of the Veronese variety is a hypersurface.

**Corollary 4.** *For $T \in S^d(V)$, if $D_{\text{eig}}(T) \neq 0$, then $T$ has a unique best rank-one approximation.*

**Corollary 5.** Let $T \in S^d(V)$ be a nonnegative tensor. If $D_{\text{eig}}(T) \neq 0$, then $T$ has a unique best rank-one approximation, which is nonnegative and symmetric.

**Example 1.** Let $T = [T_{ijk}] \in S^3(\mathbb{R}^2)$. Then $\psi_T(\lambda)$ is the resultant of the polynomials:

**Cases 2.**

$$F_0 = T_{111}x^2 + 2T_{112}xy + T_{122}y^2 - \lambda xz, \; F_1 = T_{112}x^2 + 2T_{122}xy + T_{222}y^2 - \lambda yz, \; F_2 = x^2 + y^2 - z^2.$$

In fact, $\psi_T(\lambda) = \frac{1}{512}\det(G)$, where $G$ is defined by:

$$G = \begin{bmatrix} T_{111} & T_{122} & 0 & 2T_{112} & -\lambda & 0 \\ T_{112} & T_{222} & 0 & 2T_{122} & 0 & -\lambda \\ 1 & 1 & -1 & 0 & 0 & 0 \\ 12T_{122} & 4T_{111}-8T_{122}\lambda & 4T_{111}\lambda+4T_{122}\lambda & 8T_{222}\lambda-16T_{112}\lambda & 16T_{112}^2-4\lambda^2-16T_{111}T_{122} & 8T_{112}T_{122}-8T_{111}T_{222} \\ 4T_{222}\lambda-8T_{112}\lambda & 12T_{112}\lambda & 4T_{112}\lambda+4T_{222}\lambda & 8T_{111}\lambda-16T_{122}\lambda & 8T_{112}T_{122}-8T_{111}T_{222} & 16T_{122}^2-4\lambda^2-16T_{112}T_{222} \\ 8T_{112}^2-8T_{111}T_{122}-2\lambda^2 & 8T_{122}^2-8T_{222}T_{112}-2\lambda^2 & -6\lambda^2 & 8T_{112}T_{122}-8T_{111}T_{222} & 8T_{122}\lambda+8T_{111}\lambda & 8T_{112}\lambda+8T_{222}\lambda \end{bmatrix}.$$

Thus, $\psi_T(\lambda) = p_2\lambda^6 + p_4\lambda^4 + p_6\lambda^2 + p_8$, where each $p_m$ is a homogeneous polynomial of degree $m$ in $T_{ijk}$. See also [65,71].

For a general $T \in S^3(\mathbb{R}^2)$, $\psi_T(\lambda) = \alpha(\lambda^2 - \gamma_1)(\lambda^2 - \gamma_2)(\lambda^2 - \gamma_3)$ for some $\alpha \in \mathbb{C}$, where $\gamma_1, \gamma_2, \gamma_3$ are distinct. So $D_{\text{eig}}(T) \neq 0$.

For $T \in H_{\text{disc}}$, $\psi_T(\lambda)$ has multiple roots. For example, let $A \in S^3(\mathbb{R}^2)$ be defined by $A_{111} = A_{222} = 1$ and set other $A_{ijk} = 0$. Then $D_{\text{eig}}(A) = 0$, which implies that $A$ has a nonsimple eigenpair. Here $\psi_A(\lambda) = (\lambda+1)^2(\lambda-1)^2(2\lambda^2-1)$. So $A$ has two eigenvectors $(1,0)$ and $(0,1)$ with eigenvalue 1, and two eigenvectors $(-1,0)$ and $(0,-1)$ with eigenvalue $-1$. This computation coincides with the fact that $A = a^{\otimes 3} + b^{\otimes 3}$ has two best rank-one approximations, namely $a^{\otimes 3}$ and $b^{\otimes 3}$, where $a$ and $b$ are two orthonormal vectors in $\mathbb{R}^2$.

Similarly, we can define characteristic polynomials for non-symmetric tensors. Let $W_1, \ldots, W_d$ be complex vector spaces. For $T \in W_1 \otimes \cdots \otimes W_d$, $u_i \in W_i$, and $\alpha_i \in \mathbb{C}$, we denote the resultant of the following equations by $\varphi_T(\lambda)$.

$$\begin{cases} \alpha_i \langle T, u_1 \otimes \cdots \otimes \widehat{u_i} \otimes \cdots \otimes u_d \rangle = \lambda(\prod_{j \neq i}\alpha_j)u_i, \\ \langle u_i, u_i \rangle = \alpha_i^2, \end{cases} \quad (13)$$

where $i = 1, \ldots, d$. Then, $\varphi_T(\lambda)$ vanishes if and only if (13) has a nontrivial solution.

**Definition 13.** $\varphi_T(\lambda)$ is called the singular characteristic polynomial of $T$.

The following is an analogue of Definition 10.

**Definition 14.** Let $T \in W_1 \otimes \cdots \otimes W_d$. Two normalized singular pairs $(\lambda, u_1, \ldots, u_d)$ and $(\mu, v_1, \ldots, v_d)$ of $T$ are called equivalent if $(\lambda, u_1, \ldots, u_d) = (\mu, v_1, \ldots, v_d)$, or $(-1)^{d-2}\lambda = \mu$ and $u_i = -v_i$ for $i = 1, \ldots, d$.

It follows from [72] that the subset $X \subseteq V_1 \otimes \cdots \otimes V_d$ consisting of tensors which do not have unique best rank-one approximations is contained in some hypersurface. In fact we can strengthen the result by showing that $X$ is a hypersurface.

**Theorem 15.** The following subset is an algebraic hypersurface in $V_1 \otimes \cdots \otimes V_d$ [48],

$$X := \{T \in V_1 \otimes \cdots \otimes V_d : T \text{ has non-unique best rank-one approximations}\}.$$

Besides, we have the following property.

**Proposition 8.** *Let $W_1, \ldots, W_d$ be complex vector spaces. Then for a general $T \in W_1 \otimes \cdots \otimes W_d$, the equivalence classes of normalized singular pairs of T are distinct [48].*

**Definition 15.** *The resultant of $\varphi_T$ and its derivative $\varphi'_T$ is called the singular discriminant and denoted by $D_{\mathrm{sing}}(T)$.*

**Theorem 16.** *$D_{\mathrm{sing}}(T) = 0$ is a defining equation of the hypersurface [48]*

$$X_{\mathrm{disc}} := \{T \in W_1 \otimes \cdots \otimes W_d \mid T \text{ has a non-simple normalized singular value}\},$$

*and X consists of some components of the real points of $X_{\mathrm{disc}}$.*

**Corollary 6.** *For a real tensor T, if $D_{\mathrm{sing}}(T) \neq 0$, then T has a unique best rank-one approximation.*

**Corollary 7.** *For a nonnegative tensor T, if $D_{\mathrm{sing}}(T) \neq 0$, then T has a unique best rank-one approximation, which is nonnegative.*

Theorem 16 shows that the ED discriminant $X_{\mathrm{disc}}$ of $\mathrm{Seg}(\mathbb{P}W_1 \times \cdots \times \mathbb{P}W_d)$ is a complex hypersurface when $d \geq 3$, and the set of real points of $X_{\mathrm{disc}}$ is a real hypersurface. It is worth noting that when $d = 2$, the set of real points of the ED discriminant of $\mathrm{Seg}(\mathbb{P}W_1 \times \mathbb{P}W_2)$ has codimension 2 ([50], Example 7.6).

## 8. EM Algorithm

Expectation–Maximization (EM) algorithm, as a classical technique, has been used in nonnegative matrix factorizations, and its performance and geometry has been carefully studied. See [73] and the references therein. However, to the best of our knowledge, such an analysis for nonnegative tensors has not been written down. In this section, we routinely apply the EM algorithm to nonnegative tensor decompositions and give a description of the EM fixed points.

Given a real function:

$$f(p_1, \ldots, p_n) = \sum_{i=1}^{n} u_i \log p_i$$

of $p_1, \ldots, p_n$ with parameters $u_1, \ldots, u_n$, where $p_1, \ldots, p_n, u_1, \ldots, u_n$ satisfy

$$0 \leq u_i, p_i \leq 1 \text{ for } i = 1, \ldots, n, \quad \text{and} \quad \sum_{i=1}^{n} u_i = \sum_{i=1}^{n} p_i = 1.$$

Then, the maximum of $f$ is obtained when $p_i = u_i$ for $i = 1, \ldots, n$. In fact $(p_1 = u_1, \ldots, p_n = u_n)$ is a critical point of the Lagrangian:

$$\sum_{i=1}^{n} u_i \log p_i - \lambda(1 - \sum_{i=1}^{n} p_i).$$

Hence, for a given nonnegative rank-$r$ tensor $u = (u_{i_1, \ldots, i_d})$ with:

$$u_+ = \sum_{i_1, \ldots, i_d} u_{i_1, \ldots, i_d} = 1,$$

a nonnegative rank decomposition:

$$u = \sum_{l=1}^{r} \lambda_l v_1^{(l)} \otimes \cdots \otimes v_d^{(l)},$$

in coordinates:
$$u_{i_1,\ldots,i_d} = \sum_{l=1}^{r} \lambda_l v_{1,i_1}^{(l)} \cdots v_{d,i_d}^{(l)}$$

where $v_{j,i_j}^{(l)}$ is the $i_j$th entry of the vector $v_j^{(l)}$, gives a maximum of the likelihood function:

$$\mathcal{L}(\lambda, v_i^{(l)}) = \sum_{i_1,\ldots,i_d} u_{i_1,\ldots,i_d} \log(\sum_{l=1}^{r} \lambda_l v_{1,i_1}^{(l)} \cdots v_{d,i_d}^{(l)}). \tag{14}$$

This is a hidden model in statistics, and a classical way to optimize (14) is that we first use the EM algorithm to maximize the following likelihood function

$$L(\lambda, v_i^{(l)}) = \sum_{i_1,\ldots,i_d} \sum_{l=1}^{r} w_{i_1,\ldots,i_d}^{(l)} \log(\lambda_l v_{1,i_1}^{(l)} \cdots v_{d,i_d}^{(l)}), \tag{15}$$

where:
$$u_{i_1,\ldots,i_d} = \sum_{l=1}^{r} w_{i_1,\ldots,i_d}^{(l)}.$$

By ([74], Theorem 1.15), the value of the likelihood function (15) weakly increases during every iteration of the EM algorithm, and the local maxima of $\mathcal{L}$ are among the EM fixed points (final outputs) of $L$ [73,74].

**Remark 1.** *EM algorithm and its analogues have been widely used in nonnegative matrix factorizations, by maximizing different likelihood functions, i.e., finding critical points of different divergences, for example Kullback-Leibler divergence, β-divergence and so on. Similarly we can obtain other algorithms for nonnegative tensor decompositions as well by using different divergences. Usually the fixed points of these algorithms contain critical points, and the local maxima are among the critical points.*

A fixed point of EM algorithm need satisfy the following equations:

$$\lambda_k = \frac{1}{u_+} \sum_{i_1,\ldots,i_d} \frac{\lambda_k v_{1,i_1}^{(k)} \cdots v_{d,i_d}^{(k)}}{\sum_{l=1}^{r} \lambda_l v_{1,i_1}^{(l)} \cdots v_{d,i_d}^{(l)}} u_{i_1,\ldots,i_d} \tag{16}$$

$$v_{j,i_j}^{(k)} = \frac{1}{\lambda_k u_+} \sum_{i_1,\ldots,i_{j-1},i_{j+1},\ldots,i_d} \frac{\lambda_k v_{1,i_1}^{(k)} \cdots v_{d,i_d}^{(k)}}{\sum_{l=1}^{r} \lambda_l v_{1,i_1}^{(l)} \cdots v_{d,i_d}^{(l)}} u_{i_1,\ldots,i_d} \tag{17}$$

Since $\sum_{i_j} v_{j,i_j}^{(l)} = 1$, $\lambda_k > 0$. By canceling $\lambda_k$ in (16) and (17), we have:

$$1 = \frac{1}{u_+} \sum_{i_1,\ldots,i_d} \frac{v_{1,i_1}^{(k)} \cdots v_{d,i_d}^{(k)}}{\sum_{l=1}^{r} \lambda_l v_{1,i_1}^{(l)} \cdots v_{d,i_d}^{(l)}} u_{i_1,\ldots,i_d}, \tag{18}$$

$$v_{j,i_j}^{(k)} = \frac{1}{u_+} \sum_{i_1,\ldots,i_{j-1},i_{j+1},\ldots,i_d} \frac{v_{1,i_1}^{(k)} \cdots v_{d,i_d}^{(k)}}{\sum_{l=1}^{r} \lambda_l v_{1,i_1}^{(l)} \cdots v_{d,i_d}^{(l)}} u_{i_1,\ldots,i_d}, \tag{19}$$

and (18) can be obtained from (19). By (19) we have:

$$v_{j,i_j}^{(k)}\Big(\sum_{i_1,\ldots,i_{j-1},i_{j+1},\ldots,i_d}(u_+ - \frac{u_{i_1,\ldots,i_d}}{p_{i_1,\ldots,i_d}})v_{1,i_1}^{(k)}\cdots\widehat{v_{j,i_j}^{(k)}}\cdots v_{d,i_d}^{(k)}\Big) = 0, \tag{20}$$

where $p_{i_1,\ldots,i_d}$ is the output in Algorithm 1. Let:

$$R_{i_1,\ldots,i_d} = u_+ - \frac{u_{i_1,\ldots,i_d}}{p_{i_1,\ldots,i_d}},$$

then (20) is equivalent to:

$$v_j^{(k)} \circ \langle R, v_1^{(k)} \otimes \cdots \otimes \widehat{v_j^{(k)}} \otimes \cdots \otimes v_d^{(k)}\rangle = 0, \tag{21}$$

where $\circ$ denotes the Hadamard product. Since we require $\sum_{i_1,\ldots,i_d} p_{i_1,\ldots,i_d} = 1$, the likelihood function

$$\mathcal{L} = \sum_{i_1,\ldots,i_d} u_{i_1,\ldots,i_d} \log p_{i_1,\ldots,i_d} - u_+ \log\Big(\sum_{i_1,\ldots,i_d} p_{i_1,\ldots,i_d}\Big),$$

then the gradient of $\mathcal{L}$ is $R$. Hence:

$$\langle R, v_1^{(k)} \otimes \cdots \otimes \widehat{v_j^{(k)}} \otimes \cdots \otimes v_d^{(k)}\rangle = 0$$

implies that $R$ is orthogonal to the tangent space of $\hat{\sigma}_r(\mathrm{Seg}(\mathbb{P}^{n_1-1} \times \cdots \times \mathbb{P}^{n_d-1}))$, i.e.,

$$\sum_k v_1^{(k)} \otimes \cdots \otimes v_d^{(k)}$$

is a critical point. Therefore, we arrive at the following description, which is a trivial generalization of ([73], Theorem 3).

---

**Algorithm 1** EM Algorithm

---

Step 0: Given $\epsilon > 0$, select random $v_i^{(l)} \in \Delta_{n_i-1}, \lambda \in \Delta_{r-1}$;

E-Step: Define the expected hidden data tensor $w = (w_{i_1,\ldots,i_d}^{(k)})$ by

$$w_{i_1,\ldots,i_d}^{(k)} = \frac{\lambda_k v_{1,i_1}^{(k)} \cdots v_{d,i_d}^{(k)}}{\sum_{l=1}^r \lambda_l v_{1,i_1}^{(l)} \cdots v_{d,i_d}^{(l)}} u_{i_1,\ldots,i_d};$$

M-Step: Compute $v_i^{(l)}, \lambda$ to maximize the likelihood function:

$$\lambda_k^* = \frac{\sum_{i_1,\ldots,i_d} w_{i_1,\ldots,i_d}^{(k)}}{u_+},$$

$$v_{j,i_j}^{(k)*} = \frac{\sum_{i_1,\ldots,i_{j-1},i_{j+1},\ldots,i_d} w_{i_1,\ldots,i_d}^{(k)}}{\lambda_k^* u_+};$$

Step 3: If $|L(\lambda^*, v_i^{(l)*}) - L(\lambda, v_i^{(l)})| > \epsilon$ then set $(\lambda, v_i^{(l)}) = (\lambda^*, v_i^{(l)*})$ and go to the E-Step;

Step 4: Output $p_{i_1,\ldots,i_d} = \sum_k \lambda_k v_{1,i_1}^{(k)} \cdots v_{d,i_d}^{(k)}$.

---

**Proposition 9.** *The variety of EM fixed points is defined by:*

$$v_j^{(k)} \circ \langle R, v_1^{(k)} \otimes \cdots \otimes \widehat{v_j^{(k)}} \otimes \cdots \otimes v_d^{(k)} \rangle = 0, \quad j = 1, \ldots, d.$$

*The subset that are critical is defined by:*

$$\langle R, v_1^{(k)} \otimes \cdots \otimes \widehat{v_j^{(k)}} \otimes \cdots \otimes v_d^{(k)} \rangle = 0, \quad j = 1, \ldots, d.$$

A semialgebraic characterization of the set of nonnegative matrices with nonnegative ranks no greater than 3 was given in [73]. A semialgebraic characterization of the set of nonnegative tensors with nonnegative ranks no greater than 2 was given in [75].

**Question 3.** *Give a semialgebraic characterization of the set of nonnegative tensors with nonnegative ranks no greater than 3.*

## 9. Conclusions

In this short note, we give a very brief introduction to nonnegative tensors, mainly from the geometric perspective. More precisely, we review the generic uniqueness of rank decompositions of subgeneric nonnegative tensors and nonnegative typical ranks, and thus see the difference among the nonnegative, real, and complex settings. We review the generic uniqueness of nonnegative low-rank approximations. In particular, the rank-one approximation problem leads us to the spectral theory of nonnegative tensors. Finally, we describe the semialgebraic geometry of EM algorithm. Most of the results we present are obtained by studying the corresponding geometric properties of nonnegative tensors, and we have seen there are many open problems and unknown properties in this direction, which we hope would be understood better when more geometries are unveiled.

**Funding:** This research received no external funding.

**Acknowledgments:** The author is grateful to L. Chiantini and L.-H. Lim for their kind encouragements and useful discussions. Special thanks go to the anonymous referees for their careful reading and invaluable suggestions that significantly improved the presentation of this article.

**Conflicts of Interest:** The authors declare no conflict of interest.

## References

1. Bocci, C.; Chiantini, L.; Ottaviani, G. Refined methods for the identifiability of tensors. *Ann. Mat. Pura Appl.* **2014**, *193*, 1691–1702. [CrossRef]
2. Chiantini, L.; Ottaviani, G.; Vannieuwenhoven, N. On generic identifiability of symmetric tensors of subgeneric rank. *Trans. Am. Math. Soc.* **2017**, *369*, 4021–4042. [CrossRef]
3. Comon, P.; Golub, G.; Lim, L.H.; Mourrain, B. Symmetric tensors and symmetric tensor rank. *SIAM J. Matrix Anal. Appl.* **2008**, *30*, 1254–1279. [CrossRef]
4. Brachat, J.; Comon, P.; Mourrain, B.; Tsigaridas, E. Symmetric tensor decomposition. *Linear Algebra Appl.* **2010**, *433*, 1851–1872. [CrossRef]
5. Lim, L.H.; Comon, P. Nonnegative approximations of nonnegative tensors. *J. Chemom.* **2009**, *23*, 432–441. [CrossRef]
6. Shashua, A.; Hazan, T. Non-negative tensor factorization with applications to statistics and computer vision. In Proceedings of the 22nd International Conference on Machine Learning, Bonn, Germany, 7–11 August 2005; pp. 792–799.
7. Smilde, A.; Bro, R.; Geladi, P. *Multi-Way Analysis*; Wiley: Chichester, UK, 2004.
8. Zhang, Q.; Wang, H.; Plemmons, R.J.; Pauca, V.P. Tensor methods for hyperspectral data analysis: A space object material identification study. *J. Opt. Soc. Am. A* **2008**, *25*, 3001–3012. [CrossRef]
9. Garcia, L.D.; Stillman, M.; Sturmfels, B. Algebraic geometry of Bayesian networks. *J. Symb. Comput.* **2005**, *39*, 331–355. [CrossRef]

10. Jordan, M.I. Graphical models. *Stat. Sci.* **2004**, *19*, 140–155. [CrossRef]
11. Koller, D.; Friedman, N. *Probabilistic Graphical Models: Principles and techniques*; Adaptive Computation and Machine Learning; MIT Press: Cambridge, MA, USA, 2009; p. xxxvi+1231.
12. Zhou, J.; Bhattacharya, A.; Herring, A.H.; Dunson, D.B. Bayesian factorizations of big sparse tensors. *J. Am. Stat. Assoc.* **2015**, *110*, 1562–1576. [CrossRef]
13. Shashua, A.; Zass, R.; Hazan, T. Multi-way Clustering Using Super-Symmetric Non-negative Tensor Factorization. In *Computer Vision—ECCV 2006*; Leonardis, A., Bischof, H., Pinz, A., Eds.; Springer: Berlin/Heidelberg, Germany, 2006; pp. 595–608.
14. Veganzones, M.; Cohen, J.; Farias, R.; Chanussot, J.; Comon, P. Nonnegative tensor CP decomposition of hyperspectral data. *IEEE Trans. Geosci. Remote Sens.* **2016**, *54*, 2577–2588. [CrossRef]
15. Bartoň, M.; Calo, V.M. Optimal quadrature rules for odd-degree spline spaces and their application to tensor-product-based isogeometric analysis. *Comput. Methods Appl. Mech. Energy* **2016**, *305*, 217–240. [CrossRef]
16. Mantzaflaris, A.; Jüttler, B.; Khoromskij, B.N.; Langer, U. Low rank tensor methods in Galerkin-based isogeometric analysis. *Comput. Methods Appl. Mech. Energy* **2017**, *316*, 1062–1085. [CrossRef]
17. Scholz, F.; Mantzaflaris, A.; Jüttler, B. Partial tensor decomposition for decoupling isogeometric Galerkin discretizations. *Comput. Methods Appl. Mech. Energy* **2018**, *336*, 485–506. [CrossRef]
18. Lee, D.D.; Seung, H.S. Algorithms for non-negative matrix factorization. *Adv. Neural Inf. Process. Syst.* **2001**, 556–562.
19. Chu, M.T.; Lin, M.M. Low-dimensional polytope approximation and its applications to nonnegative matrix factorization. *SIAM J. Sci. Comput.* **2008**, *30*, 1131–1155. [CrossRef]
20. Kim, H.; Park, H. Nonnegative matrix factorization based on alternating nonnegativity constrained least squares and active set method. *SIAM J. Matrix Anal. Appl.* **2008**, *30*, 713–730. [CrossRef]
21. Kim, J.; Park, H. Fast nonnegative matrix factorization: an active-set-like method and comparisons. *SIAM J. Sci. Comput.* **2011**, *33*, 3261–3281. [CrossRef]
22. Ho, N.D. Nonnegative Matrix Factorization Algorithms and Applications. Ph.D. Thesis, Ecole Polytechnique de Louvain, Ottignies-Louvain-la-Neuve, Belgium, 2008.
23. Zhou, G.; Cichocki, A.; Xie, S. Fast nonnegative matrix/tensor factorization based on low-rank approximation. *IEEE Trans. Signal Process.* **2012**, *60*, 2928–2940. [CrossRef]
24. Vavasis, S.A. On the complexity of nonnegative matrix factorization. *SIAM J. Optim.* **2009**, *20*, 1364–1377. [CrossRef]
25. Arora, S.; Ge, R.; Kannan, R.; Moitra, A. Computing a nonnegative matrix factorization—provably. In *STOC'12—Proceedings of the 2012 ACM Symposium on Theory of Computing*; ACM: New York, NY, USA, 2012; pp. 145–161, doi:10.1145/2213977.2213994.
26. Bro, R. Multi-Way Analysis in the Food Industry: Models, Algorithms, and Applications. Ph.D. Thesis, Universiteit van Amsterdam, Amsterdam, The Netherlands, 1998.
27. Friedlander, M.P.; Hatz, K. Computing non-negative tensor factorizations. *Optim. Methods Softw.* **2008**, *23*, 631–647. [CrossRef]
28. Cohen, J.; Farias, R.C.; Comon, P. Fast Decomposition of Large Nonnegative Tensors. *IEEE Signal Process. Lett.* **2015**, *22*, 862–866. [CrossRef]
29. Cichocki, A.; Zdunek, R.; Phan, A.H.; Amari, S.I. *Nonnegative Matrix and Tensor Factorizations: Applications to Exploratory Multi-Way Data Analysis and Blind Source Separation*; Wiley Publishing: Hoboken, NJ, USA, 2009.
30. Kim, J.; He, Y.; Park, H. Algorithms for nonnegative matrix and tensor factorizations: A unified view based on block coordinate descent framework. *J. Glob. Optim.* **2014**, *58*, 285–319. [CrossRef]
31. Zhou, G.; Cichocki, A.; Zhao, Q.; Xie, S. Nonnegative matrix and tensor factorizations: An algorithmic perspective. *IEEE Signal Process. Mag.* **2014**, *31*, 54–65. [CrossRef]
32. Strassen, V. Rank and optimal computation of generic tensors. *Linear Algebra Appl.* **1983**, *52*, 645–685. [CrossRef]
33. Lickteig, T. Typical tensorial rank. *Linear Algebra Appl.* **1985**, *69*, 95–120. [CrossRef]
34. Abo, H.; Ottaviani, G.; Peterson, C. Induction for secant varieties of Segre varieties. *Trans. Am. Math. Soc.* **2009**, *361*, 767–792. [CrossRef]
35. Kruskal, J.B. Three-way arrays: Rank and uniqueness of trilinear decompositions, with application to arithmetic complexity and statistics. *Linear Algebra Appl.* **1977**, *18*, 95–138. [CrossRef]

36. Stegeman, A. On uniqueness conditions for Candecomp/Parafac and Indscal with full column rank in one mode. *Linear Algebra Appl.* **2009**, *431*, 211–227. [CrossRef]
37. Chiantini, L.; Ottaviani, G. On generic identifiability of 3-tensors of small rank. *SIAM J. Matrix Anal. Appl.* **2012**, *33*, 1018–1037. [CrossRef]
38. Domanov, I.; De Lathauwer, L. On the uniqueness of the canonical polyadic decomposition of third-order tensors—Part I: Basic results and uniqueness of one factor matrix. *SIAM J. Matrix Anal. Appl.* **2013**, *34*, 855–875. [CrossRef]
39. Domanov, I.; De Lathauwer, L. On the uniqueness of the canonical polyadic decomposition of third-order tensors—Part II: Uniqueness of the overall decomposition. *SIAM J. Matrix Anal. Appl.* **2013**, *34*, 876–903. [CrossRef]
40. Chiantini, L.; Ottaviani, G.; Vannieuwenhoven, N. An algorithm for generic and low-rank specific identifiability of complex tensors. *SIAM J. Matrix Anal. Appl.* **2014**, *35*, 1265–1287. [CrossRef]
41. Domanov, I.; De Lathauwer, L. Generic uniqueness conditions for the canonical polyadic decomposition and INDSCAL. *SIAM J. Matrix Anal. Appl.* **2015**, *36*, 1567–1589. [CrossRef]
42. Qi, Y.; Comon, P.; Lim, L.H. Semialgebraic geometry of nonnegative tensor rank. *SIAM J. Matrix Anal. Appl.* **2016**, *37*, 1556–1580. [CrossRef]
43. Alexander, J.; Hirschowitz, A. Polynomial interpolation in several variables. *J. Algebr. Geom.* **1995**, *4*, 201–222.
44. Chiantini, L.; Ciliberto, C. On the concept of *k*-secant order of a variety. *J. Lond. Math. Soc.* **2006**, *73*, 436–454. [CrossRef]
45. Ballico, E. On the weak non-defectivity of Veronese embeddings of projective spaces. *Cent. Eur. J. Math.* **2005**, *3*, 183–187. [CrossRef]
46. Mella, M. Singularities of linear systems and the Waring problem. *Trans. Am. Math. Soc.* **2006**, *358*, 5523–5538. [CrossRef]
47. Landsberg, J.M. *Tensors: Geometry and Applications*; Graduate Studies in Mathematics; American Mathematical Society: Providence, RI, USA, 2012; Volume 128, p. xx+439.
48. Qi, Y.; Comon, P.; Lim, L.H. Uniqueness of nonnegative tensor approximations. *IEEE Trans. Inform. Theory* **2016**, *62*, 2170–2183. [CrossRef]
49. Friedland, S.; Ottaviani, G. The number of singular vector tuples and uniqueness of best rank-one approximation of tensors. *Found. Comput. Math.* **2014**, *14*, 1209–1242. [CrossRef]
50. Draisma, J.; Horobeţ, E.; Ottaviani, G.; Sturmfels, B.; Thomas, R.R. The Euclidean distance degree of an algebraic variety. *Found. Comput. Math.* **2016**, *16*, 99–149. [CrossRef]
51. Draisma, J.; Horobeţ, E. The average number of critical rank-one approximations to a tensor. *Linear Multilinear Algebra* **2016**, *64*, 2498–2518. [CrossRef]
52. Berman, A.; Plemmons, R.J. *Nonnegative Matrices in the Mathematical Sciences*; Classics in Applied Mathematics; Society for Industrial and Applied Mathematics (SIAM): Philadelphia, PA, USA, 1994; Volume 9, p. xx+340, doi:10.1137/1.9781611971262.
53. Chang, K.C.; Pearson, K.; Zhang, T. Perron-Frobenius theorem for nonnegative tensors. *Commun. Math. Sci.* **2008**, *6*, 507–520. [CrossRef]
54. Friedland, S.; Gaubert, S.; Han, L. Perron-Frobenius theorem for nonnegative multilinear forms and extensions. *Linear Algebra Appl.* **2013**, *438*, 738–749. [CrossRef]
55. Lim, L.H. Singular Values and Eigenvalues of tensors: A Variational Approach. In Proceedings of the 1st IEEE International Workshop on Computational Advances in Multi-Sensor Adaptive Processing, Puerto Vallarta, Mexico, 13–15 December 2005; pp. 129–132.
56. Yang, Y.; Yang, Q. Further results for Perron-Frobenius theorem for nonnegative tensors. *SIAM J. Matrix Anal. Appl.* **2010**, *31*, 2517–2530. [CrossRef]
57. Hillar, C.J.; Lim, L.H. Most tensor problems are NP-hard. *J. ACM* **2013**, *60*. [CrossRef]
58. Stegeman, A.; Comon, P. Subtracting a best rank-1 approximation may increase tensor rank. *Linear Algebra Appl.* **2010**, *433*, 1276–1300. [CrossRef]
59. Vannieuwenhoven, N.; Nicaise, J.; Vandebril, R.; Meerbergen, K. On generic nonexistence of the Schmidt-Eckart-Young decomposition for complex tensors. *SIAM J. Matrix Anal. Appl.* **2014**, *35*, 886–903. [CrossRef]
60. Chang, K.; Qi, L.; Zhang, T. A survey on the spectral theory of nonnegative tensors. *Numer. Linear Algebra Appl.* **2013**, *20*, 891–912. [CrossRef]
61. Qi, L. Eigenvalues of a real supersymmetric tensor. *J. Symb. Comput.* **2005**, *40*, 1302–1324. [CrossRef]

62. Banach, S. Über homogene Polynome in ($L^2$). *Stud. Math.* **1938**, *7*, 36–44. [CrossRef]
63. Friedland, S. Best rank one approximation of real symmetric tensors can be chosen symmetric. *Front. Math. China* **2013**, *8*, 19–40. [CrossRef]
64. Breiding, P. The expected number of eigenvalues of a real Gaussian tensor. *SIAM J. Appl. Algebra Geom.* **2017**, *1*, 254–271. [CrossRef]
65. Cartwright, D.; Sturmfels, B. The number of eigenvalues of a tensor. *Linear Algebra Appl.* **2013**, *438*, 942–952. [CrossRef]
66. Oeding, L.; Ottaviani, G. Eigenvectors of tensors and algorithms for Waring decomposition. *J. Symb. Comput.* **2013**, *54*, 9–35. [CrossRef]
67. Breiding, P. The average number of critical rank-one-approximations to a symmetric tensor. *arXiv* **2017**, arXiv:1701.07312.
68. Gel'fand, I.M.; Kapranov, M.M.; Zelevinsky, A.V. *Discriminants, Resultants, and Multidimensional Determinants*; Mathematics: Theory & Applications; Birkhäuser Boston, Inc.: Boston, MA, USA, 1994; p. x+523.
69. Cox, D.A.; Little, J.; O'Shea, D. *Using Algebraic Geometry*, 2nd ed.; Graduate Texts in Mathematics; Springer: New York, NY, USA, 2005; Volume 185, p. xii+572.
70. Qi, L. Eigenvalues and invariants of tensors. *J. Math. Anal. Appl.* **2007**, *325*, 1363–1377. [CrossRef]
71. Li, A.M.; Qi, L.; Zhang, B. E-characteristic polynomials of tensors. *Commun. Math. Sci.* **2013**, *11*, 33–53. [CrossRef]
72. Friedland, S.; Stawiska, M.G. Some approximation problems in semi-algebraic geometry. *Constr. Approx. Funct.* **2015**, *107*, 133–147. [CrossRef]
73. Kubjas, K.; Robeva, E.; Sturmfels, B. Fixed points EM algorithm and nonnegative rank boundaries. *Ann. Stat.* **2015**, *43*, 422–461. [CrossRef]
74. Pachter, L.; Sturmfels, B. *Algebraic Statistics for Computational Biology*; Cambridge University Press: New York, NY, USA, 2005.
75. Allman, E.S.; Rhodes, J.A.; Sturmfels, B.; Zwiernik, P. Tensors of nonnegative rank two. *Linear Algebra Appl.* **2015**, *473*, 37–53. [CrossRef]

© 2018 by the author. Licensee MDPI, Basel, Switzerland. This article is an open access article distributed under the terms and conditions of the Creative Commons Attribution (CC BY) license (http://creativecommons.org/licenses/by/4.0/).

# Article
# Set Evincing the Ranks with Respect to an Embedded Variety (Symmetric Tensor Rank and Tensor Rank

**Edoardo Ballico**

Department of Mathematics, University of Trento, 38123 Povo, Italy; edoardo.ballico@unitn.it

Received: 11 July 2018; Accepted: 8 August 2018; Published: 14 August 2018

**Abstract:** Let $X \subset \mathbb{P}^r$ be an integral and non-degenerate variety. We study when a finite set $S \subset X$ evinces the $X$-rank of the general point of the linear span of $S$. We give a criterion when $X$ is the order $d$ Veronese embedding $X_{n,d}$ of $\mathbb{P}^n$ and $|S| \leq \binom{n+\lfloor d/2 \rfloor}{n}$. For the tensor rank, we describe the cases with $|S| \leq 3$. For $X_{n,d}$, we raise some questions of the maximum rank for $d \gg 0$ (for a fixed $n$) and for $n \gg 0$ (for a fixed $d$).

**Keywords:** $X$-rank; symmetric tensor rank; tensor rank; veronese variety; segre variety

## 1. Introduction

Let $X \subset \mathbb{P}^r$ be an integral and non-degenerate variety. For any $q \in \mathbb{P}^r$, the $X$-rank $r_X(q)$ of $q$ is the minimal cardinality of a finite set $S \subset X$ such that $q \in \langle S \rangle$, where $\langle \ \rangle$ denotes the linear span. The definition of $X$-ranks captures the notion of tensor rank (take as $X$ the Segre embedding of a multiprojective space) of rank decomposition of a homogeneous polynomial (take as $X$ a Veronese embedding of a projective space) of partially symmetric tensor rank (take a complete linear system of a multiprojective space) and small variations of it may be adapted to cover other applications. See [1] for many applications and [2] for many algebraic insights. For the pioneering works on the applied side, see, for instance, [3–7]. The paper [7] proved that $X$-rank is not continuous and showed why this has practical importance. The dimensions of the secant varieties (i.e., the closure of the set of all $q \in \mathbb{P}^r$ with a prescribed rank) has a huge theoretical and practical importance. The Alexander–Hirschowitz theorem computes in all cases the dimensions of the secant varieties of the Veronese embeddings of a projective space ([8–14]). For the dimensions of secant varieties, see [15–17] for tensors and [18–27] for partially symmetric tensors (i.e., Segre–Veronese embeddings of multiprojective spaces). For the important problem of the uniqueness of the set evincing a rank (in particular for the important case of tensors) after the classical [28], see [29–38]. See [39–47] for other theoretical works.

Let $S \subset X$ be a finite set and $q \in \mathbb{P}^r$. We say that $S$ *evinces the $X$-rank of $q$* if $q \in \langle S \rangle$ and $|S| = r_X(q)$. We say that $S$ *evinces an $X$-rank* if there is $q \in \mathbb{P}^r$ such that $S$ evinces the $X$-rank of $q$. Obviously, $S$ may evince an $X$-rank only if it is linearly independent, but this condition is not a sufficient one, except in very trivial cases, like when $r_X(q) \leq 2$ for all $q \in \mathbb{P}^r$. Call $r_{X,\max}$ the maximum of all integers $r_X(q)$. An obvious necessary condition is that $|S| \leq r_{X,\max}$ and this is in very special cases a sufficient condition (see Propositions 1 for the rational normal curve). If $S$ evinces the $X$-rank of $q \in \mathbb{P}^r$, then $q \in \langle S \rangle$ and $q \notin \langle S' \rangle$ for any $S' \subsetneq S$. For any finite set $S \subset \mathbb{P}^r$, set $\langle S \rangle' := \langle S \rangle \setminus (\cup_{S' \subsetneq S} \langle S' \rangle)$. Note that $\langle S \rangle' = \emptyset$ if and only if either $S = \emptyset$ or $S$ is linearly dependent (when $|S| = 1$, $\langle S \rangle' = S$ and $S$ evinces itself). In some cases, it is possible to show that some finite $S \subset X$ evinces the $X$-rank of all points of $\langle S \rangle'$. We say that $S$ *evinces generically the*

X-ranks if there is a non-empty Zariski open subset $U$ of $\langle S \rangle$ such that $S$ evinces the X-ranks of all $q \in U$. We say that $S$ totally evinces the X-ranks if $S$ evinces the X-ranks of all $q \in \langle S \rangle'$. We first need an elementary and well-known bound to compare it with our results.

Let $\rho(X)$ be the maximal integer such that each subset of $X$ with cardinality $\rho(X)$ is linearly independent. See ([43] Lemma 2.6, Theorem 1.18) and ([42] Proposition 2.5) for some uses of the integer $\rho(X)$. Obviously, $\rho(X) \leq r+1$ and it is easy to check and well known that equality holds if and only if $X$ is a Veronese embedding of $\mathbb{P}^1$ (Remark 1). If $|S| \leq \lfloor (\rho(X)+1)/2 \rfloor$, then $S$ totally evinces the X-ranks (as in [43] Theorem 1.18) while, for each integer $t > \lfloor (\rho(X)+1)/2 \rfloor$ with $t \leq r+1$, there is a linearly independent subset of $X$ with cardinality $t$ and not totally evincing the X-ranks ( Lemma 3). Thus, to say something more, we need to make some assumptions on $S$ and these assumptions must be related to the geometry of $X$ or the reasons for the interest of the X-ranks. We do this in Section 3 for the Veronese embeddings and in Section 4 for the tensor rank. For tensors, we only have results for $|S| \leq 3$ (Propositions 3 and 4).

For all positive integers $n, d$ let $v_{d,n} : \mathbb{P}^n \to \mathbb{P}^r$, $r = \binom{n+d}{n} - 1$, denote the Veronese embedding of $\mathbb{P}^n$, i.e., the embedding of $\mathbb{P}^n$ induced by the complete linear system $|\mathcal{O}_{\mathbb{P}^n}(d)|$. Set $X_{n,d} := v_{d,n}(\mathbb{P}^n)$. At least over an algebraically closed base field of characteristic 0 (i.e., in the set-up of this paper), for any $q \in \mathbb{P}^r$, the integer $r_{X_{n,d}}(q)$ is the minimal number of $d$-powers of linear forms in $n+1$ variables whose sum is the homogeneous polynomial associated to $q$.

We prove the following result, whose proof is elementary (see Section 3 for the proof). In its statement, the assumption "$h^1(\mathcal{I}_A(\lfloor d/2 \rfloor)) = 0$" just means that the vector space of all degree $\lfloor d/2 \rfloor$ homogeneous polynomials in $n+1$ variables vanishing on $A$ has dimension $\binom{n+\lfloor d/2 \rfloor}{n} - |A|$, i.e., $A$ imposes $|A|$ independent conditions to the homogeneous polynomials of degree $\lfloor d/2 \rfloor$ in $n+1$ variables.

**Theorem 1.** *Fix integers $n \geq 2$, $d > k > 2$ and a finite set $A \subset \mathbb{P}^n$ such that $h^1(\mathcal{I}_A(\lfloor d/2 \rfloor)) = 0$. Set $S := v_{d,n}(A)$. Then, $S$ totally evinces the ranks for $X_{n,d}$.*

A general $A \subset \mathbb{P}^n$ satisfies the assumption of Theorem 1 if and only if $|A| \leq \binom{n+\lfloor d/2 \rfloor}{n}$. For much smaller $|A|$, one can check the condition $h^1(\mathcal{I}_A(\lfloor d/2 \rfloor)) = 0$ if $A$ satisfies some geometric conditions (e.g., if $A$ is in linearly general position, it is sufficient to assume $|A| \leq n \lfloor d/2 \rfloor + 1$).

We conclude the paper with some questions related to the maximum of the X-ranks when $X$ is a Veronese embedding of $\mathbb{P}^n$.

## 2. Preliminary Lemmas

**Remark 1.** *Let $X \subset \mathbb{P}^r$ be an integral and non-degenerate variety. Since any $r+2$ points of $\mathbb{P}^r$ are linearly dependent, we have $\rho(X) \leq r+1$. If $X$ is a rational normal curve, then $\rho(X) = r+1$ because any $r+1$ points of $X$ spans $\mathbb{P}^r$. Now, we check that, if $\rho(X) = r+1$, then $X$ is a rational normal curve. This is well known, but usually stated in the set-up of Veronese embeddings or the X-ranks of curves. Set $n := \dim X$ and $d := \deg(X)$. Assume $\rho(X) = r+1$. Let $H \subset \mathbb{P}^r$ be a general hyperplane. If $n > 1$, then $X \cap H$ has dimension $n-1 > 0$ and in particular it has infinitely many points. Any $r+1$ points of $X \cap H$ are linearly dependent. Now, assume $n = 1$. Since $X$ is non-degenerate, we have $d \geq n$. By Bertini's theorem, $X \cap H$ contains $d$ points of $X$. Since $\rho(X) = r+1$, $\dim H = r-1$ and $H \cap X \subset H$, we have $d \leq r$. Hence, $d = r$, i.e., $X$ is a rational normal curve.*

The following example shows, that in many cases, there are are sets evincing X-ranks, but not totally evincing X-ranks or even generically evincing X-ranks.

**Example 1.** *Let $X \subset \mathbb{P}^r$, $r \geq 3$, be a rational normal curve. Take $q \in \mathbb{P}^r$ with $r_X(q) = r$, i.e., take $q \in \tau(X) \setminus X$, where $\tau(X)$ is the tangential variety of $X$ ([48]). Take $S \subset X$ evincing the X-rank of $q$. Thus, $|S| = r$ and $S$ spans a hyperplane $\langle S \rangle$. Since $\dim \tau(X) = 2$ and $\tau(X)$ spans $\mathbb{P}^r$, $\langle S \rangle \cap \tau(X)$ is a* proper closed algebraic subset of $\langle S \rangle$. Thus, for a general $p \in \langle S \rangle$, we have $r_X(p) < |S|$ and hence $S$ does not generically evinces X-ranks.

**Lemma 1.** *If $S \subset X$ is a finite set evincing the rank of some $q \in \mathbb{P}^r$, then each $S' \subset S$, $S' \neq \emptyset$, evinces the X-rank of some $q' \in \mathbb{P}^r$.*

**Proof.** We may assume $S' \neq S$. Write $S'' := S \setminus S'$. Since $S$ evinces the rank of $q$, $S$ is linearly independent, but $S \cup \{q\}$ is not linearly independent. Since $S' \neq \emptyset$ and $S'' \neq \emptyset$, there are unique $q' \in \langle S' \rangle$ and $q'' \in \langle S'' \rangle$ such that $q \in \langle \{q', q''\} \rangle$. Since $S$ evinces the rank of $q$, $S'$ evinces the rank of $q'$. □

**Lemma 2.** *Every non-empty subset of a set evincing generically (resp. totally) X-ranks evinces generically (resp. totally) the X-ranks.*

**Proof.** Assume that $S$ evinces generically the X-ranks and call $U$ a non-empty open subset of $\langle S \rangle'$ such that $r_X(q) = |S|$ for all $q \in U$; if $S$ evinces totally the X-ranks, take $U := \langle S \rangle'$. Fix $S' \subsetneq S$, $S' \neq 0$ and set $S'' := S \setminus S'$. Let $E$ be the set of all $q \in \langle S \rangle'$ such that $\langle \{q\} \cup S'' \rangle \cap U \neq \emptyset$. If $q \in E$, then $r_X(q) = |S'|$ because $r_X(q') = |S|$ for each $q' \in \langle \{q\} \cup S'' \rangle \cap U$. Since $S' \cap S'' = \emptyset$ and $S' \cup S'' = S$ is linearly independent, $E$ is a non-empty open subset of $\langle S \rangle'$ (a general element of $\langle S \rangle$ is contained in the linear span of a general element of $\langle S' \rangle$ and a general element of $\langle S' \rangle$). Now, assume $U = \langle S \rangle'$. Every element of $\langle S \rangle'$ is in the linear span of an element of $\langle S' \rangle'$ and an element of $\langle S'' \rangle'$. □

**Lemma 3.** *Take a finite set $S \subset X$, $S \neq \emptyset$.*

(a) *If $|S| \leq \lfloor (\rho(X)+1)/2 \rfloor$, then $S$ totally evinces the X-ranks.*
(b) *For each integer $t > \lfloor (\rho(X)+1)/2 \rfloor$, there is $A \subset X$ such that $|A| = t$ and $A$ does not totally evince the X-ranks.*

**Proof.** Take $q \in \langle S \rangle'$ and assume $r_X(q) < |S|$. Take $B \subset X$ evincing the X-rank of $q$. Since $|B| < |S|$, we have $B \neq S$. Since $q \in \langle S \rangle \cap \langle B \rangle$, but no proper subset of either $B$ or $S$ spans $q$, $S \cup B$ is linearly dependent. Since $|B| \leq |S| - 1$, we have $|B \cup S| \leq \rho(X)$, contradicting the definition of $\rho(X)$.

Now, we prove part (b). By Lemma 1, it is sufficient to do the case $t = \lfloor (\rho(X)+1)/2 \rfloor + 1$. By the definition of the integer $\rho(X)$, there is a subset $D \subset X$ with $|D| = \rho(X) + 1$ and $D$ linearly dependent. Write $D = A \sqcup E$ with $|A| = \lfloor (\rho(X)+1)/2 \rfloor + 1$ and $|E| = \rho(X) + 1 - |A|$. Note that $|A| > |E|$. Since $|A| \leq \rho(X)$ (remember that $\rho(X) \geq 2$), both $A$ and $E$ are linearly independent. Since $A \cup E$ is linearly dependent, there is $q \in \langle A \rangle \cap \langle E \rangle$. Since $|D| = \rho(X) + 1$, every proper subset of $D$ is linearly independent. Hence, $\langle A' \rangle \cap \langle E \rangle = \emptyset$ for all $A' \subsetneq A$. Thus, $q \in \langle A \rangle'$. Since $|E| < |A|$, $A$ does not evince the X-rank of $q$. □

**Remark 2.** *Take $X \subset \mathbb{P}^r$ such that $r_X(q) \leq 2$ for all $q \in \mathbb{P}^r$ (e.g., by [49], we may take most space curves). Any set $S \subset X$ with $|S| = 2$ evinces its X-ranks if and only if $X$ contains no line.*

## 3. The Veronese Embeddings of Projective Spaces

Let $v_{d,n} : \mathbb{P}^n \to \mathbb{P}^r$, $r := -1 + \binom{n+d}{n}$, denote the Veronese embedding of $\mathbb{P}^n$. Set $X_{n,d} := v_{d,n}(\mathbb{P}^n)$.

**Proposition 1.** *Let $X \subset \mathbb{P}^d$, $d \geq 2$, be the rational normal curve.*

(a) *A non-empty finite set $S \subset X$ evinces some rank of $\mathbb{P}^d$ if and only if $|S| \leq d$.*
(b) *A non-empty finite set $A \subset X$ totally evinces the X-ranks if and only if $|A| \leq \lfloor (d+2)/2 \rfloor$.*

**Proof.** By a theorem of Sylvester's ([48]), every $q \in \mathbb{P}^d$ has $X$-rank at most $d$. Thus, the condition $|S| \leq d$ is a necessary condition for evincing some rank. By Lemma 1 to prove part (a), it is sufficient to prove it when $|S| = d$. Take any connected zero-dimensional scheme $Z \subset X$ with $\deg(Z) = 2$ and $S \cap Z = \emptyset$. Thus, $\deg(Z \cup S) = d + 2$. Since $X \cong \mathbb{P}^1$, $\deg(\mathcal{O}_X(1)) = d$ and $X$ is projectively normal, we have $h^1(\mathcal{I}_{S \cup Z}(1)) = 1$ and $h^1(\mathcal{I}_W(1)) = 0$ for each $W' \subsetneq S \cup Z$. This is equivalent to say that the line $\langle Z \rangle$ meets $\langle S \rangle$ at a unique point, $q$ and $q \neq Z_{\text{red}}$. By Sylvester's theorem, $r_X(q) = d$ ([48]). Since $q \in \langle S \rangle$ and $|S| = d$, $S$ evinces the $X$-rank of $q$.

If $A \neq \emptyset$ and $|A| \leq \lfloor (d+2)/2 \rfloor$, then $A$ totally evinces the $X$-ranks by part (a) of Lemma 3 and the fact that $\rho(X) = d + 1$. Now, assume $d \geq |A| > \lfloor (d+2)/2 \rfloor$. Fix a set $E \subset X \setminus A$ with $|E| = d + 2 - |A|$. Adapt the proof of part (b) of Lemma 3. □

**Proposition 2.** *Fix a set $S \subset X_{n,d}$, $n \geq 2$, with $|S| = d + 1$. The following conditions are equivalent:*

1. *there is a line $L \subset \mathbb{P}^n$ such that $|S \cap L| > \lfloor (d+2)/2 \rfloor$;*
2. *$S$ evinces no $X_{n,d}$-rank;*
3. *there is $q \in \langle S \rangle'$ such that $S$ does not evince the $X_{n,d}$-rank of $q$.*

**Proof.** Obviously, (2) implies (3). If $X' \subset X$ is a subvariety and $q \in \langle X' \rangle$, we have $r_{X'}(q) \geq r_X(q)$. Thus, Sylvester's theorem ([48]) and Lemma 2 show that (1) implies (2).

Now, assume the existence of $q \in \langle S \rangle'$ such that $S$ does not evince the $X$-rank of $q$, i.e., $r_X(q) \leq d$. Take $A \subset \mathbb{P}^n$ such that $\nu(A) = S$ and take $B \subset \mathbb{P}^n$ such that $\nu_d(B)$ evinces the $X$-rank of $q$. Since $q \in \langle S \rangle'$, (Ref. [50] Lemma 1) gives $h^1(\mathbb{P}^n, \mathcal{I}_{A \cup B}(d)) > 0$. Since $|A \cup B| \leq 2d + 1$, (Ref. [51] Lemma 34) gives the existence of a line $L \subset \mathbb{P}^n$ such that $|L \cap (A \cup B)| \geq d + 2$. Let $H \subset \mathbb{P}^n$ be a general hyperplane containing $L$. Since $H$ is general and $A \cup B$ is a finite set, we have $H \cap (A \cup B) = L \cap (A \cup B)$. Since $|L \cap (A \cup B)| \geq d + 2$, we have $|A \cup B \setminus H \cap (A \cup B)| \leq d - 1$ and hence $h^1(\mathbb{P}^n, \mathcal{I}_{A \cup B \setminus H \cap (A \cup B)}(d - 1)) = 0$. By ([52] Lemma 5.2), we have $A \setminus A \cap H = B \setminus B \cap H$. □

See [53,54] for some results on the geometry of sets $S \subset X_{n,d}$ with controlled Hilbert function and that may be useful to extend Proposition 2.

**Proof of Theorem 1:** Set $k := \lfloor d/2 \rfloor$. Note that $h^1(\mathcal{I}_A(x)) = 0$ for all $x \geq k$ and in particular $h^1(\mathcal{I}_A(d-k)) = 0$. Fix $q \in \langle \nu_{d,n}(A) \rangle'$ and assume $r_{X_{n,d}}(q) < |A|$. Fix $B \subset \mathbb{P}^n$ such that $\nu_{d,n}(B)$ evinces the $X_{n,d}$-rank of $q$. Since $h^1(\mathcal{I}_A(k)) = 0$ and $|A| > |B|$, we have $h^0(\mathcal{I}_B(k)) > h^0(\mathcal{I}_A(k))$. Thus, there is $M \in |\mathcal{O}_{\mathbb{P}^n}(k)|$ containing $B$, but with $A \not\subset M$, i.e., $A \setminus A \cap M \neq \emptyset$, while $B \setminus B \cap M = \emptyset$. Since $h^1(\mathcal{I}_A(d-k)) = 0$, we have $h^1(\mathcal{I}_{A \setminus A \cap M}(d-k)) = 0$. Since $h^1(\mathcal{I}_A(d)) = 0$, $\nu_{d,n}(A)$ is linearly independent. Since $\nu_{d,n}(B)$ evinces a rank, it is linearly independent. Grassmann's formula gives $\dim \langle \nu_{d,k}(A) \rangle \cap \langle \nu_{d,b}(B) \rangle = |A \cap B| + h^1(\mathcal{I}_{A \cup B}(d)) - 1$. We have $A \cup B = ((A \cup B) \cap M) \cup (A \setminus A \cap M)$. Since $A \setminus A \cap B$ is a finite set, we have $h^2(\mathcal{I}_{A \setminus A \cap B}(d-k)) = h^2(\mathcal{O}_{\mathbb{P}^n}(d-k)) = 0$. Since $h^1(\mathcal{I}_{A \setminus A \cap M}(d-k)) = 0$, the residual exact sequence (also known as the Castelnuovo's sequence)

$$0 \to \mathcal{I}_{A \setminus A \cap B}(d-k) \to \mathcal{I}_{A \cup B}(d) \to \mathcal{I}_{M \cap (A \cup B), M}(d) \to 0$$

gives $h^1(\mathcal{I}_{A \cup B}(d)) = h^1(M, \mathcal{I}_{M \cap (A \cup B)}(d))$. Since $M$ is projectively normal, $h^1(M, \mathcal{I}_{M \cap (A \cup B)}(d)) = h^1(\mathcal{I}_{A \cup B}(d))$. Thus, the Grassmann's formula gives $\dim \langle \nu_{d,n}(A \cap M) \rangle \cap \langle \nu_{d,n}(B \cap M) \rangle = |A \cap B \cap M| + h^1(\mathcal{I}_{A \cup B}(d)) - 1$. Since $B \subset M$, we get $\langle \nu_{d,n}(A \cap M) \rangle \cap \langle \nu_{d,n}(B \cap M) \rangle = \langle \nu_{d,k}(A) \rangle \cap \langle \nu_{d,b}(B) \rangle$. Since $A \cap M \supsetneq A$, we get $q \notin \langle \nu_{d,n}(A) \rangle'$, a contradiction. □

## 4. Tensors, i.e., the Segre Varieties

Fix an integer $k \geq 2$ and positive integers $n_1, \ldots, n_k$. Set $Y := \prod_{i=1}^k \mathbb{P}^{n_i}$ (the Segre variety) and $N := -1 + \prod_{i=1}^k (n_i + 1)$. Let $\nu : Y \to \mathbb{P}^N$ denote the Segre embedding. Let $\pi_i : Y \to \mathbb{P}^{n_i}$ denote the projection on the $i$-th factor. For any $i \in \{1, \ldots, k\}$, set $Y[i] := \prod_{h \neq i} \mathbb{P}^{n_h}$ and call $\eta_i : Y \to Y[i]$ the projection which forgets the $i$-th component. Let $v[i] : Y[i] \to \mathbb{P}^{N_i}$, $N_i := -1 + \prod_{h \neq i}(n_h + 1)$ denote the Segre embedding of $Y[i]$. A key difficulty is that $\rho(v(Y)) = 2$ because $v(Y)$ contains lines.

**Lemma 4.** *Let $S \subset Y$ be any finite set such that there is $i \in \{1, \ldots k\}$ with $\eta_{i|S}$ not injective. Then, $v(S)$ evinces no rank.*

**Proof.** By Lemma 1, we reduce to the case $|S| = 2$, say $S = \{a, b\}$ with $a = (a_1, \ldots, a_k)$, $b = (b_1, \ldots, b_k)$ with $a_i = b_i$ if and only if $i > 1$. Since all lines of $Y$ are contained in one of the factors of $Y$ and all lines of $v(Y)$ are images of lines of $Y$, we get $S \subset v(Y)$. Thus, each element of $\langle v(S) \rangle$ is contained in $v(Y)$ and hence it has rank 1. Since $|S| > 1$, $v(S)$ evinces no rank. □

**Lemma 5.** *Let $S \subset Y$ such that there are $S' \subseteq S$ and $i \in \{1, \ldots, k\}$ with $|S'| = 3$, $v_i(\eta_i(S'))$ linearly dependent and $\pi_i(S') \subset \mathbb{P}^{n_i}$ linearly dependent. Then, $v(S)$ evinces no rank.*

**Proof.** Let $Q \subset \mathbb{P}^3$ be a smooth quadric surface. $Q$ is projectively equivalent to the Segre embedding of $\mathbb{P}^1 \times \mathbb{P}^1$ and each point of $\mathbb{P}^3$ has at most $Q$-rank 2 by [47] (Proposition 5.1). By Lemma 1, we may assume $S' = S$. By Lemma 4, we may assume that $\eta_{i|S}$ is injective. Thus, $|\eta_{i|S}| = 3$. Since $v_i(\eta_i(S))$ is not linearly independent and it has cardinality 3, it is contained in a line of $v_i(Y[i])$. Thus, $\eta_i(S)$ is contained in a line of one of the factors of $Y[i]$. By assumption, $\pi_i(S)$ is contained in a line of $\mathbb{P}^{n_i}$. Thus, $S$ is contained in a subscheme of $Y$ isomorphic to $\mathbb{P}^1 \times \mathbb{P}^1$. Since each point of $\mathbb{P}^3$ has $Q$-rank $\leq 2$ and $|S| = 3$, $v(S)$ evinces no rank. □

**Remark 3.** *Fix a finite set $A \subset Y$ such that $S := v(A)$ is linearly independent. $S$ evinces no tensor rank if there is a multiprojective subspace $Y' \subset Y$ such that $A \subset Y'$ and $|S|$ is larger than the maximum tensor rank of $v(Y')$.*

Note that Lemmas 4 and 5 may be restated as a way to check for very low $|S|$ if there is some $Y'$ as in Lemma 3 exists.

**Proposition 3.** *Take $S \subset v(Y)$ with $|S| = 2$. Let $Y'$ be the minimal multiprojective subspace of $Y$ containing $S$. The following conditions are equivalent:*

1. *$S$ evinces no rank;*
2. *$S$ does not generically evince ranks;*
3. *$S$ does not totally evince ranks;*
4. *$Y' \cong \mathbb{P}^1$.*

**Proof.** Since any two distinct points of $\mathbb{P}^N$ are linearly independent (i.e., $\langle S \rangle$ is a line) and $v(Y)$ is the set of all points with $v(Y)$-rank 1, $S$ evinces no rank if and only if $\langle S \rangle \subset v(Y)$. Use the fact that the lines of $v(Y)$ are contained in one of the factors of $v(Y)$. Since $v(Y)$ is cut out by quadrics, if $\langle S \rangle \not\subset v(Y)$, then $|\langle S \rangle \cap v(Y)| \leq 2$. Since $S \subset \langle S \rangle \cap v(Y)$, we see that all points of $\langle S \rangle \setminus S$ have rank 2. □

**Proposition 4.** *Take $S \subset v(Y)$ with $|S| = 3$ and $v(S)$ linearly independent. Write $S = v(A)$ with $A \subset Y'$. Let $Y'$ be the minimal multiprojective subspace of $Y$ containing $A$. Write $Y' = \mathbb{P}^{m_1} \times \cdots \mathbb{P}^{m_s}$ with $s \geq 1$ and $m_1 \geq \cdots \geq m_s > 0$. We have $m_1 \leq 2$.*

*If $\eta_{i|A}$ is injective for all $i$ and either $m_2 = 2$ or $s \geq 4$ or $m_1 = 2$ and $s = 3$, then $S$ totally evinces its ranks. In all other cases for a general $E \in Y'$ with $|E| = 3$, $v(E)$ does not generically evince its ranks.*

**Proof.** If $\eta_{i|A}$ is not injective for some $i$, then $S$ evinces no rank by Lemma 4. Thus, we may assume that each $\eta_{i|A}$ is injective for all $i$. Each factor of $Y'$ is the linear span of $\pi_i(A)$ in $\mathbb{P}^{n_i}$. Hence, $m_1 \leq 2$.

Omitting all factors which are points, we get the form of $Y'$ we use. If $Y' = \mathbb{P}^1$ (resp. $\mathbb{P}^2$, resp. $\mathbb{P}^1 \times \mathbb{P}^1$), then each point of $\langle S \rangle$ has rank 1 (resp. 1, resp. $\leq 2$). Thus, in these cases, $S$ evinces no rank. If either $Y' = \mathbb{P}^2 \times \mathbb{P}^1$ or $Y' = (\mathbb{P}^1)^3$, then $\sigma_2(\mathbb{P}^2 \times \mathbb{P}^1) = \mathbb{P}^5$ and $\sigma_2((\mathbb{P}^1)^3) = \mathbb{P}^7$ ([23,26]). Thus, the last assertion of the proposition is completed.

(a) Assume $s \geq 2$ and $m_2 = 2$. Taking a projection onto the first two factors, we reduce to the case $s = 2$ (this reduction step is used only to simplify the notation). Take a $H \in |\mathcal{O}_{Y'}(1,0)|$ containing $B$ (this is possible because $h^0(\mathcal{O}_{\mathbb{P}^2}(1)) = 3 > |B|$). Since $Y'$ is the minimal multiprojective subspace of $Y$ containing $A$, we have $A \setminus A \cap H \neq \emptyset$. Since $B \setminus B \cap H = \emptyset$, (Ref. [52] Lemma 5.1) gives $h^1(\mathcal{I}_{A \setminus A \cap H}(0,1)) > 0$. Thus, either there is $A' \subset A$ with $|A'| = 2$ and $\eta_{1|A'}$ not injective (we excluded this possibility) or $|A \setminus A \cap H| = 3$ (i.e., $A \cap H = \emptyset$) and $\eta_1(A) \subset \mathbb{P}^2$ is contained in a line $R$. Set $M := \mathbb{P}^2 \times R$. We get $A \subset M$ and hence $A$ is a contained in a proper multiprojective subspace, contradicting the definition of $Y'$.

(b) Assume $s \geq 3$ and $m_1 = 2$. By part (a), we may assume $m_2 = 1$. Taking a projection, we reduce to the case $s = 3$, i.e., $Y' = \mathbb{P}^2 \times \mathbb{P}^1 \times \mathbb{P}^1$. Take $H$ as in step (a). As in step (a), we get $A \cap H = \emptyset$ and $\eta_1(A)$ contained in a line $R$ of the Segre embedding of $\mathbb{P}^1 \times \mathbb{P}^1$, contradicting the definition of $Y'$.

(c) Assume $s \geq 4$. By step (b), we may assume $m_1 = 1$. Taking a projection onto the first four factors of $Y'$, we reduce to the case $Y' = (\mathbb{P}^1)^4$. Fix any $H \in |\mathcal{O}_{Y'}(1,1,0,0)|$ containing $B$. Assume for the moment $A \not\subseteq H$. By ([52] Lemma 5.1), we have $h^1(\mathcal{I}_{A \setminus A \cap H}(0,0,1,1)) > 0$, i.e., either there are $a = (a_1, a_2, a_3, a_4) \in A$, $b = (b_1, b_2, b_3, b_4) \in A$ with $a \neq b$ and $(a_3, a_4) = (b_3, b_4)$ of $A \cap H = \emptyset$ and the projection of $A$ onto the last 2 factors of $Y'$ is contained in a line. The last possibility is excluded by the minimality of $Y'$. Thus, $a, b \in A$ exists. Set $A := \{a, b, c\}$ and write $c = (c_1, c_2, c_3, c_4)$. Permuting the factors of $Y'$, we see that, for each $E \subset \{1, 2, 3, 4\}$, there is $A_E \subset A$ with $|A_E| = 2$ and $\pi_E(A_E)$ is a singleton, where $\pi_E : Y' \to \mathbb{P}^1 \times \mathbb{P}^1$ denote the projection onto the factors of $Y'$ corresponding to $E$. Since the cardinality of the set $\mathcal{S}$ of all subset of $\{1, 2, 3, 4\}$ with cardinality 2 is larger than the cardinality of the set of all subsets of $A$ with cardinality 3, there are $E, F \in \mathcal{S}$ such that $E \neq F$ and $A_E = A_F$. If $E \cap F \neq \emptyset$, say $E \cap F = \{i\}$, then $\eta_{i|A}$ is not injective, contradicting our assumption. If $E \cap F = \emptyset$, we have $E \cup F = \{1, 2, 3, 4\}$. Since $A_E = A_F$, we get $|A_E| = 1$, a contradiction. □

**Remark 4.** *Take a finite $S \subset \nu(Y)$ and fix $q \in \langle \nu(S) \rangle'$. Let $A \subset Y$ be the subset with $\nu(A) = S$. It is easier to prove that $S$ evinces the rank of $q$ if we know that the minimal multiprojective subspace of $Y$ containing $A$ is the minimal multiprojective subspace $Y''$ of $Y$ with $q \in \langle \nu(Y'') \rangle$. Note that this is always true if $Y'' = Y$, i.e., if the tensor $q$ is concise.*

## 5. Questions on the Case of Veronese Varieties

Let $r_{\max}(n, d)$ denote the maximum of all $X_{n,d}$-ranks (in [55,56] it is denoted with $r_{\max}(n+1, d)$). The integer $r_{\max}(n, d)$ depends on two variables, $n$ and $d$. In this section, we ask some question on the asymptotic behavior of $r_{\max}(n, d)$ when we fix one variable, while the other one goes to $+\infty$.

Let $r_{\text{gen}}(n, d)$ denote the $X_{n,d}$-rank of a general $q \in \mathbb{P}^r$. These integers do not depend on the choice of the algebraically closed base field $\mathbb{K}$ with characteristic 0. The diagonalization of quadratic forms gives $r_{\max}(n, 2) = r_{\text{gen}}(n, 2) = n + 1$. The integers $r_{\text{gen}}(n, d)$, $d > 2$, are known by an important theorem of Alexander and Hirschowitz ([8–13]); with four exceptional cases, we have $r_{\text{gen}}(n, d) = \lceil \binom{n+d}{n}/(n+1) \rceil$. An important theorem of Blekherman and Teitler gives $r_{\max}(n, d) \leq 2 r_{\text{gen}}(n, d)$ (and even $r_{\max}(n, d) \leq 2 r_{\text{gen}}(n, d) - 1$ with a few obvious exceptions) ([57,58]). In particular, for a fixed $n$, we have

$$\frac{1}{(n+1)!} \leq \liminf_{d \to +\infty} r_{\max}(n, d)/d^n \leq \limsup_{d \to +\infty} r_{\max}(n, d)/d^n \leq \frac{2}{(n+1)!}.$$

It is reasonable to ask if $\liminf_{d \to +\infty} r_{\max}(n,d)/d^n$ exists and its value. Of course, it is tempting also to ask a more precise information about $r_{\max}(n,d)$ for $d \gg 0$. In the case $n = 2$, De Paris proved in [55,56] that $r_{\max}(2,d) \geq \lfloor (d^2 + 2d + 5)/4 \rfloor$ ([56] Theorem 3), which equality holds if $d$ is even ([56] (Proposition 2.4)) and suggested that equality holds for all $d$. Since $r_{\max}(2, d+1) \geq r_{\max}(2,d)$ even for odd $d$, the integer $r_{\max}(2,d)$ grows like $d^2/4$. Thus, there is an interesting interval between the general upper bound of [57] (which, in this case, has order $d^2/3$) and $r_{\max}(2,d)$. There are very interesting upper bounds for the dimensions of the set of all points with rank bigger than the generic one ([59]).

What are
$$\limsup_{n \to +\infty} \frac{(n+1)! r_{\max}(n,d)}{d^n} \text{ and } \liminf_{n \to +\infty} \frac{(n+1)! r_{\max}(n,d)}{d^n} ?$$

For all $d \geq 3$, study $r_{\max}(n,d) - r_{\max}(n, d-1)$ and compare for $d \gg 0$ $r_{\max}(n,d) - r_{\max}(n, d-1)$ with $r_{\max}(n-1,d)$ and $r_{\text{gen}}(n-1,d)$. Of course, this is almost exactly known when $n = 2$ by Sylvester's theorem ([48]) and De Paris ([55,56]), but $r_{\max}(2,d) - r_{\max}(2, d-1)$ for $d \gg 0$ is both $\sim r_{\text{gen}}(1,d)$ and $\sim r_{\max}(1,d)/2$ and so we do not have any suggestion for the case $n > 2$.

**Funding:** The author was partially supported by MIUR and GNSAGA of INdAM (Italy).

**Conflicts of Interest:** The author declares no conflict of interest.

## References

1. Landsberg, J.M. *Tensors: Geometry and Applications Graduate Studies in Mathematics*; American Mathematical Society: Kingston, ON, Canada; New York, NY, USA, 2012; Volume 128.
2. Iarrobino, A.; Kanev, V. Lecture Notes in Mathematics. In *Power Sums, Gorenstein Algebras, and Determinantal Loci*; Springer-Verlag: Berlin, Germany; New York, NY, USA, 1999; Volume 1721.
3. Comon, P. Tensor decompositions: state of the art and applications, Mathematics. In *Signal Processing, V (Coventry, 2000)*; McWhirter, J.G., Proudler, I.K., Eds.; Clarendon Press: Oxford, UK, 2002; Volume 71, pp. 1–24.
4. Kolda, T.G.; Bader, B.W. Tensor decomposition and applications. *SIAM Rev.* **2009**, *51*, 455–500. [CrossRef]
5. Lickteig, T. Typical tensorial rank. *Linear Algebra Appl.* **1985**, *69*, 95–120. [CrossRef]
6. Lim, L.H.; Comon, P. Multiarray signal processing: Tensor decomposition meets compressed sensing. *C. R. Mecanique* **2010**, *338*, 311–320. [CrossRef]
7. Lim, L.H.; de Silva, V. Tensor rank and the ill-posedness of the best low-rank approximation problem. *SIAM J. Matrix Anal. Appl.* **2008**, *30*, 1084–1127.
8. Alexander, J. Singularités imposables en position générale aux hypersurfaces de $\mathbb{P}^n$. *Composi. Math.* **1988**, *68*, 305–354.
9. Alexander, J.; Hirschowitz, A. Un lemme d'Horace différentiel: application aux singularité hyperquartiques de $\mathbb{P}^5$. *J. Algebr. Geom.* **1992**, *1*, 411–426.
10. Alexander, J.; Hirschowitz, A. La méthode d'Horace éclaté: application à l'interpolation en degré quatre. *Invent. Math.* **1992**, *107*, 585–602. [CrossRef]
11. Alexander, J.; Hirschowitz, A. Polynomial interpolation in several variables. *J. Algebr. Geom.* **1995**, *4*, 201–222.
12. Brambilla, M.C.; Ottaviani, G. On the Alexander–Hirschowitz Theorem. *J. Pure Appl. Algebra* **2008**, *212*, 1229–1251. [CrossRef]
13. Chandler, K. A brief proof of a maximal rank theorem for generic double points in projective space. *Trans. Am. Math. Soc.* **2000**, *353*, 1907–1920. [CrossRef]
14. Postinghel, E. A new proof of the Alexander–Hirschowitz interpolation theorem. *Ann. Mat. Pura Appl.* **2012**, *191*, 77–94. [CrossRef]
15. Abo, H.; Ottaviani, G.; Peterson, C. Induction for secant varieties of Segre varieties. *Trans. Am. Math. Soc.* **2009**, *361*, 67–792. [CrossRef]

16. Aladpoosh, T.; Haghighi, H. On the dimension of higher secant varieties of Segre varieties $\mathbb{P}^n \times \cdots \times \mathbb{P}^n$. *J. Pure Appl. Algebra* **2011**, *215*, 1040–1052. [CrossRef]
17. Catalisano, M.V.; Geramita, A.V.; Gimigliano, A. Secant varieties of $\mathbb{P}^1 \times \cdots \times \mathbb{P}^1$ (*n*-times) are NOT defective for $n \geq 5$. *J. Algebr. Geom.* **2011**, *20*, 295–327. [CrossRef]
18. Abo, H. On non-defectivity of certain Segre-Veronese varieties. *J. Symb. Comput.* **2010**, *45*, 1254–1269. [CrossRef]
19. Abo, H.; Brambilla, M.C. Secant varieties of Segre-Veronese varieties $\mathbb{P}^m \times \mathbb{P}^n$ embedded by $\mathcal{O}(1,2)$. *Exp. Math.* **2009**, *18*, 369–384. [CrossRef]
20. Ab, H.; Brambilla, M.C. New examples of defective secant varieties of Segre-Veronese varieties. *Collect. Math.* **2012**, *63*, 287–297. [CrossRef]
21. Abo, H.; Brambilla, M.C. On the dimensions of secant varieties of Segre-Veronese varieties. *Ann. Mat. Pura Appl.* **2013**, *192*, 61–92. [CrossRef]
22. Baur, K.; Draisma, J.; de Graaf, W. Secant dimensions of minimal orbits: computations and conjectures. *Exp. Math.* **2007**, *16*, 239–250. [CrossRef]
23. Baur, K.; Draisma, J. Secant dimensions of low-dimensional homogeneous varieties. *Adv. Geom.* **2010**, *10*, 1–29. [CrossRef]
24. Catalisano, M.V.; Geramita, A.V.; Gimigliano, A. On the rank of tensors, via secant varieties and fat points. In *Zero-Dimensional Schemes and Applications (Naples, 2000)*; Landsberg, J.M., Ed.; American Mathematical Society: Kingston, ON, Canada; New York, NY, USA, 2002; Volume 123, pp. 133–147.
25. Catalisano, M.V.; Geramita, A.V.; Gimigliano, A. Ranks of tensors, secant varieties of Segre varieties and fat points. *Linear Algebra Appl.* **2002**, *355*, 263–285. [CrossRef]
26. Catalisano, M.V.; Geramita, A.V.; Gimigliano, A. Segre-Veronese embeddings of $\mathbb{P}^1 \times \mathbb{P}^1 \times \mathbb{P}^1$ and their secant varieties. *Collect. Math.* **2007**, *58*, 1–24.
27. Laface, A.; Postinghel, E. Secant varieties of Segre-Veronese embeddings of $(\mathbb{P}^1)^r$. *Math. Ann.* **2013**, *356*, 1455–1470. [CrossRef]
28. Kruskal, J.B. Three-way arrays: rank and uniqueness of trilinear decompositions, with application to arithmetic complexity and statistics. *Linear Algebra Appl.* **1977**, *18*, 95–138. [CrossRef]
29. Bocci, C.; Chiantini, L.; Ottaviani, G. Refined methods for the identifiability of tensors. *Ann. Mat. Pura Appl.* **2014**, *193*, 1691–1702. [CrossRef]
30. Chiantini, L.; Ottaviani, G. On generic identifiability of 3-tensors of small rank. *SIAM J. Matrix Anal. Appl.* **2012**, *33*, 1018–1037. [CrossRef]
31. Chiantini, L.; Ottaviani, G.; Vanniuwenhoven, N. An algorithm for generic and low-rank specific identifiability of complex tensors. *SIAM J. Matrix Anal. Appl.* **2014**, *35*, 1265–1287. [CrossRef]
32. Chiantini, L.; Ottaviani, G.; Vanniuwenhoven, N. On identifiability of symmetric tensors of subgeneric rank. *Trans. Amer. Math. Soc.* **2017**, *369*, 4021–4042. [CrossRef]
33. Chiantini, L.; Ottaviani, G.; Vanniuwenhoven, N. Effective criteria for specific identifiability of tensors and forms. *SIAM J. Matrix Anal. Appl.* **2017**, *38*, 656–681. [CrossRef]
34. Domanov, I.; De Lathauwer, L. On the uniqueness of the canonical polyadic decomposition of third-order tensors–part I: Basic results and unique- ness of one factor matrix. *SIAM J. Matrix Anal. Appl.* **2013**, *34*, 855–875. [CrossRef]
35. Domanov, I.; De Lathauwer, L. On the uniqueness of the canonical polyadic decomposition of third-order tensors—Part II: Uniqueness of the overall decomposition. *SIAM J. Matrix Anal. Appl.* **2013**, *34*, 876–903. [CrossRef]
36. Domanov, I.; De Lathauwer, L. Generic uniqueness conditions for the canonical polyadic decomposition and INDSCAL. *SIAM J. Matrix Anal. Appl.* **2015**, *36*, 1567–1589. [CrossRef]
37. Massarenti, A.; Mella, M.; Staglian'o, G. Effective identifiability criteria for tensors and polynomials. *J. Symbolic Comput.* **2018**, *87*, 227–237. [CrossRef]
38. Sidiropoulos, N.D.; Bro, R. On the uniqueness of multilinear decomposition of N-way arrays. *J. Chemom.* **2000**, *14*, 229–239. [CrossRef]
39. Abo, H.; Ottaviani, G.; Peterson, C. Non-defectivity of Grassmannians of planes. *J. Algebr. Geom.* **2012**, *21*, 1–20. [CrossRef]

40. Araujo, C.; Massarenti, A.; Rischter, R. On nonsecant defectivity of Segre-Veronese varieties. *arXiv* **2016**, arXiv:1611.01674
41. Boralevi, A. A note on secants of Grassmannians. *Rend. Istit. Mat. Univ. Trieste* **2013**, *45*, 67–72.
42. Buczyńska, W.; Buczyński, J. Secant varieties to high degree Veronese reembeddings, catalecticant matrices and smoothable Gorenstein schemes. *J. Algebr. Geom.* **2014**, *23*, 63–90. [CrossRef]
43. Buczyński, J.; Ginensky, A.; Landsberg, J.M. Determinantal equations for secant varieties and the Eisenbud–Koh–Stillman conjecture. *J. Lond. Math. Soc.* **2013**, *88*, 1–24. [CrossRef]
44. Buczyński, J.; Landsberg, J.M. Ranks of tensors and a generalization of secant varieties. *Linear Algebra Appl.* **2013**, *438*, 668–689. [CrossRef]
45. Chiantini, L.; Ciliberto, C. On the dimension of secant varieties. *J. Europ. Math. Soc.* **2006**, *73*, 436–454. [CrossRef]
46. Draisma, J. A tropical approach to secant dimension. *J. Pure Appl. Algebra* **2008**, *212*, 349–363. [CrossRef]
47. Landsberg, J.M.; Teitler, Z. On the ranks and border ranks of symmetric tensors. *Found. Comput. Math.* **2010**, *10*, 339–366. [CrossRef]
48. Comas, G.; Seiguer, M. On the rank of a binary form. *Found. Comp. Math.* **2011**, *11*, 65–78. [CrossRef]
49. Piene, R. Cuspidal projections of space curves. *Math. Ann.* **1981**, *256*, 95–119. [CrossRef]
50. Ballico, E.; Bernardi, A. Decomposition of homogeneous polynomials with low rank. *Math. Z.* **2012**, *271*, 1141–1149. [CrossRef]
51. Bernardi, A.; Gimigliano, A.; Idà, M. Computing symmetric rank for symmetric tensors. *J. Symb. Comput.* **2011**, *46*, 34–53. [CrossRef]
52. Ballico, E.; Bernardi, A. Stratification of the fourth secant variety of Veronese variety via the symmetric rank. *Adv. Pure Appl. Math.* **2013**, *4*, 215–250. [CrossRef]
53. Ballico, E. Finite subsets of projective spaces with bad postulation in a fixed degree. *Beitrage zur Algebra und Geometrie* **2013**, *54*, 81–103 [CrossRef]
54. Ballico, E. Finite defective subsets of projective spaces. *Riv. Mat. Univ. Parma* **2013**, *4*, 113–122.
55. De Paris, A. The asymptotic leading term for maximum rank for ternary forms of a given degree. *Linear Algebra Appl.* **2016**, *500*, 15–29. [CrossRef]
56. De Paris, A. High-rank ternary forms of even degree. *Arch. Math.* **2017**, *109*, 505–510. [CrossRef]
57. Blekherman, G.; Teitler, Z. On maximum, typical and generic ranks. *Math. Ann.* **2015**, *362*, 1021–1031. [CrossRef]
58. Blekherman, G.; Teitler, Z. Some examples of forms of high rank. *Collect. Math.* **2016**, *67*, 431–441.
59. Buczyński, J.; Han, K.; Mella, M.; Teitler, Z. On the locus of points of high rank. *Eur. J. Math.* **2018**, *4*, 113–136. [CrossRef]

© 2018 by the author. Licensee MDPI, Basel, Switzerland. This article is an open access article distributed under the terms and conditions of the Creative Commons Attribution (CC BY) license (http://creativecommons.org/licenses/by/4.0/).

Article
# On Comon's and Strassen's Conjectures

Alex Casarotti [1], Alex Massarenti [1,2,*] and Massimiliano Mella [1]

[1] Department of Mathematics and Informatics, University of Ferrara, 44121 Ferrara, Italy; csrlxa@unife.it (A.C.); mll@unife.it (M.M.)
[2] Institute of Mathematics and Statistics, Universidade Federal Fluminense, Niterói 24210-200, Brazil
* Correspondence: alexmassarenti@id.uff.br

Received: 18 September 2018; Accepted: 22 October 2018; Published: 25 October 2018

**Abstract:** Comon's conjecture on the equality of the rank and the symmetric rank of a symmetric tensor, and Strassen's conjecture on the additivity of the rank of tensors are two of the most challenging and guiding problems in the area of tensor decomposition. We survey the main known results on these conjectures, and, under suitable bounds on the rank, we prove them, building on classical techniques used in the case of symmetric tensors, for mixed tensors. Finally, we improve the bound for Comon's conjecture given by flattenings by producing new equations for secant varieties of Veronese and Segre varieties.

**Keywords:** Strassen's conjecture; Comon's conjecture; tensor decomposition; Waring decomposition

**MSC:** Primary 15A69, 15A72, 11P05; Secondary 14N05, 15A69

## 1. Introduction

Let $X \subset \mathbb{P}^N$ be an irreducible and reduced non-degenerate variety. The *rank* $\operatorname{rank}_X(p)$ with respect to $X$ of a point $p \in \mathbb{P}^N$ is the minimal integer $h$ such that $p$ lies in the linear span of $h$ distinct points of $X$. In particular, if $Y \subseteq X$ we have that $\operatorname{rank}_X(p) \leqslant \operatorname{rank}_Y(p)$.

Since the $h$-*secant variety* $\mathbb{S}ec_h(X)$ of $X$ is the subvariety of $\mathbb{P}^N$ obtained as the closure of the union of all $(h-1)$-planes spanned by $h$ general points of $X$, for a general point $p \in \mathbb{S}ec_h(X)$ we have $\operatorname{rank}_X(p) = h$.

When the ambient projective space is a space parametrizing tensors we enter the area of tensor decomposition. A tensor rank decomposition expresses a tensor as a linear combination of simpler tensors. More precisely, given a tensor $T$, lying in a given tensor space over a field $k$, a tensor rank-1 decomposition of $T$ is an expression of the form

$$T = \lambda_1 U_1 + \dots + \lambda_h U_h \tag{1}$$

where the $U_i$'s are linearly independent rank one tensors, and $\lambda_i \in k^*$. The rank of $T$ is the minimal positive integer $h$ such that $T$ admits such a decomposition.

Tensor decomposition problems come out naturally in many areas of mathematics and applied sciences. For instance, in signal processing, numerical linear algebra, computer vision, numerical analysis, neuroscience, graph analysis, control theory and electrical networks [1–7]. In pure mathematics tensor decomposition issues arise while studying the additive decompositions of a general tensor [8–14].

Comon's conjecture [3], which states the equality of the rank and symmetric rank of a symmetric tensor, and Strassen's conjecture on the additivity of the rank of tensors [15] are two of the most important and guiding problems in the area of tensor decomposition.

More precisely, Comon's conjecture predicts that the rank of a homogeneous polynomial $F \in k[x_0,\ldots,x_n]_d$ with respect to the Veronese variety $\mathcal{V}_d^n$ is equal to its rank with respect to the Segre variety $\mathcal{S}^{\underline{n}} \cong (\mathbb{P}^n)^d$ into which $\mathcal{V}_d^n$ is diagonally embedded, that is $\mathrm{rank}_{\mathcal{V}_d^n}(F) = \mathrm{rank}_{\mathcal{S}^{\underline{n}}}(F)$.

Strassen's conjecture was originally stated for triple tensors and then generalized to several different contexts. For instance, for homogeneous polynomials it says that if $F \in k[x_0,\ldots,x_n]_d$ and $G \in k[y_0,\ldots,y_m]_d$ are homogeneous polynomials in distinct sets of variables then $\mathrm{rank}_{\mathcal{V}_d^{n+m+1}}(F+G) = \mathrm{rank}_{\mathcal{V}_d^n}(F) + \mathrm{rank}_{\mathcal{V}_d^m}(G)$.

In Sections 3 and 4, while surveying the state of the art on Comon's and Strassen's conjectures, we push a bit forward some standard techniques, based on catalecticant matrices and more generally on flattenings, to extend some results on these conjectures, known in the setting of Veronese and Segre varieties, for Segre-Veronese and Segre-Grassmann varieties that is to the context of mixed tensors.

In Section 5 we introduce a method to improve a classical result on Comon's conjecture. By standard arguments involving catalecticant matrices it is not hard to prove that Comon's conjecture holds for the general polynomial in $k[x_0,\ldots,x_n]_d$ of symmetric rank $h$ as soon as $h < \binom{n+\lfloor \frac{d}{2}\rfloor}{n}$, see Proposition 1. We manage to improve this bound looking for equations for the $(h-1)$-secant variety $\mathrm{Sec}_{h-1}(\mathcal{V}_d^n)$, not coming from catalecticant matrices, that are restrictions to the space of symmetric tensors of equations of the $(h-1)$-secant variety $\mathrm{Sec}_{h-1}(\mathcal{S}^{\underline{n}})$. We will do so by embedding the space of degree $d$ polynomials into the space of degree $d+1$ polynomials by mapping $F$ to $x_0 F$ and then considering suitable catalecticant matrices of $x_0 F$ rather than those of $F$ itself.

Implementing this method in Macaulay2 we are able to prove for instance that Comon's conjecture holds for the general cubic polynomial in $n+1$ variables of rank $h = n+1$ as long as $n \leqslant 30$. Please note that for cubics the usual flattenings work for $h \leqslant n$.

## 2. Notation

Let $\underline{n} = (n_1,\ldots,n_p)$ and $\underline{d} = (d_1,\ldots,d_p)$ be two $p$-uples of positive integers. Set

$$d = d_1 + \cdots + d_p, \; n = n_1 + \cdots + n_p, \text{ and } N(\underline{n},\underline{d}) = \prod_{i=1}^{p} \binom{n_i+d_i}{n_i}$$

Let $V_1,\ldots,V_p$ be vector spaces of dimensions $n_1+1 \leqslant n_2+1 \leqslant \cdots \leqslant n_p+1$, and consider the product

$$\mathbb{P}^{\underline{n}} = \mathbb{P}(V_1^*) \times \cdots \times \mathbb{P}(V_p^*).$$

The line bundle

$$\mathcal{O}_{\mathbb{P}^{\underline{n}}}(d_1,\ldots,d_p) = \mathcal{O}_{\mathbb{P}(V_1^*)}(d_1) \boxtimes \cdots \boxtimes \mathcal{O}_{\mathbb{P}(V_1^*)}(d_p)$$

induces an embedding

$$\sigma\nu_{\underline{d}}^{\underline{n}}: \mathbb{P}(V_1^*) \times \cdots \times \mathbb{P}(V_p^*) \longrightarrow \mathbb{P}(\mathrm{Sym}^{d_1} V_1^* \otimes \cdots \otimes \mathrm{Sym}^{d_p} V_p^*) = \mathbb{P}^{N(\underline{n},\underline{d})-1},$$
$$([v_1],\ldots,[v_p]) \longmapsto [v_1^{d_1} \otimes \cdots \otimes v_p^{d_p}]$$

where $v_i \in V_i$. We call the image

$$\mathcal{SV}_{\underline{d}}^{\underline{n}} = \sigma\nu_{\underline{d}}^{\underline{n}}(\mathbb{P}^{\underline{n}}) \subset \mathbb{P}^{N(\underline{n},\underline{d})-1}$$

a Segre-Veronese variety. It is a smooth variety of dimension $n$ and degree $\frac{(n_1+\cdots+n_p)!}{n_1!\cdots n_p!} d_1^{n_1}\cdots d_p^{n_p}$ in $\mathbb{P}^{N(\underline{n},\underline{d})-1}$.

When $p = 1$, $\mathcal{SV}_d^n$ is a Veronese variety. In this case, we write $\mathcal{V}_d^n$ for $\mathcal{SV}_d^n$, and $\nu_d^n$ for the Veronese embedding. When $d_1 = \cdots = d_p = 1$, $\mathcal{SV}_{1,\ldots,1}^{\underline{n}}$ is a Segre variety. In this case, we write $\mathcal{S}^{\underline{n}}$ for $\mathcal{SV}_{1,\ldots,1}^{\underline{n}}$, and $\sigma^{\underline{n}}$ for the Segre embedding. Please note that

$$\sigma\nu_{\underline{d}}^{\underline{n}} = \sigma^{\underline{n}'} \circ \left(\nu_{d_1}^{n_1} \times \cdots \times \nu_{d_p}^{n_p}\right),$$

where $\underline{n}' = (N(n_1, d_1) - 1, \ldots, N(n_p, d_p) - 1)$.

Similarly, given a $p$-uple of $k$-vector spaces $(V_1^{n_1}, \ldots, V_p^{n_p})$ and $p$-uple of positive integers $\underline{d} = (d_1, \ldots, d_p)$ we may consider the Segre-Plücker embedding

$$\sigma\pi_{\underline{d}}^{\underline{n}}: Gr(d_1, n_1) \times \cdots \times Gr(d_p, n_p) \longrightarrow \mathbb{P}(\bigwedge^{d_1} V_1^{n_1} \otimes \cdots \otimes \bigwedge^{d_p} V_p^{n_p}) = \mathbb{P}^{N(\underline{n},\underline{d})-1},$$
$$([H_1], \ldots, [H_p]) \longmapsto [H_1 \otimes \cdots \otimes H_p]$$

where $N(\underline{n}, \underline{d}) = \prod_{i=1}^{p} \binom{n_i}{d_i}$. We call the image

$$SG_{\underline{d}}^{\underline{n}} = \sigma\pi_{\underline{d}}^{\underline{n}}(Gr(d_1, n_1) \times \cdots \times Gr(d_p, n_p)) \subset \mathbb{P}^{N(\underline{n},\underline{d})}$$

a *Segre-Grassmann variety*.

### 2.1. Flattenings

Let $V_1, \ldots, V_p$ be $k$-vector spaces of finite dimension, and consider the tensor product $V_1 \otimes \ldots \otimes V_p = (V_{a_1} \otimes \ldots \otimes V_{a_s}) \otimes (V_{b_1} \otimes \ldots \otimes V_{b_{p-s}}) = V_A \otimes V_B$ with $A \cup B = \{1, \ldots, p\}$, $B = A^c$. Then we may interpret a tensor

$$T \in V_1 \otimes \ldots \otimes V_p = V_A \otimes V_B$$

as a linear map $\widetilde{T}: V_A^* \to V_{A^c}$. Clearly, if the rank of $T$ is at most $r$ then the rank of $\widetilde{T}$ is at most $r$ as well. Indeed, a decomposition of $T$ as a linear combination of $r$ rank one tensors yields a linear subspace of $V_{A^c}$, generated by the corresponding rank one tensors, containing $\widetilde{T}(V_A^*) \subseteq V_{A^c}$. The matrix associated with the linear map $\widetilde{T}$ is called an $(A, B)$-*flattening* of $T$.

In the case of mixed tensors we can consider the embedding

$$\text{Sym}^{d_1} V_1 \otimes \ldots \otimes \text{Sym}^{d_p} V_p \hookrightarrow V_A \otimes V_B$$

where $V_A = \text{Sym}^{a_1} V_1 \otimes \ldots \otimes \text{Sym}^{a_p} V_p$, $V_B = \text{Sym}^{b_1} V_1 \otimes \ldots \otimes \text{Sym}^{b_p} V_p$, with $d_i = a_i + b_i$ for any $i = 1, \ldots, p$. In particular, if $n = 1$ we may interpret a tensor $F \in \text{Sym}^{d_1} V_1$ as a degree $d_1$ homogeneous polynomial on $\mathbb{P}(V_1^*)$. In this case, the matrix associated with the linear map $\widetilde{F}: V_A^* \to V_B$ is nothing but the $a_1$-th *catalecticant matrix* of $F$, that is the matrix whose rows are the coefficient of the partial derivatives of order $a_1$ of $F$.

Similarly, by considering the inclusion

$$\bigwedge^{d_1} V_1 \otimes \ldots \otimes \bigwedge^{d_p} V_p \hookrightarrow V_A \otimes V_B$$

where $V_A = \bigwedge^{a_1} V_1 \otimes \ldots \otimes \bigwedge^{a_p} V_p$, $V_B = \bigwedge^{b_1} V_1 \otimes \ldots \otimes \bigwedge^{b_p} V_p$, with $d_i = a_i + b_i$ for any $i = 1, \ldots, p$, we get the so called *skew-flattenings*. We refer to [16] for details on the subject.

**Remark 1.** *The partial derivatives of an homogeneous polynomials are particular flattenings. The partial derivatives of a polynomial $F \in k[x_0, \ldots, x_n]_d$ are $\binom{n+s}{n}$ homogeneous polynomials of degree $d - s$ spanning a linear space $H_{\partial^s F} \subseteq \mathbb{P}(k[x_0, \ldots, x_n]_{d-s})$.*

*If $F \in k[x_0, \ldots, x_n]_d$ admits a decomposition as in (1) then $F \in \text{Sec}_h(V_d^n)$, and conversely a general $F \in \text{Sec}_h(V_d^n)$ can be written as in (1). If $F = \lambda_1 L_1^d + \ldots + \lambda_h L_h^d$ is a decomposition then the partial derivatives of order $s$ of $F$ can be decomposed as linear combinations of $L_1^{d-s}, \ldots, L_h^{d-s}$ as well. Therefore, the linear space $\langle L_1^{d-s}, \ldots, L_h^{d-s} \rangle$ contains $H_{\partial^s F}$.*

## 2.2. Rank and Border Rank

Let $X \subset \mathbb{P}^N$ be an irreducible and reduced non-degenerate variety. We define the rank $\mathrm{rank}_X(p)$ with respect to $X$ of a point $p \in \mathbb{P}^N$ as the minimal integer $h$ such that there exist $h$ points in linear general position $x_1, \ldots, x_h \in X$ with $p \in \langle x_1, \ldots, x_h \rangle$. Clearly, if $Y \subseteq X$ we have that

$$\mathrm{rank}_X(p) \leqslant \mathrm{rank}_Y(p) \tag{2}$$

The border rank $\underline{\mathrm{rank}}_X(p)$ of $p \in \mathbb{P}^N$ with respect to $X$ is the smallest integer $r > 0$ such that $p$ is in the Zariski closure of the set of points $q \in \mathbb{P}^N$ such that $\mathrm{rank}_X(q) = r$. In particular $\underline{\mathrm{rank}}_X(p) \leqslant \mathrm{rank}_X(p)$.

Recall that given an irreducible and reduced non-degenerate variety $X \subset \mathbb{P}^N$, and a positive integer $h \leqslant N$ the $h$-secant variety $\mathrm{Sec}_h(X)$ of $X$ is the subvariety of $\mathbb{P}^N$ obtained as the Zariski closure of the union of all $(h-1)$-planes spanned by $h$ general points of $X$.

In other words $\underline{\mathrm{rank}}_X(p)$ is computed by the smallest secant variety $\mathrm{Sec}_h(X)$ containing $p \in \mathbb{P}^N$.

Now, let $Y, Z$ be subvarieties of an irreducible projective variety $X \subset \mathbb{P}^N$, spanning two linear subspaces $\mathbb{P}^{N_Y} := \langle Y \rangle, \mathbb{P}^{N_Z} := \langle Z \rangle \subseteq \mathbb{P}^N$. Fix two points $p_Y \in \mathbb{P}^{N_Y}, p_Z \in \mathbb{P}^{N_Z}$, and consider a point $p \in \langle p_Y, p_Z \rangle$. Clearly

$$\mathrm{rank}_X(p) \leqslant \mathrm{rank}_Y(p_Y) + \mathrm{rank}_Z(p_Z) \tag{3}$$

## 3. Comon's Conjecture

It is natural to ask under which assumptions (2) is indeed an equality. Consider the Segre-Veronese embedding $\sigma \nu_{\underline{d}}^n : \mathbb{P}(V_1^*) \times \cdots \times \mathbb{P}(V_p^*) \rightarrow \mathbb{P}(\mathrm{Sym}^{d_1} V_1^* \otimes \cdots \otimes \mathrm{Sym}^{d_p} V_p^*) = \mathbb{P}^{N(n,\underline{d})-1}$ with $V_1 \cong \cdots \cong V_p \cong V$ $k$-vector spaces of dimension $n+1$. Its composition with the diagonal embedding $i : \mathbb{P}(V^*) \rightarrow \mathbb{P}(V_1^*) \times \cdots \times \mathbb{P}(V_p^*)$ is the Veronese embedding of $\nu_d^n$ of degree $d = d_1 + \cdots + d_p$. Let $\mathcal{V}_{\underline{d}}^n \subseteq \mathcal{SV}_{\underline{d}}^n$ be the corresponding Veronese variety. We will denote by $\Pi_{n,\underline{d}}$ the linear span of $\mathcal{V}_{\underline{d}}^n$ in $\mathbb{P}^{N(n,\underline{d})-1}$.

In the notations of Section 2.2 set $X = \mathcal{SV}_{\underline{d}}^n$ and $Y = \mathcal{V}_d^n$. For any symmetric tensor $T \in \Pi_{n,\underline{d}}$ we may consider its symmetric rank $\mathrm{srk}(T) := \mathrm{rank}_{\mathcal{V}_d^n}(T)$ and its rank $\mathrm{rank}(T) := \mathrm{rank}_{\mathcal{SV}_{\underline{d}}^n}(T)$ as a mixed tensor. Comon's conjecture predicts that in this particular setting the inequality (2) is indeed an equality [3].

**Conjecture 1** (Comon's). *Let $T$ be a symmetric tensor. Then $\mathrm{rank}(T) = \mathrm{srk}(T)$.*

Conjecture 1 has been generalized in several directions for complex border rank, real rank and real border rank, see Section 5.7.2 in [16] for a full overview.

Please note that when $d = 2$ Comon's conjecture is true. Indeed, $\mathrm{Sec}_h(S^n)$ is cut out by the size $(h+1) \times (h+1)$ minors of a general square matrix and $\mathrm{Sec}_h(\mathcal{V}_2^n)$ is cut out by the size $(h+1) \times (h+1)$ minors of a general symmetric matrix, that is $\mathrm{Sec}_h(\mathcal{V}_2^n) = \mathrm{Sec}_h(S^n) \cap \Pi_{n,2}$.

Conjecture 1 has been proved in several special cases. For instance, when the symmetric rank is at most two [3], when the rank is less than or equal to the order [17], for tensors belonging to tangential varieties to Veronese varieties [18], for tensors in $\mathbb{C}^2 \otimes \mathbb{C}^n \otimes \mathbb{C}^n$ [19], when the rank is at most the flattening rank plus one [20], for the so called Coppersmith–Winograd tensors [21], for symmetric tensors in $\mathbb{C}^4 \otimes \mathbb{C}^4 \otimes \mathbb{C}^4$ and also for symmetric tensors of symmetric rank at most seven in $\mathbb{C}^n \otimes \mathbb{C}^n \otimes \mathbb{C}^n$ [22].

On the other hand, a counter-example to Comon's conjecture has recently been found by Y. Shitov [23]. The counter-example consists of a symmetric tensor $T$ in $\mathbb{C}^{800} \times \mathbb{C}^{800} \times \mathbb{C}^{800}$ which can be written as a sum of 903 rank one tensors but not as a sum of 903 symmetric rank one tensors. It is important to stress that for this tensor $T$ rank and border rank are quite different. Comon's conjecture for border ranks is still completely open (Problem 25 in [23]).

Even though it has been recently proven false in full generality, we believe that Comon's conjecture is true for a general symmetric tensor, perhaps it is even true for those tensor for which rank $T = $ rank $T$.

In what follows we use simple arguments based on flattenings to give sufficient conditions for Comon's conjecture, recovering a known result, and its skew-symmetric analogue.

**Lemma 1.** *The tensors $T \in \text{Sec}_h(SV_{\underline{d}}^n)$ such that $\dim(\widetilde{T}(V_A^*)) \leq h - 1$ for a given flattening $\widetilde{T}$ form a proper closed subset of $\text{Sec}_h(SV_{\underline{d}}^n)$. Furthermore, the same result holds if we replace the Segre-Veronese variety $SV_{\underline{d}}^n$ with the Segre-Grassmann variety $SG_{\underline{d}}^n$.*

**Proof.** Let $T \in \text{Sec}_h(SV_{\underline{d}}^n)$ be a general point. Assume that $\dim(\widetilde{T}(V_A^*)) \leq h - 1$. This condition forces the $(A, B)$-flattening matrix to have rank at most $h - 1$. On the other hand, by Proposition 4.1 in [24] these minors do not vanish on $\text{Sec}_h(SV_{\underline{d}}^n)$, and therefore define a proper closed subset of $\text{Sec}_h(SV_{\underline{d}}^n)$. In the Segre-Grassmann setting we argue in the same way by using skew-flattenings. □

**Proposition 1.** [25] *For any integer $h < \binom{n+\lfloor \frac{d}{2} \rfloor}{n}$ there exists an open subset $\mathcal{U}_h \subseteq \text{Sec}(V_n^d)$ such that for any $T \in \mathcal{U}_h$ the rank and the symmetric rank of $T$ coincide, that is*

$$\text{rank}(T) = \text{srk}(T)$$

**Proof.** First of all, note that we always have $\text{rank}(T) \leq \text{srk}(T)$. Furthermore, Section 2.1 yields that for any $(A, B)$-flattening $\widetilde{T} : V_A^* \to V_B$ the inequality $\text{rank}(T) \geq \dim(\widetilde{T}(V_A^*))$ holds. Since $T$ is symmetric and its catalecticant matrices are particular flattenings we get that $\text{rank}(T) \geq \dim(H_{\partial^s T})$ for any $s \geq 0$.

Now, for a general $T \in \text{Sec}_h(V_d^n)$ we have $\text{srk}(T) = h$, and if $h < \binom{n+\bar{s}}{n}$, where $\bar{s} = \lfloor \frac{d}{2} \rfloor$, then Lemma 1 yields $\dim(H_{\partial^{\bar{s}} T}) = h$. Therefore, under these conditions we have the following chain of inequalities

$$\dim(H_{\partial^{\bar{s}} T}) \leq \text{rank}(T) \leq \text{srk}(T) = \dim(H_{\partial^{\bar{s}} T})$$

and hence $\text{rank}(T) = \text{srk}(T)$. □

Now, consider the Segre-Plücker embedding $\mathbb{P}(V_1) \times \ldots \times \mathbb{P}(V_p) \to \mathbb{P}(\bigwedge^{d_1} V_1 \otimes \cdots \otimes \bigwedge^{d_p} V_p) = \mathbb{P}^{N(\underline{n},\underline{d})-1}$ with $V_1 \cong \ldots \cong V_p \cong V$ $k$-vector spaces of dimension $n+1$. Its composition with the diagonal embedding $i : \mathbb{P}(V) \to \mathbb{P}(V_1) \times \cdots \times \mathbb{P}(V_p)$ is the Plücker embedding of $Gr(d, n)$ with $d = d_1 + \ldots + d_p$. Let $Gr(d, n) \subseteq SG_{\underline{d}}^n$ be the corresponding Grassmannian and let us denote by $\Pi_{n,\underline{d}}$ its linear span in $\mathbb{P}^{N(\underline{n},\underline{d})-1}$.

For any skew-symmetric tensor $T \in \Pi_{n,\underline{d}}$ we may consider its skew rank $\text{skrk}(T)$ that is its rank with respect to the Grassmannian $Gr(d, n) \subseteq \Pi_{n,\underline{d}}$, and its rank $\text{rank}(T)$ as a mixed tensor. Playing the same game as in Proposition 1 we have the following.

**Proposition 2.** *For any integer $h < \binom{n}{\lfloor \frac{d}{2} \rfloor}$ there exists an open subset $\mathcal{U}_h \subseteq \text{Sec}_h(Gr(d, n))$ such that for any $T \in \mathcal{U}_h$ the rank and the skew rank of $T$ coincide, that is*

$$\text{rank}(T) = \text{skrk}(T)$$

**Proof.** As before for any tensor $T$ we have $\text{rank}(T) \leq \text{skrk}(T)$. For any $(A, B)$-skew-flattening $\widetilde{T} : V_A^* \to V_B$ we have $\text{skrk}(T) \geq \dim(\widetilde{T}(V_A^*))$. Furthermore, since $\widetilde{T}$ is in particular a flattening also the inequality $\text{rank}(T) \geq \dim(\widetilde{T}(V_A^*))$ holds.

Now, for a general $T \in \text{Sec}_h(Gr(d, n))$ we have $\text{skrk}(T) = h$, and if $h < \binom{n}{\bar{s}}$, where $\bar{s} = \lfloor \frac{d}{2} \rfloor$, Lemma 1 yields $\text{skrk}(T) = \dim(\widetilde{T}_{\bar{s}}(V_A^*))$, where $\widetilde{T}_{\bar{s}}$ is the skew-flattening corresponding to the partition $(\bar{s}, d - \bar{s})$ of $d$. Therefore, we deduce that

$$\dim(\widetilde{T}_{\bar{s}}(V_A^*)) \leq \text{rank}(T) \leq \text{skrk}(T) = \dim(\widetilde{T}_{\bar{s}}(V_A^*))$$

and hence rank($T$) = skrk($T$). □

**Remark 2.** *Propositions 1 and 2 suggest that whenever we are able to write determinantal equations for secant varieties we are able to verify Comon's conjecture. We conclude this section suggesting a possible way to improve the range where the general Comon's conjecture holds giving a conjectural way to produce determinantal equations for some secant varieties.*

Set $\underline{n} = (n, \ldots, n)$, $(d+1)$-times, $\underline{n}_1 = (n, \ldots, n)$, $d$-times, and consider the corresponding Segre varieties $X := S^{\underline{n}}$, $X_1 := S^{\underline{n}_1}$ and Veronese varieties $Y = V^n_{d+1}$, $Y_1 := V^n_d$. Fix the polynomial $x_0^{d+1} \in Y$ and let $\Pi$ be the linear space spanned by the polynomials of the form $x_0 F$, where $F$ is a polynomial of degree $d$. This allow us to see $Y_1 \subseteq \Pi$. Please note that polynomials of the form $x_0 L_1^d$ lie in the tangent space of $Y$ at $L_1^{d+1}$, and therefore $\text{rank}_Y(x_0 L^{\otimes d}) = 2$.

Hence for a polynomial $F$ of degree $d$ we have $\text{rank}_Y(x_0 F) \leq 2 \text{rank}_{Y_1}(F)$. Our aim is to understand when the equality holds.

We may mimic the same construction for the Segre varieties $X$ and $X_1$, and use determinantal equations for the secant varieties of $X_1$ to give determinantal equations of the secant varieties of $X$ and henceforth conclude Comon's conjecture. In particular, as soon as $d$ is odd and $d < n$, this produces new determinantal equations for $\text{Sec}_h(X_1)$ and $\text{Sec}_h(Y_1)$ with $2h < \binom{n+\frac{d+1}{2}}{n}$. Therefore, this would give new cases in which the general Comon's conjecture holds. Unfortunately, we are only able to successfully implement this procedure in very special cases, see Section 5.

## 4. Strassen's Conjecture

Another natural problem consists in giving hypotheses under which in Equation (3) equality holds. Consider the triple Segre embedding $\sigma^{\underline{n}} : \mathbb{P}(V_1^*) \times \mathbb{P}(V_2^*) \times \mathbb{P}(V_3^*) = \mathbb{P}^a \times \mathbb{P}^b \times \mathbb{P}^c \to \mathbb{P}(V_1^* \otimes V_2^* \otimes V_1^*) = \mathbb{P}^{N(\underline{n},d)-1}$, and let $S^{\underline{n}}$ be the corresponding Segre variety. Now, take complementary subspaces $\mathbb{P}^{a_1}, \mathbb{P}^{a_2} \subset \mathbb{P}^a$, $\mathbb{P}^{b_1}, \mathbb{P}^{b_2} \subset \mathbb{P}^b$, $\mathbb{P}^{c_1}, \mathbb{P}^{c_2} \subset \mathbb{P}^c$, and let $S^{(a_1,b_1,c_1)}, S^{(a_2,b_2,c_2)}$ be the Segre varieties associated respectively to $\mathbb{P}^{a_1} \times \mathbb{P}^{b_1} \times \mathbb{P}^{c_1}$ and $\mathbb{P}^{a_2} \times \mathbb{P}^{b_2} \times \mathbb{P}^{c_2}$.

In the notations of Section 2.2 set $X = S^{\underline{n}}$, $Y = S^{(a_1,b_1,c_1)}$ and $Z = S^{(a_1,b_1,c_1)}$. Strassen's conjecture states that the additivity of the rank holds for triple tensors, or in onther words that in this setting the inequality (3) is indeed an equality [15].

**Conjecture 2** (Strassen's). *In the above notation let $T_1 \in \left\langle S^{(a_1,b_1,c_1)} \right\rangle$, $T_2 \in \left\langle S^{(a_2,b_2,c_2)} \right\rangle$ be two tensors. Then $\text{rank}(T_1 \oplus T_2) = \text{rank}(T_1) + \text{rank}(T_2)$.*

Even though Conjecture 2 was originally stated in the context of triple tensors that is bilinear forms, with particular attention to the complexity of matrix multiplication, several generalizations are immediate. For instance, we could ask the same question for higher order tensors, symmetric tensors, mixed tensors and skew-symmetric tensors. It is also natural to ask for the analogue of Conjecture 2 for border rank. This has been answered negatively [26].

Conjecture 2 and its analogues have been proven when either $T_1$ or $T_2$ has dimension at most two, when rank($T_1$) can be determined by the so called substitution method [21], when dim($V_1$) = 2 both for the rank and the border rank [27], when $T_1, T_2$ are symmetric that is homogeneous polynomials in disjoint sets of variables, either $T_1, T_2$ is a power, or both $T_1$ and $T_2$ have two variables, or either $T_1$ or $T_2$ has small rank [28], and also for other classes of homogeneous polynomials [29,30].

As for Comon's conjecture a counterexample to Strassen's conjecture has recently been given by Y. Shitov [31]. In this case Y. Shitov proved that over any infinite field there exist tensors $T_1, T_2$ such that the inequality in Conjecture 2 is strict.

In what follows, we give sufficient conditions for Strassen's conjecture, recovering a known result, and for its mixed and skew-symmetric analogues.

**Proposition 3.** [25] Let $V_1, V_2$ be k-vector spaces of dimensions $n+1, m+1$, and consider $V = V_1 \oplus V_2$. Let $F \in \operatorname{Sym}^d(V_1) \subset \operatorname{Sym}^d(V)$ and $G \in \operatorname{Sym}^d(V_2) \subset \operatorname{Sym}^d(V)$ be two homogeneous polynomials. If there exists an integer $s > 0$ such that
$$\dim(H_{\partial^s F}) = \operatorname{srk}(F), \quad \dim(H_{\partial^s G}) = \operatorname{srk}(G)$$
then $\operatorname{srk}(F + G) = \operatorname{srk}(F) + \operatorname{srk}(G)$.

**Proof.** Clearly, $\operatorname{srk}(F + G) \leq \operatorname{srk}(F) + \operatorname{srk}(G)$ holds in general. On the other hand, our hypothesis yields
$$\operatorname{srk}(F) + \operatorname{srk}(G) = \dim(H_{\partial^s F}) + \dim(H_{\partial^s F}) = \dim(H_{\partial^s F+G}) \leq \operatorname{srk}(F+G)$$
where the last inequality follows from Remark 1. □

**Remark 3.** The argument used in the proof of Proposition 3 works for $F \in \mathbb{P}^{N(n,d)}$ general only if for the generic rank we have $\lfloor \frac{\binom{n+d}{d}}{n+1} \rfloor \leq \binom{n+\lfloor \frac{d}{2} \rfloor}{n}$. For instance, when $n = 3, d = 6$ the generic rank is 21 while the maximal dimension of the spaces spanned by partial derivatives is 20.

**Proposition 4.** Let $V_1, \ldots, V_p$ and $W_1, \ldots, W_p$ be k-vector spaces of dimension $n_1 + 1, \ldots, n_p + 1$ and $m_1 + 1, \ldots, m_p + 1$ respectively. Consider $U_i = V_i \oplus W_i$ for every $1 \leq i \leq p$. Let $T_1 \in \operatorname{Sym}^{d_1} V_1 \otimes \cdots \otimes \operatorname{Sym}^{d_p} V_p \subset \operatorname{Sym}^{d_1} U_1 \otimes \cdots \otimes \operatorname{Sym}^{d_p} U_p$ and $T_2 \in \operatorname{Sym}^{d_1} W_1 \otimes \cdots \otimes \operatorname{Sym}^{d_p} W_p \subset \operatorname{Sym}^{d_1} U_1 \otimes \cdots \otimes \operatorname{Sym}^{d_p} U_p$ be two mixed tensors.

If for any $i \in \{1, ..., p\}$ there exists a pair $(a_i, b_i)$ with $a_i + b_i = d_i$ and $(A, B)$-flattenings $\tilde{T}_1 : V_A^* \to V_B$, $\tilde{T}_2 : V_A^* \to V_B$ as in (Section 2.1) such that
$$\dim(\tilde{T}_1(V_A^*)) = \operatorname{rank}(T_1), \quad \dim(\tilde{T}_2(V_A^*)) = \operatorname{rank}(T_2)$$
then $\operatorname{rank}(T_1 + T_2) = \operatorname{rank}(T_1) + \operatorname{rank}(T_2)$.

**Proof.** Clearly, $\operatorname{rank}(T_1 + T_2) \leq \operatorname{rank}(T_1) + \operatorname{rank}(T_2)$. On the other hand, our hypothesis yields
$$\operatorname{rank}(T_1) + \operatorname{rank}(T_2) = \dim(\tilde{T}_1(V_A^*)) + \dim(\tilde{T}_2(V_A^*)) = \dim(\widetilde{T_1 + T_2}(V_A^*)) \leq \operatorname{rank}(T_1 + T_2)$$
where $\widetilde{T_1 + T_2}$ denotes the $(A, B)$-flattening of the mixed tensor $T_1 + T_2$. □

Arguing as in the proof of Proposition 4 with skew-symmetric flattenings we have an analogous statement in the Segre-Grassmann setting.

**Proposition 5.** Let $V_1, \ldots, V_p$ and $W_1, \ldots, W_p$ be k-vector spaces of dimension $n_1 + 1, \ldots, n_p + 1$ and $m_1 + 1, \ldots, m_p + 1$ respectively. Consider $U_i = V_i \oplus W_i$ for every $1 \leq i \leq p$, and let $T_1 \in \bigwedge^{d_1} V_1 \otimes \cdots \otimes \bigwedge^{d_p} V_p \subset \bigwedge^{d_1} U_1 \otimes \cdots \otimes \bigwedge^{d_p} U_p$ and $T_2 \in \bigwedge^{d_1} W_1 \otimes \cdots \otimes \bigwedge^{d_p} W_p \subset \bigwedge^{d_p} U_1 \otimes \cdots \otimes \bigwedge^{d_p} U_p$ be two skew-symmetric tensors with $d_i \leq \min\{n_i + 1, m_i + 1\}$.

If for any $i \in \{1, \ldots, p\}$ there exists a pair $(a_i, b_i)$ with $a_i + b_i = d_i$ and $(A, B)$-skew-flattenings $\tilde{T}_1 : V_A^* \to V_B$, $\tilde{T}_2 : V_A^* \to V_B$ as in (Section 2.1) such that
$$\dim(\tilde{T}_1(V_A^*)) = \operatorname{rank}(T_1), \quad \dim(\tilde{T}_2(V_A^*)) = \operatorname{rank}(T_2)$$
then $\operatorname{rank}(T_1 + T_2) = \operatorname{rank}(T_1) + \operatorname{rank}(T_2)$.

## 5. On the Rank of $x_0 F$

In this section, building on Remark 2, we present new cases in which Comon's conjecture holds. Recall, that for a smooth point $x \in X$, the a-osculating space $\mathbb{T}_x^a X$ of $X$ at $x$ is roughly the smaller linear subspace locally approximating $X$ up to order $a$ at $x$, and the $a$-osculating variety $T^a X$ of $X$ is defined as the closure of the union of all the osculating spaces

$$T^a X = \overline{\bigcup_{x \in X} T^a_x X}$$

For any $1 \leq a \leq d - 1$ the osculating space $\mathbb{T}^a_{[L^d]} \mathcal{V}^n_d$ of order $a$ at the point $[L^d] \in \mathcal{V}_d$ can be written as

$$\mathbb{T}^a_{[L^d]} \mathcal{V}^n_d = \left\langle L^{d-a} F \mid F \in k[x_0, \ldots, x_n]_a \right\rangle \subseteq \mathbb{P}^N$$

Equivalently, $\mathbb{T}^a_{[L^d]} \mathcal{V}^n_d$ is the space of homogeneous polynomials whose derivatives of order less than or equal to $a$ in the direction given by the linear form $L$ vanish. Please note that $\dim(\mathbb{T}^a_{[L^d]} \mathcal{V}^n_d) = \binom{n+a}{n} - 1$ and $\mathbb{T}^b_{[L^d]} \mathcal{V}^n_d \subseteq \mathbb{T}^a_{[L^d]} \mathcal{V}^n_d$ for any $b \leq a$. Moreover, for any $1 \leq a \leq d$ and $[L^d] \in \mathcal{V}^n_d$ we can embed a copy of $\mathcal{V}^n_a$ into the osculating space $\mathbb{T}^a_{[L^d]} \mathcal{V}^n_d$ by considering

$$\mathcal{V}^n_a = \{L^{d-a} M^a \mid M \in k[x_0, \ldots, x_n]_1\} \subseteq \mathbb{T}^a_{[L^d]} \mathcal{V}^n_d$$

**Remark 4.** Let us expand the ideas in Remark 2. We can embed

$$\mathcal{V}^n_d = \{x_0 L^d \mid L \in k[x_0, \ldots, x_n]_1\} \subseteq \mathbb{T}^d_{[x_0^d]} \mathcal{V}^n_{d+1}$$

and Remark 2 yields that

$$\mathrm{Sec}_h(\mathcal{V}^n_d) \subseteq \mathrm{Sec}_{2h}(\mathcal{V}^n_{d+1}) \cap \mathbb{T}^d_{[L^{d+1}]} \mathcal{V}^n_{d+1} \tag{4}$$

This embedding extends to an embedding at the level of Segre varieties, and, in the notation of Remark 2, we have that $\mathrm{Sec}_h(S^{n_1}) \subseteq \mathrm{Sec}_{2h}(S^n)$.

Assume that for a polynomial $F \in \mathrm{Sec}_h(\mathcal{V}^n_d)$ we have $F \in \mathrm{Sec}_{h-1}(S^{n_1})$. Then $x_0 F \in \mathrm{Sec}_{2h-2}(S^n)$. Now, if we find a determinantal equation of $\mathrm{Sec}_{2h-2}(\mathcal{V}^n_{d+1})$ coming as the restriction to $\Pi$, the space of symmetric tensors, of a determinantal equation of $\mathrm{Sec}_{2h-2}(S^n)$, and not vanishing at $x_0 F$ then $x_0 F \notin \mathrm{Sec}_{2h-2}(S^n)$ and hence $F \notin \mathrm{Sec}_{h-1}(S^{n_1})$ proving Comon's conjecture for $F$.

This will be the leading idea to keep in mind in what follows. The determinantal equations involved will always come from minors of suitable catalecticant matrices, that can be therefore seen as the restriction to $\Pi$ of determinantal equations for the secants of the Segre coming from non symmetric flattenings.

It is easy to give examples where the inequality (4) is strict. When $n = 1$ the generic rank is $g_d = \lceil \frac{d+1}{2} \rceil$. Then for $d$ odd we have $g_d = g_{d-1}$ while for $d$ even we have $g_d = g_{d-1} + 1$. Hence $\mathrm{rank}_{\mathcal{V}_d} x_0 F < 2 \mathrm{rank}_{\mathcal{V}_{d-1}} F$ if $2 \mathrm{rank}_{\mathcal{V}_{d-1}} F > \frac{g_d}{2}$, where $\mathcal{V}_d := \mathcal{V}^1_d$ is the rational normal curve. It is natural to ask if the inequality is indeed an equality as long as the rank is subgeneric. In the case $n = 1$ we have the following result.

**Proposition 6.** Let $\mathcal{V}_d := \mathcal{V}^1_d$ be the degree $d$ rational normal curve. If $2h < g_{d+1}$ then there does not exist $k_h > 0$ such that $\mathrm{Sec}_h(\mathcal{V}_d) \subseteq \mathrm{Sec}_{2h-k_h}(\mathcal{V}_{d+1}) \cap \mathbb{T}^d_{[x^{d+1}]} \mathcal{V}_{d+1}$.

**Proof.** Clearly, it is enough to prove the statement for $k_h = 1$. Let $p \in \mathrm{Sec}_h(\mathcal{V}_d)$ be a general point. Then $p \in \left\langle [x_0 L^d_1], \ldots, [x_0 L^d_h] \right\rangle$ with $L_i$ general linear forms. In particular

$$p \in H := \left\langle \mathbb{T}^{d+1}_{[L^{d+1}_1]} \mathcal{V}_{d+1}, \ldots, \mathbb{T}^{d+1}_{[L^{d+1}_h]} \mathcal{V}_{d+1} \right\rangle$$

Please note that $\dim(H) = 2h - 1$. Now, assume that $p$ is contained also in $\mathrm{Sec}_{2h-1}(\mathcal{V}_{d+1})$. Then there exists a linear subspace $H' \subset \mathbb{P}^{d+1}$ of dimension $2h - 2$ passing through $p$ intersecting $\mathcal{V}_{d+1}$ at $2h - 1$ points $q_1, \ldots, q_r$ counted with multiplicity. Let $q_{i_1}, \ldots, q_{i_r}$ be the points among the $q_i$ coinciding with some of the $[L_i^{d+1}]$ and such that the intersection multiplicity of $H'$ and $\mathcal{V}_{d+1}$ at $q_{i_j}$ is one, and $q_{j_1}, \ldots, q_{j_r}$ be the points among the $q_i$ coinciding with some of the $[L_i^{d+1}]$ and such that the intersection multiplicity of $H'$ and $\mathcal{V}_d$ at $q_{j_k}$ is greater that or equal to two.

Set $\Pi := \langle H, H' \rangle$, then $\dim(\Pi) = 2h - 1 + 2h - 2 - i_r - 2j_r$ and $\Pi$ intersects $\mathcal{V}_{d+1}$ at $2h + (2h - 1 - i_r - 2j_r)$ points counted with multiplicity. Consider general points $b_1, \ldots, b_s \in \mathcal{V}_{d+1}$ with $s = i_r + 2j_r$, and the linear space $\Pi' = \langle \Pi, b_1, \ldots, b_s \rangle$. Therefore, $\dim(\Pi') = 4h - 3$ and $\Pi'$ intersects $\mathcal{V}_{d+1}$ at $4h - 1$ points counted with multiplicity. Since $2h \leq \frac{d+3}{2}$ adding enough general points to $\Pi'$ we may construct a hyperplane in $\mathbb{P}^{d+1}$ intersecting $\mathcal{V}_{d+1}$ at $d + 2$ points counted with multiplicity, a contradiction. □

Proposition 6 can be applied to get results on the rank of a special class of matrices called Hankel matrices.

Let $F = Z_0 x_0^d + \ldots + \binom{d}{d-i} Z_i x_0^{d-i} x_1^i + \ldots + Z_d x_1^d$ be a binary form and consider $[Z_0, \ldots, Z_d]$ as homogeneous coordinates on $\mathbb{P}(k[x_0, x_1]_d)$. Furthermore, consider the matrices

$$M_{2n} = \begin{pmatrix} Z_0 & \cdots & Z_n \\ \vdots & \ddots & \vdots \\ Z_n & \cdots & Z_d \end{pmatrix}, \quad M_{2n+1} = \begin{pmatrix} Z_0 & \cdots & Z_n \\ \vdots & \ddots & \vdots \\ Z_{n+1} & \cdots & Z_d \end{pmatrix}$$

It is well known that the ideal of $\mathrm{Sec}_h(\mathcal{V}_d)$ is cut out by the minors of $M_d$ of size $(h+1) \times (h+1)$ [4]. Now, consider a polynomial $F \in k[x_0, x_1]_d$ with homogeneous coordinates $[Z_0, \ldots, Z_d]$. Then $F' := x_0 F \in k[x_0, x_1]_{d+1}$ has homogeneous coordinates $[Z'_0, \ldots, Z'_{d+1}]$ with

$$Z'_i = \frac{d+1-i}{d+1} Z_i$$

To determine the rank of $F'$ we have to relate the rank of the matrices

$$N_{2n} = \begin{pmatrix} Z_0 & \frac{d}{d+1} Z_1 & \cdots & \frac{d+1-n}{d+1} Z_n \\ \frac{d}{d+1} Z_1 & \cdots & \cdots & \frac{d-n}{d+1} Z_{n+1} \\ \vdots & \ddots & \ddots & \vdots \\ \frac{d-n+2}{d+1} Z_{n-1} & \cdots & \cdots & \frac{1}{d+1} Z_d \\ \frac{d-n+1}{d+1} Z_n & \cdots & \frac{1}{d+1} Z_d & 0 \end{pmatrix}$$

$$N_{2n+1} = \begin{pmatrix} Z_0 & \frac{d}{d+1} Z_1 & \cdots & \frac{d-n}{d+1} Z_n \\ \frac{d}{d+1} Z_1 & \cdots & \cdots & \frac{d-n-1}{d+1} Z_{n+2} \\ \vdots & \ddots & \ddots & \vdots \\ \frac{d-n+2}{d+1} Z_n & \cdots & \cdots & \frac{1}{d+1} Z_d \\ \frac{d-n+1}{d+1} Z_{n+1} & \cdots & \frac{1}{d+1} Z_d & 0 \end{pmatrix}$$

with the rank of $M_d$.

**Definition 1.** *A matrix $A = (A_{i,j}) \in M(a, b)$ such that $A_{i,j} = A_{h,k}$ whenever $i + j = h + k$ is called a Hankel matrix.*

In particular all the matrices of the form $M_d$ and $N_d$ considered above are Hankel matrices.

Let $M(a, b)$ be the vector space of $a \times b$ matrices with coefficients in the base field $k$. For any $h \leq \min\{a, b\}$ let $\mathrm{Rank}_r(M(a, b)) \subseteq M(a, b)$ be the subvariety consisting of all matrices of rank at most $h$. Now, consider the map $\beta : \mathbb{N} \longrightarrow \mathbb{N} \times \mathbb{N}$ given by $\beta(2n) = (n+1, n+1)$ and $\beta(2n+1) = (n+2, n+1)$. For any $d \geq 1$ we can view the subspace $H_d \subseteq M(\beta(d))$ formed by matrices of the form $M_d$ as the subspace of Hankel matrices. Now, given any linear morphism $f : M(a, b) \to M(c, d)$ we can ask if for some $s \leq \min\{c, d\}$ we have $f(\mathrm{Rank}_h(M(a, b))) \subseteq \mathrm{Rank}_s(M(c, d))$.

**Corollary 1.** *Consider the linear morphism*

$$\alpha_d : M(\beta(d)) \longrightarrow M(\beta(d+1))$$
$$(A_{i,j}) \longmapsto \left(\frac{d-(i+j-3)}{d+1} A_{i,j}\right)$$

*Then $\alpha_d(H_d) \subseteq H_{d+1}$ and $\alpha_d(Rank_h(M(\beta(d)) \cap M_d)) \subseteq Rank_{2h}(M(\beta(d+1))) \cap M_{d+1}$.*

**Proof.** Since $\alpha_d(A_{i,j}) = \alpha_d(A_{h,k})$ when $i+j = h+k$ we have that $\alpha_d(H_d) \subseteq H_{d+1}$. By Proposition 6 $Rank_h(M(\beta(d)) \cap H_d = Sec_h(V_d)$, and by construction $\alpha_d(M_d)$ is the linear change of coordinates mapping a binary form $F \in k[x_0, x_1]_d$ to $F' = x_0 F \in k[x_0, x_1]_{d+1}$.

Since $Sec_h(V_d) \subseteq Sec_{2h}(V_{d+1}) \cap \mathbb{T}^d_{[x^{d+1}]} V_{d+1}$, if an $h \times h$ minor of a general matrix $B$ in $M(\beta(d))$ does not vanish, under the assumption that all the $(h+1) \times (h+1)$ minors of $B$ vanish, then there is a $2h \times 2h$ minor of $\alpha_d(B)$ that does not vanish. □

When $n \geqslant 2$ we are able to determine, via Macaulay2 [32] aided methods, the rank of $x_0 F$ in some special cases.

i  $(n,d) = (2,2)$. The variety $Sec_3(V_3^2)$ is the hypersurface in $\mathbb{P}^9$ cut out by the Aronhold invariant, see for instance (Section 1.1 in [4]). With a Macaulay2 computation we prove that if $F \in Sec_2(V_2^2)$ is general then the Aronhold invariant does not vanish at $x_0 F$, hence rank $x_0 F = 2$ rank $F$.

ii  $(n,d) = (2,3)$. The varieties $Sec_5(V_4^2)$ and $Sec_3(V_3^2)$ are both hypersurfaces, given respectively by the determinant of the catalecticant matrix of second partial derivatives and the Aronhold invariant (Section 1.1 in [4]). With Macaulay2 we prove that the determinant of the second catalecticant matrix does not vanish at $x_0 F$ for $F \in Sec_3(V_3^2)$ general, hence rank $x_0 F = 2$ rank $F$.

iii  $(n,d) = (3,3)$. The secant variety $Sec_9(V_4^3)$ is the hypersurface cut out by the second catalecticant matrix (Section 1.1 in [4]) while $Sec_5(V_3^3)$ is the entire osculating space. A Macaulay2 computation shows that $\mathbb{T}^3_{[x_0^4]} V_4^3 \subseteq Sec_9(V_4^3)$. This proves that rank $x_0 F < 2$ rank $F$, for $F$ general.

iv  $(n,d) = (4,3)$. In this case $Sec_8(V_3^4) = \mathbb{T}^3_{[x_0^4]} V_4^4$ and $Sec_{14}(V_4^4)$ is given by the determinant of the second catalecticant matrix (Section 1.1 in [4]). Again using Macaulay2 we show that $\mathbb{T}^3_{[x_0^4]} V_4^4 \subseteq Sec_{14}(V_4^4)$. This proves that rank $x_0 F < 2$ rank $F$, for $F$ general.

**Corollary 2.** *For the osculating varieties $T^3 V_4^3$ and $T^3 V_4^4$ we have*

$$T^3 V_4^3 \subseteq Sec_9(V_4^3), \quad T^3 V_4^4 \subseteq Sec_{14}(V_4^4)$$

**Proof.** The action of $PGL(n+1)$ on $\mathbb{P}^n$ extends naturally to an action on $\mathbb{P}^{N(n,d)}$ stabilizing $V_d^n$ and more generally the secant varieties $Sec_h(V_d^n)$. Since this action is transitive on $V_d^n$ we have $\mathbb{T}^a_{[x_0^d]} V_d^n \subseteq Sec_h(V_d^n)$ if and only if $\mathbb{T}^a_{[L^d]} V_d^n \subseteq Sec_h V_d^n$ for any point $[L^d] \in V_d^n$ that is $T^a V_d^n \subseteq Sec_h V_d^n$. Finally, we conclude by applying iii and iv in the list above. □

*Macaulay2 Implementation*

In the Macaulay2 file Comon-1.0.m2 we provide a function called Comon which operates as follows:

- Comon takes in input three natural numbers $n, d, h$;
- if $h < \binom{n+\lfloor\frac{d}{2}\rfloor}{n}$ then the function returns that Comon's conjecture holds for the general degree $d$ polynomial in $n+1$ variables of rank $h$ by the usual flattenings method in Proposition 1. If not, and $d$ is even then it returns that the method does not apply;
- if $d$ is odd and $\binom{n+k}{n} < 2\binom{n+k-1}{n}$, where $k = \lfloor\frac{d+1}{2}\rfloor$, then again it returns that the method does not apply;
- if $d$ is odd, $\binom{n+k}{n} \geqslant 2\binom{n+k-1}{n}$ and $2h-1 > \binom{n+k}{n}$ then it returns that the method does not apply since $2h-2$ must be smaller than the number of order $k$ partial derivatives;

- if $d$ is odd, $\binom{n+k}{n} \geq 2\binom{n+k-1}{n}$ and $2h - 1 \leq \binom{n+k}{n}$ then Comon, in the spirit of Remark 4, produces a polynomial of the form

$$F = \sum_{i=1}^{h}(a_{i,0}x_0 + \cdots + a_{i,n}x_n)^d$$

then substitutes random rational values to the $a_{i,j}$, computes the polynomial $G = x_0 F$, the catalecticant matrix $D$ of order $k$ partial derivatives of $G$, extracts the most up left $2h - 1 \times 2h - 1$ minor $P$ of $D$, and compute the determinant $\det(P)$ of $P$;
- if $\det(P) = 0$ then Comon returns that the method does not apply, otherwise it returns that Comon's conjecture holds for the general degree $d$ polynomial in $n+1$ variables of rank $h$.

Please note that since the function random is involved Comon may return that the method does not apply even though it does. Clearly, this event is extremely unlikely. Thanks to this function we are able to prove that Comon's conjecture holds in some new cases that are not covered by Proposition 1. Since the case $n = 1$ is covered by Proposition 6 in the following we assume that $n \geq 2$.

**Theorem 1.** *Assume $n \geq 2$ and set $h = \binom{n+\lfloor\frac{d}{2}\rfloor}{n}$. Then Comon's conjecture holds for the general degree $d$ homogeneous polynomial in $n+1$ variables of rank $h$ in the following cases:*

- $d = 3$ and $2 \leq n \leq 30$;
- $d = 5$ and $3 \leq n \leq 8$;
- $d = 7$ and $n = 4$.

**Proof.** The proof is based on Macaualy2 computations using the function Comon exactly as shown in Example 1 below. □

**Example 1.** *We apply the function Comon in a few interesting cases:*

```
Macaulay2, version 1.12
with packages: ConwayPolynomials, Elimination, IntegralClosure, InverseSystems,
LLLBases, PrimaryDecomposition, ReesAlgebra, TangentCone
i1 : loadPackage "Comon-1.0.m2";
i2 : Comon(5,3,4)
Lowest rank for which the usual flattenings method does not work = 6
o2 = Comon's conjecture holds for the general degree 3 homogeneous polynomial
in 6 variables of rank 4 by the usual flattenings method
i3 : Comon(5,3,6)
Lowest rank for which the usual flattenings method does not work = 6
o3 = Comon's conjecture holds for the general degree 3 homogeneous polynomial
in 6 variables of rank 6
i4 : Comon(5,3,7)
Lowest rank for which the usual flattenings method does not work = 6
o4 = The method does not apply --- The determinant vanishes
i5 : Comon(5,5,21)
Lowest rank for which the usual flattenings method does not work = 21
o5 = Comon's conjecture holds for the general degree 5 homogeneous polynomial
in 6 variables of rank 21
i6 : Comon(4,7,35)
Lowest rank for which the usual flattenings method does not work = 35
o6 = Comon's conjecture holds for the general degree 7 homogeneous polynomial
in 5 variables of rank 35
```

**Author Contributions:** The authors contributed equally to this work.

**Funding:** This research received no external funding.

**Acknowledgments:** The second and third named authors are members of the Gruppo Nazionale per le Strutture Algebriche, Geometriche e le loro Applicazioni of the Istituto Nazionale di Alta Matematica "F. Severi" (GNSAGA-INDAM). We thank the referees for helping us to improve the exposition.

**Conflicts of Interest:** The authors declare no conflict of interest.

## References

1. Kolda, T.G.; Bader, B.W. Tensor decompositions and applications. *SIAM Rev.* **2009**, *51*, 455–500. [CrossRef]
2. Comon, P.; Mourrain, B. Decomposition of quantics in sums of powers of linear forms. *Signal Process.* **1996**, *53*, 93–107. [CrossRef]
3. Comon, P.; Golub, G.; Lim, L.; Mourrain, B. Symmetric tensors and symmetric tensor rank. *SIAM J. Matrix Anal. Appl.* **2008**, *30*, 1254–1279. [CrossRef]
4. Landsberg, J.M.; Ottaviani, G. New lower bounds for the border rank of matrix multiplication. *Theory Comput.* **2015**, *11*, 285–298. [CrossRef]
5. Massarenti, A.; Raviolo, E. The rank of $n \times n$ matrix multiplication is at least $3n^2 - 2\sqrt{2}n^{\frac{3}{2}} - 3n$. *Linear Algebra Appl.* **2013**, *438*, 4500–4509. [CrossRef]
6. Massarenti, A.; Raviolo, E. Corrigendum to The rank of $n \times n$ matrix multiplication is at least $3n^2 - 2\sqrt{2}n^{\frac{3}{2}} - 3n$. *Linear Algebra Appl.* **2014**, *445*, 369–371. [CrossRef]
7. Buscarino, A.; Fortuna, L.; Frasca, M.; Xibilia, M. Positive-real systems under lossless transformations: Invariants and reduced order models. *J. Frankl. Inst.* **2017**, *354*, 4273–4288. [CrossRef]
8. Dolgachev, I.V. Dual homogeneous forms and varieties of power sums. *Milan J. Math.* **2004**, *72*, 163–187. [CrossRef]
9. Dolgachev, I.V.; Kanev, V. Polar covariants of plane cubics and quartics. *Adv. Math.* **1993**, *98*, 216–301. [CrossRef]
10. Massarenti, A.; Mella, M. Birational aspects of the geometry of varieties of sums of powers. *Adv. Math.* **2013**, *243*, 187–202. [CrossRef]
11. Massarenti, A. Generalized varieties of sums of powers. *Bull. Braz. Math. Soc. (N.S.)* **2016**, *47*, 911–934. [CrossRef]
12. Ranestad, K.; Schreyer, F.O. Varieties of sums of powers. *J. Reine Angew. Math.* **2000**, *525*, 147–181. [CrossRef]
13. Takagi, H.; Zucconi, F. Spin curves and Scorza quartics. *Math. Ann.* **2011**, *349*, 623–645. [CrossRef]
14. Massarenti, A.; Mella, M.; Staglianò, G. Effective identifiability criteria for tensors and polynomials. *J. Symbolic Comput.* **2018**, *87*, 227–237. [CrossRef]
15. Strassen, V. Vermeidung von Divisionen. *J. Reine Angew. Math.* **1973**, *264*, 184–202.
16. Landsberg, J.M. Tensors: Geometry and applications. In *Graduate Studies in Mathematics*; American Mathematical Society: Providence, RI, USA, 2012; Volume 128, p. 439.
17. Zhang, X.; Huang, Z.H.; Qi, L. Comon's conjecture, rank decomposition, and symmetric rank decomposition of symmetric tensors. *SIAM J. Matrix Anal. Appl.* **2016**, *37*, 1719–1728. [CrossRef]
18. Ballico, E.; Bernardi, A. Tensor ranks on tangent developable of Segre varieties. *Linear Multilinear Algebra* **2013**, *61*, 881–894. [CrossRef]
19. Buczyński, J.; Landsberg, J.M. Ranks of tensors and a generalization of secant varieties. *Linear Algebra Appl.* **2013**, *438*, 668–689. [CrossRef]
20. Friedland, S. Remarks on the symmetric rank of symmetric tensors. *SIAM J. Matrix Anal. Appl.* **2016**, *37*, 320–337. [CrossRef]
21. Landsberg, J.M.; Michałek, M. Abelian tensors. *J. Math. Pures Appl.* **2017**, *108*, 333–371. [CrossRef]
22. Seigal, A. Ranks and Symmetric Ranks of Cubic Surfaces. *arXiv* **2018**, arXiv:1801.05377v1.
23. Shitov, Y. A counterexample to Comon's conjecture. *arXiv* **2017**, arXiv:1705.08740v2.
24. Simis, A.; Ulrich, B. On the ideal of an embedded join. *J. Algebra* **2000**, *226*, 1–14. [CrossRef]
25. Iarrobino, A.; Kanev, V. Power Sums, Gorenstein Algebras, and Determinantal Loci. Available online: https://www.springer.com/gp/book/9783540667667 (accessed on 11 September 2018).
26. Schönhage, A. Partial and total matrix multiplication. *SIAM J. Comput.* **1981**, *10*, 434–455. [CrossRef]

27. Buczyński, J.; Ginensky, A.; Landsberg, J.M. Determinantal equations for secant varieties and the Eisenbud-Koh-Stillman conjecture. *J. Lond. Math. Soc.* **2013**, *88*, 1–24. [CrossRef]
28. Carlini, E.; Catalisano, M.V.; Chiantini, L. Progress on the symmetric Strassen conjecture. *J. Pure Appl. Algebra* **2015**, *219*, 3149–3157. [CrossRef]
29. Carlini, E.; Catalisano, M.V.; Oneto, A. Waring loci and the Strassen conjecture. *Adv. Math.* **2017**, *314*, 630–662. [CrossRef]
30. Teitler, Z. Sufficient conditions for Strassen's additivity conjecture. *Ill. J. Math.* **2015**, *59*, 1071–1085.
31. Shitov, Y. A counterexample to Strassen's direct sum conjecture. *arXiv* **2017**, arXiv:1712.08660v1.
32. Macaulay2. Macaulay2 a Software System Devoted to Supporting Research in Algebraic Geometry and Commutative Algebra. 1992. Available online: http://www.math.uiuc.edu/Macaulay2/ (accessed on 11 September 2018).

© 2018 by the authors. Licensee MDPI, Basel, Switzerland. This article is an open access article distributed under the terms and conditions of the Creative Commons Attribution (CC BY) license (http://creativecommons.org/licenses/by/4.0/).

*Article*
# Seeking for the Maximum Symmetric Rank

Alessandro De Paris

Dipartimento di Matematica e Applicazioni "Renato Caccioppoli", Università di Napoli Federico II, I-80126 Napoli, Italy; deparis@unina.it

Received: 9 October 2018 ; Accepted: 3 November 2018; Published: 12 November 2018

**Abstract:** We present the state-of-the-art on maximum symmetric tensor rank, for each given dimension and order. After a general discussion on the interplay between symmetric tensors, polynomials and divided powers, we introduce the technical environment and the methods that have been set up in recent times to find new lower and upper bounds.

**Keywords:** symmetric tensor; tensor rank; Waring rank; power sum

**MSC:** 15-02

---

## 1. Introduction

Until recently, if a scientist or an engineer were asked what tensors are good for, probably in many cases they would have organized their answer around examples such as the Cauchy stress tensor or the tensors that arise in general relativity. However, these are examples of a more complex concept: that of a *tensor field*. To give a first explanation of the difference, let us point out that a scalar field is a function defined on a portion of space (on which one can perform the various operations of Calculus, e.g., take the partial derivatives), whereas a scalar is just a real (or complex) number. To refine the explanation, one might discuss the difference between a vector field and a single vector. In the context of the present survey article, tensors are considered as mathematical objects per se. Thus, we disregard how tensors can vary in a portion of space, and look at them as single elements of a space of tensors, i.e. just the set of all tensors of the same type. In the last decades, it has become clearer and clearer that even this considerably simpler notion is of utmost applicative importance: a convincing account is provided by Landsberg's book [1], especially in Section 1.3 and the whole Part 3.

In most of the applications, vectors are either physical quantities which can be represented by arrows, or arrays of numbers. Since spaces of these two kinds of objects reveal the same fundamental structure, the twentieth century Mathematics has established the unifying concept of a *vector space*. This abstract object turns out to be suitable for a much wider class of situations, including those in which vectors can vary in real space. Vector spaces are a kind of algebraic structure, that is, they are formally defined by means of the operations that can be performed on their elements. The multiplication of numbers by vectors can be extended by allowing more general kinds of "numbers", which can be defined by means of another algebraic structure, named a *field* (not to be confused with a field of varying objects in a portion of space). These are the building blocks of modern Linear Algebra, with which even the multilinear algebra, hence the tensor theory, can be set up. The survey is written using this language. Although not unknown, such an abstract mathematical approach might be a bit demanding for some tensor practitioners (cf. [1], Subsection 0.3); therefore, it is likely they will prefer to accompany the reading with some of the many good books on abstract algebra.

Based on the work in [1], Part 3, the importance of tensor decompositions can be appreciated. The perhaps most basic kind of decomposition leads to the notion of *tensor rank*. It can be said that, to date, tensor rank is not understood well, as no general algorithm for determining it is known. The same is true for the symmetric rank of a symmetric tensor. Moreover, many of the techniques that have led to some success in tensor theory apply both to rank and to symmetric rank, and the assertion that they coincide (for a symmetric tensor) is the content of the Comon's conjecture (see [1], Exercise 2.6.6.5 and references therein). This encourages us to believe that a better understanding of the symmetric rank may shed some light on tensor rank.

When the base field is of characteristic zero, e.g., when dealing with real or complex numbers (as in most of the applicative uses of tensors), symmetric tensors are naturally identified with homogeneous polynomials. Even in positive characteristic, symmetric tensors and polynomials are "quite close" since, as a matter of facts, symmetric tensors can always be naturally identified with the so-called *divided powers*. From the polynomial viewpoint, the symmetric rank becomes the *Waring rank*, that is, the minimum number of summands that are required to express a given homogeneous polynomial as a sum of powers of linear forms. It follows that, in the characteristic zero case, to determine the maximum symmetric rank for symmetric tensors of given dimension $n$ and order $d$, is the same as to determine the maximum Waring rank for degree $d$ homogeneous polynomials in $n$ variables. This is one of the most natural variants of the classical Waring problem on natural numbers. Now, if the symmetric rank was well understood, one probably would easily determine what is its maximum for symmetric tensors of given order and dimension. This leads us to hope that the techniques invented to find the maximum Waring rank for degree $d$ forms in $n$ variables might indicate some ways to understand tensor rank.

Following Geramita [2], let us recall that, while the classical Waring problem was solved by Hilbert, its main variant remains open, i.e. determining the minimum number $G(d)$ of summands that are needed to decompose every *sufficiently large* natural number into a sum of $d$th powers. From the polynomial viewpoint, a naturally analogous problem is to determine the maximum Waring rank of a *generic* degree $d$ homogeneous polynomials in $n$ variables, where "generic" is meant in the sense that it commonly has in Algebraic Geometry. In this case, under the hypothesis that the base field is algebraic closed of characteristic zero, the answer is provided by the celebrated Alexander–Hirschowitz theorem, which deals with interpolation of sets of fat points, but which through the Terracini's lemma has a direct translation in frame of the Waring rank (see [3] for the original proof of the theorem and [4], Section 7 for a good historical account of it). Henceforth, we shall refer to the Waring rank of a form simply as its *rank*.

Interestingly, at the end of his review of the outstanding paper [3] in the MathSciNet database, Fedor Zak wrote:

> It would also be nice to know how many linear forms are required to express an arbitrary form of degree $d$ as a sum of $d$th powers of linear forms.

In the mentioned Exposé [2], Tony Geramita called this problem the *little* (and the "generic version" solved by Alexander and Hirschowitz the *big*) Waring problem for polynomials. However, as little as it can be, similar to a mouse, the problem is also escaping, and indeed it has remained open to date.

After Geramita's Exposé, Johannes Kleppe was able to prove that the maximum rank of ternary quartics is seven, in his master thesis [5] under the supervision of Kristian Ranestad. More precisely, that work contains a study of normal forms and ranks of ternary cubics and quartics, from which the maximum ranks can be obtained.

Some years later, at the opposite side of Kleppe's result, which gives a sharp upper bound in two very specific cases, Corollary 1 [6] provided us with a very mild upper bound in the general case. After about the same amount of time, that result was dramatically improved by Blekherman and Teitler (see [7], Corollary 9). From the references in [6], some information on earlier works on the subject can be obtained. For instance, the work in [8], published in 1969, deals with the problem for arbitrary fields, with a particular care to detect in which cases there is actually a maximum for the Waring rank (as it is always the case when the base field is algebraically closed of characteristic zero). We believe that any list of references one might try to work out for the problem of our interest will likely be far from complete, since it is of a natural and elementary nature and dates back to at least the first half of the past century. For instance, B. Segre at the beginning of [9], Section 96, mentioned a question raised in [10].

The introduction of Buczyński and Teitler's paper [11] is a good source of information on what is known on the maximum Waring rank, up to recent times. As tensor theory, as well as its related algebrogeometric aspects, is currently a very active research field, a good deal of new results have been discovered since then, and they can be of use to determine that maximum. However, if we strictly focus on the problem, as far as we know, there are only two new facts that improve ([11], Table 1). The first one is [12], Proposition 3.3, which sensibly improves the Blekherman and Teitler's upper bound when $n = 3$. The other is [13], Theorem 2.5, which extends to even degrees the Buczyński and Teitler's lower bound established in [11], Theorem 1 (in the sense that it raises by one the lower bound given by the maximum rank of monomials, which for an even degree $d$ is $(d^2 + 2d)/4$).

According to the aforementioned results, for ternary forms over algebraically closed fields of characteristic zero both the lower and the upper bounds are of order $O(d^2/4)$, and the gap is $d - 1$. The main purpose of the present survey article is to convey the basic ideas that led to those bounds.

## 2. Symmetric Algebra and Apolarity

In this section, we give an overview of some fundamental facts that are worthy of being recalled in the present context. The technical treatments of basic tensor theory and apolarity that can be encountered in the literature may vary considerably. Instead of fixing one of them, we only review the main statements: whatever reference the reader is willing to adopt, the basic definitions will likely be compatible with (or at least adaptable to) our assumptions. The exposition is organized so that, by adding sufficient details, one can get a coherent theoretical development.

> Henceforth, $\mathbb{K}$ denotes a field and $V$ a $\mathbb{K}$-vector space.

### 2.1. The Tensor Algebra of a Vector Space

We take the view that the *tensor algebra* $T(V)$ can be any fixed $\mathbb{K}$-algebra that contains $V$ as a $\mathbb{K}$-vector subspace and such that for each $\mathbb{K}$-vector space homomorphism of $V$ into a (commutative or not) $\mathbb{K}$-algebra $A$ there exists a unique $\mathbb{K}$-algebra homomorphism $T(V) \to A$ that extends it. The elements of $T(V)$ are called *tensors*, the multiplication in $T(V)$ is denoted by $\otimes$ and $t_0 \otimes t_1$ is the *tensor product* of $t_0$ and $t_1$. However, $V \otimes V$ does not denote the set of all $v_0 \otimes v_1$ with $v_0, v_1 \in V$, but the vector subspace of $T(V)$ generated by that set. The *tensor power* $V \otimes \cdots \otimes V =: V^{\otimes d}$ can be similarly defined; its elements are said to have *order* $d$. The multiplication $\mu_d : V \times \cdots \times V \to V^{\otimes n}$, $(v_1, \ldots, v_n) \mapsto v_1 \otimes \cdots \otimes v_n$, is a universal $d$-linear map, that is, every $d$-linear map $m : V \times \cdots \times V \to W$ factors as $\varphi \circ \mu_d$ for a uniquely determined vector space homomorphism $\varphi : V^{\otimes d} \to W$.

This approach may be considered rather abstract, but it might be said that $T(V)$ is no more abstract than $\mathbb{C}$: similar to how one extends $\mathbb{R}$ by adding a square root of $-1$ and all objects that are consequently needed to keep the operations and their properties, here one extends $V$ and $\mathbb{K}$ by adding tensor products of vectors and all objects that are consequently needed to keep the operations of $V$, and to build a multiplication with the usual properties (apart from commutativity). The above characterization of $T(V)$ (up to algebra isomorphisms) makes precise the intuitive idea, and indeed a similar characterization holds for $\mathbb{C}$: for every $\mathbb{R}$-algebra $A$ that contains a square root $x$ of $-1$ (for instance, $A$ could be the algebra of endomorphisms of vectors in the plane and $x$ a rotation by 90°), there exists a unique $\mathbb{R}$-algebra homomorphism $\mathbb{C} \to A$ that sends the imaginary unit into $x$.

### 2.2. The Symmetric Algebra of a Vector Space

The definition of the *symmetric algebra* $S(V)$ can be given in the same way, except for the fact that now the target $\mathbb{K}$-algebra $A$ in the characteristic property is required to be commutative. A very natural $\mathbb{K}$-algebra homomorphism $T(V) \to S(V)$ arises: the one which extends the identity map of $V$. Elements of $S(V)$ are not exactly symmetric tensors, though in most cases the two things can be identified, but rather *polynomials*. Indeed, if $(x_i)_{i \in I}$ denotes a (possibly infinite) basis of the vector space $V$, any element of $S(V)$ can be written as a polynomial in the $x_i$s, and two elements are equal if and only if for each monomial they carry the same coefficient.

### 2.3. Symmetric Tensors

To define symmetric tensors, first notice that for each permutation $\sigma$ of $\{1, \ldots, n\}$, by the mentioned universal property of tensor powers we have an automorphism $\sigma_V$ of $V^{\otimes n}$ such that $v_1 \otimes \cdots \otimes v_n \mapsto v_{\sigma(1)} \otimes \cdots \otimes v_{\sigma(n)}$. A tensor in $V^{\otimes n}$ is said to be *symmetric* if it is invariant under $\sigma_V$ for all permutation $\sigma$ of $\{1, \ldots, n\}$. For instance, $v \otimes w + w \otimes v$ and $v \otimes v \otimes w + v \otimes w \otimes v + w \otimes v \otimes v$ are symmetric tensors for whatever choice of $v, w \in V$. A sum of symmetric tensors of different orders can be called symmetric as well (although very seldom one encounters such sums).

### 2.4. The Symmetric Product

The set $\bar{S}(V)$ of symmetric tensors is a vector subspace, but not a subalgebra of $T(V)$ (apart from the trivial cases when $\dim V \leq 1$). However, a meaningful multiplication in $\bar{S}(V)$ can be introduced. The *symmetric product* $s_0 s_1$ of two symmetric tensors of given orders $d_0, d_1$ can be defined as the sum of $\sigma_V(s_0 \otimes s_1)$ with $\sigma$ varying on the $(d_0, d_1)$-shuffles, that are the permutations of $\{1, \ldots, d_0 + d_1\}$ such that $\sigma(1) < \cdots < \sigma(d_0)$ and $\sigma(d_0 + 1) < \cdots < \sigma(d_0 + d_1)$. For instance,

$$u(v \otimes w + w \otimes v) = u \otimes v \otimes w + u \otimes w \otimes v + v \otimes u \otimes w + w \otimes u \otimes v + v \otimes w \otimes u + w \otimes v \otimes u = uvw \quad (1)$$

for whatever choice of $u, v, w \in V$.

There is a unique way to extend this operation on the whole of $\bar{S}(V)$, that preserves the distributive law. Note that with this definition we have $v^d = d! v^{\otimes d} = d! v \otimes \cdots \otimes v$ ($d$ factors).

### 2.5. A Variant of the Symmetric Product in Characteristic Zero

We mention that there is a much more popular variant of the symmetric product, which is in use when $\mathbb{K}$ is $\mathbb{C}$, or more generally is of characteristic zero (see, e.g., [1], 2.6.3). It can be presented as follows. Every $\mathbb{K}$-vector space automorphism $\alpha$ of any $\mathbb{K}$-algebra $A$ induces another multiplication on $A$: the one that makes $\alpha$ a $\mathbb{K}$-algebra automorphism, when it replaces the original one on the target (but not on the domain). In particular, if a sequence $(a_d)_{d \geq 0}$ of nonzero scalars is given, the multiplication by $a_d$ on order $d$ symmetric tensors gives a $\mathbb{K}$-vector space automorphism of $\bar{S}(V)$, hence a multiplication in it. The mentioned symmetric product in $\bar{S}(V)$ is given by the sequence $a_d := 1/d!$. For instance, this definition gives

$$u(v \otimes w + w \otimes v) = \frac{1}{3}(u \otimes v \otimes w + u \otimes w \otimes v + v \otimes u \otimes w + w \otimes u \otimes v + v \otimes w \otimes u + w \otimes v \otimes u) = u(2vw)$$

(compare with Equation (1)); note also that now $v^d = v^{\otimes d}$.

The reason for the popularity of the "modified" symmetric product is explained by the relationship between symmetric tensors and polynomials: it makes the restriction $\bar{S}(V) \to S(V)$ of the natural map $T(V) \to S(V)$ a $\mathbb{K}$-algebra isomorphism. In positive characteristic, a multiplication with such a desirable property can not be hoped for, simply because in this case the restriction $\bar{S}(V) \to S(V)$ is not bijective (unless $\dim V \leq 1$). It also worthy of being remarked that in characteristic zero, $\bar{S}(V)$ with the previous symmetric product is naturally isomorphic to $S(V)$ as well, but in this case through the $\mathbb{K}$-algebra homomorphism $S(V) \to \bar{S}(V)$ that arises simply because $\bar{S}(V)$ is a commutative $\mathbb{K}$-algebra containing $V$.

Although the subject of the present work is the maximum symmetric rank in the characteristic zero case, we prefer to assume that $\bar{S}(V)$ is equipped with the general symmetric product (which is defined regardless of the characteristic of $\mathbb{K}$), not with the modified one.

## 2.6. Tensors on the Dual and Multilinear Forms

Let us consider the dual space $V^*$ and denote by $V^{*d}$ the space of $d$-linear forms. A $d$-linear form $\mu : V^d \to \mathbb{K}$ and a $d'$-linear form $\mu' : V^{d'} \to \mathbb{K}$ give rise to the $(d+d')$-linear form

$$(v_1, \ldots, v_{d+d'}) \mapsto \mu(v_1, \ldots, v_d)\mu'(v_{d+1}, \ldots, v_{d+d'}).$$

This allows us to define a $\mathbb{K}$-algebra structure on

$$M(V) := \bigoplus_d V^{*d}.$$

Since $V^*$ embeds (as a summand) in $M(V)$, one comes with a natural $\mathbb{K}$-algebra homomorphism $T(V^*) \to M(V)$, which is always injective and turns out to be an isomorphism if and only if $\dim V < \infty$. On the other hand, to give a $d$-linear form on $V$ is the same as to give a linear form on $V^{\otimes d}$, and $T(V)$, as a matter of facts, is a *graded* algebra whose degree $d$ component is $V^{\otimes d}$ (regardless of the finiteness assumption). Therefore, $M(V)$ is isomorphic as a vector space to the graded dual of $T(V)$. In conclusion, when $\dim V < \infty$, the graded dual of $T(V)$ is isomorphic, as a graded vector space, to $T(V^*)$.

Let us also point out that, given $l_1, \ldots, l_d \in V^*$, the image of the tensor product $l_1 \otimes \cdots \otimes l_d$ in $M(V)$ takes value

$$l_1(v_1) \cdots l_d(v_d) \tag{2}$$

on $(v_1, \ldots, v_d)$, and this is also the value taken on $v_1 \otimes \cdots \otimes v_d$ by the image in the graded dual of $T(V)$.

For the symmetric algebra, we have the following facts. To begin with, $S(V)$ is a graded algebra as well, and the natural homomorphism $T(V) \to S(V)$ is a graded one, and as a matter of facts is surjective on each graded component. Hence, the graded dual of $S(V)$ embeds into the graded dual of $T(V)$, which is isomorphic to $M(V)$. However, similarly to what happens for $V^{\otimes d}$, to give a $d$-linear form on the symmetric power $S^d(V)$ (the subspace of degree $d$ elements of $S(V)$) is the same as to give a *symmetric* $d$-linear form on $V$. This easily implies that the image of the dual of $S^d(V)$ through the embedding into $M(V)$ is precisely the subspace of $d$-linear *symmetric forms*. Next, it is not difficult to show that an order $d$ tensor in $T(V^*)$ is symmetric if and only if its image through the embedding $T(V^*) \hookrightarrow M(V)$ is a symmetric $d$-linear map. In the finite dimensional case, the converse is also true: a $d$-linear form is symmetric if and only if it corresponds to a symmetric tensor. In conclusion, when $\dim V < \infty$, the graded dual of $S(V)$ is isomorphic, as a graded $\mathbb{K}$-vector space, to $\overline{S}(V^*)$.

According to the definition of the symmetric product, given $l_1, \ldots, l_n \in V^*$, the image of $l_1 \cdots l_d$ in $M(V)$ takes value

$$\sum_\sigma l_1(v_{\sigma(1)}) \cdots l_d(v_{\sigma(d)}) \tag{3}$$

on $(v_1, \ldots, v_d)$, that is, the *permanent* of the matrix $(l_i(v_j))_{i,j}$. This is also the value taken on $v_1 \cdots v_d$ by the image of $l_1 \cdots l_d$ in the graded dual of $S(V)$.

## 2.7. Evaluation

The algebra $\mathbb{K}^V$ of all functions $V \to \mathbb{K}$ is commutative and contains $V^*$. Hence, there is a canonical $\mathbb{K}$-algebra homomorphism $S(V^*) \to \mathbb{K}^V$. Note also that the value of the image of $f \in S(V^*)$ on $v$ is the image of through the evaluation homomorphism $S(V^*) \to \mathbb{K}$ that extends the evaluation at $v$ map $V^* \to \mathbb{K}$ ($l \mapsto l(v)$). If $f \in S(V^*)$ is considered as a polynomial, the image in $\mathbb{K}^V$ is the corresponding polynomial function.

For a form $f \in S^d(V^*)$, the corresponding symmetric $d$-linear form can be considered as a *polarization* of the corresponding polynomial function (though it is customary to affect that $d$-linear form by a factor $1/d!$).

## 2.8. Symmetric Tensors on the Dual and Divided Powers

The symmetric product may be preferred over its variant in characteristic zero, not only because it is defined regardless to the characteristic of $\mathbb{K}$, but also because it makes $\bar{S}(V)$ an algebra of divided powers, in a quite simple way: it suffices to set $v^{[d]} := v^{\otimes d}$. A perhaps more familiar way to introduce divided powers is in the context of duality, as done, e.g., in [14], Appendix A. To link the discussion earlier to the Iarrobino and Kanev's approach:

$$\boxed{\text{henceforth, we assume dim } V < \infty.}$$

Since an isomorphic image of a tensor algebra is a tensor algebra as well, we can assume that $T(V^*)$ is chosen so that $\bar{S}(V^*)$ actually *is* the graded dual of $S(V)$. Similarly, if $R = \mathbb{K}[x_1, \ldots, x_r] = \oplus_{d \geq 0} R_d$ is a (graded) ring of polynomials, we can assume $S(R_1) = R$. Let us also assume $V := R_1$ (under a suitable definition of polynomial rings, this assumption imposes no restriction on $V$). Setting $\mathcal{D} := \bar{S}(V^*)$ and denoting by $\mathcal{D}_d$ the subspace of order $d$ symmetric tensors, we get exactly in the situation at the beginning of [14], Appendix A. Let us employ a few lines below to check that setting $l^{[d]} := l^{\otimes d}$, subsequent definitions of [14], Appendix A, are automatically fulfilled.

Let $(X_1, \ldots, X_r)$ be the base of $\mathcal{D}_1$, dual to $(x_1, \ldots, x_r)$ (that is, $X_i(x_j)$ is 0 when $i \neq j$ and 1 when $i = j$). Note that $X_i X_j$ takes value 1 on $x_i x_j$ when $i \neq j$ (it is indeed given by Equation (3), and $l_i(x_j) = l_j(v_i) = 0$), but the value is 2 when $i = j$, that is, $X_i^2(x_i^2) = 2$. Instead, we have $X_i^{[2]}(x_i) = X_i^{\otimes 2}(x_i) = 1$, as it follows from the evaluation rule Equation (2) (or also from $X_i^2 = 2! X_i^{\otimes 2}$). More generally, we have that $X_1^{[d_1]} \cdots X_r^{[d_r]}$ takes value 0 on $x_1^{d_1'} \cdots x_r^{d_r'}$ when $(d_1, \ldots, d_r) \neq (d_1', \ldots, d_r')$ and value 1 when $(d_1, \ldots, d_r) = (d_1', \ldots, d_r')$. This agrees with [14], Definition A.1. Since the number of $(d, d')$-shuffles is $(d + d')!/(d! d'!)$, we have

$$X_i^{[d]} X_i^{[d']} = \frac{(d+d')!}{d! d'!} X_i^{[d+d']},$$

hence

$$X_1^{[d_1]} \cdots X_r^{[d_r]} X_1^{[d_1']} \cdots X_r^{[d_r']} = \frac{(d_1 + d_1')!}{d_1! d_1'!} \cdots \frac{(d_r + d_r')!}{d_r! d_r'!} X_1^{[d_1 + d_1']} \cdots X_r^{[d_r + d_r']},$$

in agreement with [14], (A.0.5). Similar calculations lead to

$$(a_1 X_1 + \cdots + a_r X_r)^{[d]} = \sum_{d_1 + \cdots + d_r = d} a_1^{d_1} \cdots a_r^{d_r} X_1^{[d_1]} \cdots X_r^{[d_r]},$$

in agreement with [14], Definition A.8.

## 2.9. The Contraction Map

Given $p \in R_d$, $f \in \mathcal{D}_{d+d'} = \text{Hom}(R_{d+d'}, \mathbb{K})$ and denoting by $\mu_p : R_{d'} \to R_{d+d'}$ the vector space homomorphism given by the multiplication by $p$, the composition $f \circ \mu_p$ belongs to $\mathcal{D}_{d'}$ and is called the *contraction of $f$ by $p$*. The bilinear operations $R_d \times \mathcal{D}_{d+d'} \to \mathcal{D}_{d'}$ (assuming $\mathcal{D}_{d'} = \{0\}$ when $d' < 0$) extend to a unique bilinear operation $R \times \mathcal{D} \to \mathcal{D}$, that can be called *contraction map* (in agreement with [14], Definition A.2). If $l_1, \ldots, l_d \in R_1$ and $\varphi \in M(V)$ is the $(d + d')$-linear form corresponding to $f$, the contraction of $f$ by $l_1 \cdots l_d$ corresponds to the $d'$-linear form obtained by fixing $d$ arguments of $\varphi$ equal to $l_1, \ldots, l_d$: $(v_1, \ldots, v_{d'}) \mapsto \varphi(l_1, \ldots, l_d, v_1, \ldots, v_{d'})$. This property may also be used to give an alternative definition of the contraction, because of the characteristic property of the symmetric powers (the $d$-linear assignment on all $(l_1, \ldots, l_d) \in R_1^d$ determines a homomorphism $R_d = S^d(R_1) \to V^{*d'}$, whose image is canonically isomorphic to $\mathcal{D}_{d'}$ and which depends on $f \in \mathcal{D}_{d+d'}$ in a linear way). For this reason, the contraction can also be called *insertion* and denoted by $\lrcorner$ (this attitude is perhaps more common in the context of alternating forms).

## 2.10. Contraction and Derivatives

The ordinary directional derivative of differentiable (real valued) functions fulfills the Leibnitz rule $\partial_v(fg) = (\partial_v f)g + f\partial_v g$ and on linear forms is nothing but the evaluation on $v$. These two properties can be used to characterize the derivative of polynomials in $S(V^*)$ along $v \in V$. Indeed, for each $d$, we can define $\partial_v$ on $S^d V^*$ as the unique operator into $S^{d-1} V^*$ such that

$$\partial_v (l_1 \cdots l_d) = \sum_{i=1}^{d} l_i(v) l_1 \cdots \widehat{l_i} \cdots l_d \in S^{d-1} V^*$$

for all $l_1, \ldots, l_d \in V^*$ (with the hat denoting omission). Then, $\partial_v$ extends to the whole of $S(V^*)$ by additivity. Using again the fact that linear operators on $S^d V^*$ are characterized by their values on the products $l_1 \cdots l_d$, one can easily check that the Leibnitz rule holds in $S(V^*)$ and that $\partial_v$ is the unique extension of the evaluation on $v$ in $V^*$ with this property. A partial derivative is obviously a directional derivative along a basis vector $x_i$.

Let $l_1, \ldots, l_d \in V^*$, $v_1, \ldots, v_d \in V$, and $f$ be the image in $\mathcal{D}_d$ of $l_1 \cdots l_d \in S^d V^*$ (that is, the product $l_1 \cdots l_d$ in the ring $\mathcal{D}$). From the evaluation rule Equation (3) follows that the contraction of $f$ by $v_1$ take the same value on $v_2 \cdots v_d$ as

$$\sum_{i=1}^{d} l_i(v_1) l_1 \cdots \widehat{l_i} \cdots l_d \in \mathcal{D}_{d-1}$$

(is basically a Laplace-like expansion of the permanent along the first row), which is the image of $\partial_{v_1}(l_1 \cdots l_d) \in S^{d-1} V^*$. Using additivity and again the fact that operators on symmetric powers are determined by their values on products of vectors, we conclude that, for every $v \in V$, the partial derivative $\partial_v$ and the contraction by $v$ are compatible via the canonical homomorphism $S(V^*) \to \bar{S}(V^*) = \mathcal{D}$.

A constant coefficient linear partial differential operator on $S(V^*)$ is a linear combination of compositions of directional (or partial) derivatives. The set $D$ of such operators is a commutative $\mathbb{K}$-algebra with multiplication given by composition. Hence, $v \mapsto \partial_v$ extends in a unique way to a $\mathbb{K}$-algebra homomorphism $S(V) = R \to D$. The image of each $p \in R$ in $D$ can be denoted by $\partial_p$.

## 2.11. Apolarity

When $\mathbb{K}$ is of characteristic zero, both canonical homomorphisms $S(V^*) \to \mathcal{D}$ and $R \to D$ are isomorphisms. In this case, we can assume that $S(V^*) = \mathcal{D}$ (and under a suitable definition of polynomials also $R = D$ could be assumed). This way the contraction map becomes a bilinear map $S(V) \times S(V^*) \to S(V^*)$ such that each $p$ acts as the constant coefficient linear differential operator $\partial_p$ (e.g., the contraction of $X_1 X_2^2$ by $3x_1 x_2 + x_2^2$ is $3\partial_{x_1}\partial_{x_2}(X_1 X_2^2) + \partial_{x_2}\partial_{x_2}(X_1 X_2^2) = 2X_1 + 6X_2$).

From the coordinate-free definitions is quite easy to recognize that the contraction map is invariant with respect to the canonical actions of $GL(V)$ on $V$ and on $V^*$ (cf. also [14], Proposition A3(i)). As Ehrenborg and Rota reported in [15], Introduction, for each fixed degree there is a unique invariant bilinear form $S^d(V) \times S^d(V^*) \to \mathbb{K}$, which has been much used since the nineteenth century by classical invariant theorists. They also say that this form can be called *apolar bilinear form*, that the subject of apolarity has been related with the symbolic method in classical invariant theory, and that an efficient treatment can be given in frames of Hopf algebras.

Apolarity for univariate polynomials of the same degree can also be disguised as an explicit formula with alternating signs. To understand why, it may be useful to have a quick look on what happens for the homogeneous version of such polynomials, that is, for binary forms. From a geometric, projective viewpoint, (nonzero) vectors can be viewed as points and linear forms as hyperplanes. For binary forms, that is, when $\dim V = 2$ and the projective picture is a line, hyperplanes are (singletons of) points. From the algebraic viewpoint, this amounts to the existence of an isomorphism $V \to V^*$, unique up to scalar factors (or, equivalently, to the existence of a unique, up to scalar factors, nondegenerate alternating form on $V$. In coordinates, $a_0 x_0 + a_1 x_1$ corresponds to $a_1 X_0 - a_0 X_1$ (or to a scalar multiple of it). The extension $S(V) \tilde{\to} S(V^*)$ of this isomorphism allows us to equivalently describe apolarity as a bilinear form on $S(V)$ alone. Restricting the attention on (homogeneous) polynomials of the same degree, we get a bilinear form on $S^d(V)$, which is alternating for $d$ odd and symmetric for $d$ even. In coordinates:

$$\left( \sum_i a_i x_0^{d-i} x_1^i , \sum_i a_i' x_0^{d-i} x_1^i \right) \mapsto \sum_i (-1)^i (d-i)! i! a_{d-i} a_i' .$$

In terms of univariate polynomials of an assigned degree $d$:

$$\left( \sum_i a_i x^i , \sum_i a_i' x^i \right) \mapsto \sum_i (-1)^i (d-i)! i! a_{d-i} a_i' .$$

Some classical results deal with this kind of apolarity, with $\mathbb{K} = \mathbb{C}$. For instance, Grace's theorem is sometimes named Grace's apolarity theorem for this reason. Nowadays, there is a basic result, which is largely referred to as the *apolarity lemma*. It plays a fundamental role in proving the bounds on Waring ranks we aim to present in this article. We end this section by setting up the technical environment of our presentation. In the next section, we state a version of the apolarity lemma.

2.12. *Standing Assumptions*

From the usual geometric viewpoint, forms are regarded as hypersurfaces. To begin with, for every element of $S(V^*)$, the vanishing locus of the corresponding polynomial function is an affine hypersuperface in $V$. However, a projective viewpoint is perhaps more fruitful. We define the projective space $\mathbb{P}(V)$ as the set of one-dimensional subspace of $V$, $\langle v \rangle$ with $v \neq 0$, and the hypersurface corresponding to a form $f \in S^d(V^*)$ as the zero locus in $\mathbb{P}(V)$ of the corresponding degree $d$ homogeneous function on $V$. Sometimes geometric features of the hypersurface defined by a form $f$, such as the singular locus, can give information on the rank of $f$ (cf. [1], Theorem 9.2.1.4).

However, rank determination often uses the dual viewpoint, from which forms are considered as points in a space where powers of linear forms constitute a Veronese variety (see [1], 4.3.7). Sometimes, having a simultaneous look on both viewpoint has been fruitful, and perhaps a systematic investigation with this double viewpoint could be worthy of being pursued. However, that it is not our goal here, we prefer to facilitate the interchange between $V$ and $V^*$ by working on an arbitrary pair of (finite dimensional) vector spaces, with a given perfect pairing between one another. To denote such spaces, we follow a kind of *abstract index notation*, using upper indices for one of the two spaces and lower indices for the other. In a purely algebraic context, especially one in which powers play an important role, this might be considered bad practice. However, if one takes care of not assigning an independent meaning to $x$, the use of $x^i$ causes no ambiguities. An advantage of this choice is to have a notation that can be more promptly translated (and provide insight) in physics contexts where tensors are widely used.

As we have anticipated, we are interested in the case when $\mathbb{K}$ is algebraically closed of characteristic zero (e.g., $\mathbb{K} = \mathbb{C}$). Hence, apolarity will be assumed on a (dual) pair of symmetric algebras, which we denote by $S^\bullet$ and $S_\bullet$. From our preferred geometric viewpoint, elements of $S^\bullet$ are considered as polynomial functions on the degree 1 component $S_1$ of $S_\bullet$ (in accordance with the abstract index notation, where forms take upper indices and vectors lower indices), but elements of $S^\bullet$ act also as constant coefficient linear differential operators on $S_\bullet$. Note that, to fit the first description into the previously outlined technical treatment, one needs to identify $S_\bullet$ with $S(V)$ and $S^\bullet$ with $S(V^*)$, whereas, to fit the other, one needs the converse. Polynomials whose rank is to be studied will live in $S_d$.

Following from the above said, now we set up more formally our ground technical framework for the subsequent sections.

- We assume that $\mathbb{K}$ is algebraically closed and of characteristic zero.
- The span of a subset $X \subseteq V$ is denoted by $\langle X \rangle$ (or also $\langle v_1, \ldots, v_n \rangle$ when $X = \{v_1, \ldots, v_n\}$).
- We define the projective space $\mathbb{P}(V)$ as the set $\{\langle v \rangle : v \in V \smallsetminus \{0\}\}$ of one-dimensional subspaces of $V$.
- We fix two symmetric algebras $S_\bullet$ and $S^\bullet$, whose degree $d$ components are denoted by $S_d$ and $S^d$.
- We assume $\dim S_1 < \infty$, $\dim S^1 < \infty$, $S_\bullet = S(S_1)$, $S^\bullet = S(S^1)$.
- We assume that a perfect pairing $S^1 \times S_1 \to \mathbb{K}$ is given.
- By the value $f(v)$ of $f \in S^\bullet$ at $v \in S_1$, we mean the value at $v$ of the image of $f$ through $S^\bullet \xrightarrow{\sim} S(S_1^*) \to \mathbb{K}^{S_1}$, where the first isomorphism is induced by the given perfect pairing.
- The *ideal* $I(X) \subseteq S^\bullet$ of a subset $X \subseteq \mathbb{P}(S_1)$ is the homogenous ideal with degree $d$ components given, for each $d$, by all $f \in S^d$ such that $f(v) = 0$ for all $\langle v \rangle \in X$.
- The *apolar bilinear map* $S^\bullet \times S_\bullet \to S_\bullet$ is the map induced by the earlier defined bilinear map $S(V) \times S(V^*) \to S(V^*)$, when $V = S^1$, through the isomorphism $S(S^{1*}) \xrightarrow{\sim} S_\bullet$ induced by the given perfect pairing.
- $\partial_f x$ denotes the value of the apolar map on $(f, x) \in S^\bullet \times S_\bullet$.
- $f$ and $x$ are said to be *apolar* to each other when $\partial_f x = 0$, and the ideal Ann $x \subseteq S^\bullet$ of all $f$ apolar to $x$ is called the *apolar ideal of $x$*.
- If $W \le S_d$ is a subspace, $W^\perp$ denotes the set of all $f \in S^d$ that are apolar to all elements of $W$. Similarly, for a subspace $W \le S^d$, $W^\perp$ denotes the set of all $x \in S_d$ that are apolar to all elements of $W$.

From the discussion above, it follows that apolarity induces an isomorphism $S^d \to S_d^*$ for each $d$, therefore gives a perfect pairing in each degree. Hence, $W^\perp$ denotes nothing but the orthogonal complement with respect such a perfect pairing. The notation Ann $x$ for the apolar ideal complies with the notion of the annihilator of an element of a module, because $S_\bullet$ is structured as an $S^\bullet$-module by apolarity. We prefer not to use the quite common notation $x^\perp$ for the apolar ideal (we speak about orthogonality only in a fixed degree).

When one needs schemes (for which we assume the definitions in [16]), it turns out that a point of **Proj** $S(V^*)$ rational over $\mathbb{K}$ is a maximal non-irrelevant homogeneous ideal in $S(V^*)$ such that its intersection with $S^1(V^*) = V^*$ is a hyperplane. However, every hyperplane in $V^*$ is the hyperplane of forms that vanish on some point in $\mathbb{P}(V)$. This gives the canonical identification of $\mathbb{P}(V)$ with the set of $\mathbb{K}$-points of **Proj** $S(V^*)$.

## 3. Basic Results

### 3.1. Apolarity Lemma

Let us preliminary point out that

$$f(v) = \frac{1}{d!}\partial_f v^d, \qquad \forall f \in S^d, v \in S_1, d > 0. \tag{4}$$

**Remark 1.** *From Equation (4), it follows that, given $f \in S^d$ and $v \in S_1$ such that $f(v) = 0$, for every $g \in S^{d'}$, we have*

$$\partial_g \partial_f v^{d+d'} = \partial_{gf} v^{d+d'} = (d+d')!(gf)(v) = (d+d')!g(v)f(v) = 0.$$

*However, $\partial_g \partial_f v^{d+d'} = 0$ for all $g \in S^{d'}$ implies that $\partial_f v^{d+d'} = 0$ (because apolarity gives a perfect pairing in degree $d'$). Of course, $\partial_f$ also vanishes on $v^{d'}$ when $d' < d$. Therefore, if $f(v) = 0$ then $\partial_f$ vanishes on all powers of $v$.*

*By additivity, we conclude that every $f \in I(\{\langle v_1 \rangle, \ldots, \langle v_r \rangle\})$ is apolar to every power sum $v_1^d + \cdots + v_r^d$, $d > 0$ and, more generally, to every linear combination of $v_1^d, \ldots, v_r^d$, $d > 0$.*

Now, we prove the following version of the apolarity lemma.

**Lemma 1.** *Let $x \in S_d$ with $d > 0$, and $X := \{\langle v_1 \rangle, \ldots, \langle v_r \rangle\} \subset \mathbb{P}S_1$. Then,*

$$x \in \langle v_1^d, \ldots, v_r^d \rangle \iff I(X) \subseteq \text{Ann } x.$$

**Proof.** Suppose that $x \in \langle v_1^d, \ldots, v_r^d \rangle$, that is, $x$ is a linear combination of $v_1^d, \ldots, v_r^d$. By Remark 1, every $f \in I(X)$ is apolar to such linear combination, hence is apolar to $x$. Therefore, $I(X) \subseteq \text{Ann } x$.

Conversely, let us suppose that $I(X) \subseteq \text{Ann } x$. By the evaluation of Equation (4), it follows that $f \in S^d$ vanishes on $v \in S_1$ if and only if it is apolar to $v^d$. In other terms, the set of all $f \in S^d$ that vanish on $v \in S_1$ is the orthogonal complement of $v^d$ with respect to the perfect pairing given by apolarity in degree $d$. Hence

$$I(X) \cap S^d = \langle v_1^d, \ldots, v_r^d \rangle^\perp.$$

Since $I(X) \subseteq \text{Ann } x$, we have in particular that $x$ is orthogonal to $I(X) \cap S^d$. Hence, $x \in \langle v_1^d, \ldots, v_r^d \rangle$. □

A general and detailed version of the apolarity lemma can be found in [14], Lemma 1.15. It holds in every characteristic and uses divided powers. Lemma 1 is basically equivalent to [14], Lemma 1.15(i) and (ii), restricted to the characteristic zero case.

To illustrate the lemma with a simple (nearly trivial) example, let $S_\bullet := \mathbb{C}[x_0, x_1]$, $S^\bullet := \mathbb{C}[x^0, x^1]$, with $(x^0, x^1)$ being the dual basis of $(x_0, x_1)$. The evaluation of a polynomial $p = p(x^0, x^1) \in S^\bullet$ on $a^0 x_0 + a^1 x_1 \in S_1$ is just $p(a^0, a^1)$. When $p$ is homogeneous of degree $d$ and $p(v) = 0$ for a $v \in S_1$, then $p$ vanishes on all scalar multiples of $v$, that is, on all elements of $\langle v \rangle$. If $v \neq 0$, we can say that $\langle v \rangle \in \mathbb{P}S_1$ is a root of $p$.

Let us find the sum of squares decompositions of $f := x_0 x_1 \in S_2$ using the apolarity lemma. We have $\partial_{x^0} f = x_1$ and $\partial_{x^1} f = x_0$, hence Ann $f$ has no degree 1 homogeneous nonzero elements. In degree 2 we have Ann $f \cap S^2 = \langle {x^0}^2, {x^1}^2 \rangle$. Obviously, $S^d \subset \text{Ann } f$ for all $d \geq 3$. Now, for a finite set $X = \{\langle v_1 \rangle, \ldots, \langle v_r \rangle\} \subset \mathbb{P}S_1$ of $r$ (distinct) points, $I(X)$ is the set of all (polynomial) multiples in $S^\bullet$ of the polynomial $p \in S^r$ with roots precisely $\langle v_1 \rangle, \ldots, \langle v_r \rangle$. Hence, by the apolarity lemma, for every homogeneous $p \in \text{Ann } f \cap S^r$ that has $r$ (distinct) roots $\langle v_1 \rangle, \ldots, \langle v_r \rangle$, we have $f \in \langle v_1^2, \ldots, v_r^2 \rangle$. It easily follows that $f$ can be decomposed as a sum of squares of appropriate scalar multiples of $v_1, \ldots v_r$. For instance, $p := {x^0}^2 - {x^1}^2 \in \text{Ann } f \cap S^2$ gives rise to the decomposition

$$x_0 x_1 = \left(\frac{1}{2}(x_0 + x_1)\right)^2 + \left(\frac{i}{2}(x_0 - x_1)\right)^2.$$

### 3.2. The Classically Known Results on Maximum Rank

From the elementary theory of quadratic forms, known since long time, follows that the rank of a quadratic form $f \in S_2$, equals the rank of its representing matrix with respect to whatever given (ordered) basis. Hence, the maximum rank equals $\dim S_1$, that is, the number of indeterminates (if $S_\bullet$ is considered as a ring of polynomials).

To find the maximum rank of binary forms of given degree, apolarity is very effective. Indeed, let us consider the following simple description of the apolar ideal of a binary form.

**Proposition 1.** *Let $f \in S_d \setminus \{0\}$, with $\dim S_1 = 2$. Then, Ann $f$ is generated by a form $a \in S^s$ and a form $b \in S^{d+2-s}$ for some integer $s \leq (d+2)/2$.*

**Proof.** See [14], Theorem 1.44(iv). □

As reported in [14], this classical result is due to Macaulay. When $\dim S_1 = 2$, the ideal of a set of $r$ distinct points in $\mathbb{P}S_1$ is generated by a homogeneous form. If the form $a$ in the statement of Proposition 1 is squarefree, from Lemma 1 follows that the rank of $f$ is $s$, and some appropriate pairwise non-proportional roots of $a$ in $S_1$ give the linear forms of a $d$th power sum decomposition. Taking also into account that $a$ and $b$ must be coprime, since they generate an ideal that contains $S^{d+1}$, we also deduce that if $a$ is not squarefree, then the rank is $d + 2 - s$ and for every finite subset $X \subset \mathbb{P}S_1$ there exists a $d$th power sum decomposition such that $\langle v \rangle \notin X$ for every linear form $v$ in it.

From the above, it is clear that the rank of $f$ is at most $d$, and can be $d$ only if $\operatorname{Ann} f$ contains the square of a linear form. Given a basis $(x_0, x_1)$ of $S_1$, the apolar ideal of $x_0^{d-1} x_1$ contains $x_1^2$, with $(x^0, x^1)$ being the dual basis. We conclude that the maximum rank of binary forms of degree $d > 0$ is $d$.

In [9], Sections 96 and 97, one finds that the maximum ranks of ternary and quaternary cubics are 5 and 7, respectively. As we mentioned in the Introduction, beyond these classical results, only two new cases have been recently worked out: the maximum rank is 7 for ternary quartics (which has been determined in [5,17]) and 10 for ternary quintics (see [11,18]).

To our knowledge, no other values for the maximum rank have been determined to date; the known values can therefore be summarized in the following Table 1.

**Table 1.** Known maximum Waring ranks in $S_d$, with $\dim S_1 = n$.

|   | 1 | 2 | 3 | 4 | 5 | d |
|---|---|---|---|---|---|---|
| 1 | 1 | 1 | 1 | 1 | 1 | 1 |
| 2 | 1 | 2 | 3 | 4 | 5 | d |
| 3 | 1 | 3 | 5 | 7 | 10 | - |
| 4 | 1 | 4 | 7 | - | - | - |
| n | 1 | n | - | - | - | - |

### 3.3. Elementary Bounds on Maximum Waring Rank

Let us recall a common geometric viewpoint on Waring rank. Given $f \in S_d \setminus \{0\}$, we have a point $\langle f \rangle \in \mathbb{P}S_d$ and we have to express $f$ as sum of $d$th powers. The set of (spans of) $d$th powers of linear forms is an algebraic variety in $\mathbb{P}S_d$: the image of the embedding $\mathbb{P}S_1 \to \mathbb{P}S_d$, $\langle v \rangle \mapsto \langle v^d \rangle$. This embedding turns out to be equivalent to a much studied embedding: the Veronese embedding (also called $d$-uple embedding: see, e.g., [16], Chapter I, Exercise 2,12). Its image is sometimes called the Veronese variety (see [1] 4.3.7). The problem of finding a power sum decomposition of $f$ is equivalent to the problem of finding a set of points $X$ in the Veronese variety such that $f \in \langle X \rangle$, that is, such that the point $\langle f \rangle$ lies in the projective span of $X$ (one has to take into account that a scalar multiple of a $d$th power is a $d$th power as well, since $\mathbb{K}$ is algebraically closed). Since the Veronese variety spans $\mathbb{P}S_d$, the Waring rank is well defined for all forms, and it is at most

$$\dim S_d = \binom{d + n - 1}{n - 1},$$

where $n := \dim S_1$ and the parenthesized notation stands for the binomial coefficient. This gives an elementary upper bound on rank (which could easily be slightly lowered, but we are not interested in doing this here).

Let us now consider the union $U$ of all projective spans of $r$ distinct points in the Veronese variety. Clearly, for every $\langle f \rangle \notin U$, the rank of $f$ is greater than $r$. Elementary tools of algebraic geometry allow one to estimate the dimension of the Zariski closure $\overline{U}$ of $U$ (see [1], 4.9.5 or [16], Chapter I, Section 2, p. 10), which is called the $(r-1)$th *secant variety* (of the Veronese variety). Roughly speaking, the set of all groups of $r$ points in the Veronese variety, which has dimension $n - 1 = \dim \mathbb{P}S_1$, is of dimension $r(n - 1)$. Since most of these groups spans a subspace of dimension $r - 1$, the expected dimension of $\overline{U}$ is $rn - 1$. When this number does not reach the dimension $\binom{d+n-1}{n-1} - 1$ of the entire space $\mathbb{P}S_d$, there exist forms with rank greater than $r$. Hence, the maximum rank in $S_d$ is at least

$$\left\lceil \frac{1}{n} \binom{d + n - 1}{n - 1} \right\rceil$$

(where the external parentheses denote the upper integer part; a similar notation $\lfloor \ldots \rfloor$ will be used for the lower integer part).

Note that when $d = 2$, the lower bound is $\lceil (n+1)/2 \rceil$, meanwhile the maximum rank is $n$. It is also worthy of being mentioned that the estimate of $\dim \overline{U}$ fails for $d = 2$ (and $n, r \geq 2$). More generally, that estimate fails when most points in $\overline{U}$ lie on infinitely many spans. In this case, the dimension of the secant variety drops, and the now classical theorem by Alexander and Hirschowitz gives the complete list of $n, d, r$ for which this happens (we refer the reader to the exposition in [4]). It turns out that for $d \geq 3$ the above lower bound can be raised by one in exactly four cases: $(n, d) \in \{(3,4), (4,4), (5,3), (5,4)\}$.

Given $n, d$, if $r$ is the least value for which $\overline{U} = \mathbb{P}S_d$ then, by some basic algebrogeometric considerations which we skip here, for all $\langle f \rangle$ in a nonempty open Zariski subset of $\mathbb{P}S_d$, $f$ is actually of rank $r$. In this situation, it is customary to say that $r$ is the rank of a *generic form* in $S_d$. In the context of tensor rank, the striking outcome of the Alexander–Hirschowitz theorem is indeed the exact value of the rank of a generic form (the lower bounds in the exceptional cases are only some of the many consequences). In the following sections, we review the enhanced lower and upper bounds that have been found recently.

## 4. Lower Bounds

To find a good lower bound on the set of the symmetric ranks of all symmetric tensors over $\mathbb{K}$ of given order $d$ and dimension $n$, it suffices to find a form in $S_d$, when $\dim S_1 = n$, with high Waring rank.

The (few) lower bounds which we are aware of have been obtained by finding some special forms of high rank. Since the rank of a generic form gives a lower bound, the challenge is to exceed it. In this section, we present the special forms of high rank that have given the best known lower bounds.

### 4.1. What Monomials Tell Us

To begin with, let us consider binary forms, for which ranks are quite well understood. The maximum rank of degree $d$ binary forms is $d$, while the rank of a generic degree $d$ binary form is $\lfloor (d+2)/2 \rfloor$ (it can be deduced from Proposition 1). Moreover, a degree $d$ binary form of maximum rank can be turned into a monomial by a change of coordinates. Hence, for binary forms, the maximum rank is reached by monomials. For quadrics, whose rank is obviously well understood, the maximum rank of monomials is two (unless $\dim S_1 \leq 1$), meanwhile the maximum rank is reached by generic forms (and equals $\dim S_1$, that from a polynomial viewpoint is the number of indeterminates).

The rank of all monomials has been determined by Carlini, Catalisano e Geramita in [19]. In dimension three, it turns out that the monomial $xy^s z^s$ is of rank $(s+1)^2$ and $xy^{s-1}z^s$ of rank $s(s+1)$. This gives a lower bound that asymptotically approaches $d^2/4$ for the rank of ternary forms of degree $d$, while a generic form has rank asymptotically approaching $d^2/6$. According to [12], Proposition 3.4, the asymptotic estimate of maximum rank for ternary forms is actually $d^2/4$. When the number of variables is four or greater, the maximum rank of monomials does not exceed the rank of a generic form of the same degree.

In view of the above, a first guess on maximum rank could be that the maximum rank is reached either by monomials or by generic forms. However, for ternary quartics, for which the maximum rank is known from Kleppe's master thesis [5,17], it exceeds by one the maximum rank of both the monomials and the generic forms. The maximum rank exceeds both the maximum rank, of the monomials and of the generic forms, for ternary quintics too (see [11,18]). Buczyński and Teitler in [11] also found forms in more than three variables with rank exceeding by one the rank of generic forms. An improvement by one might seem not too exciting but, at least for ternary forms, one cannot hope to go much farther. Indeed, the upper bound given in [12], Proposition 3.3 shows that the maximum rank of a degree $d$ ternary form can exceed the maximum rank of monomials by at most $d$. Thus, the initial guess may be modified by expecting that the maximum rank could only slightly exceed the maximum rank of either monomials or generic forms.

For a detailed discussion on maximum rank of monomials, we refer the reader to [20]. Let us now outline how the rank of monomials has been bounded from below, and how that technique has been enhanced by Buczyński and Teitler, to exceed the previously known lower bounds on maximum ranks.

### 4.2. What Hilbert Functions Tell Us

The *Hibert function* of a graded module $\oplus_{d \in \mathbb{Z}} M_d$ over the graded $\mathbb{K}$-algebra $S^\bullet$ can be simply defined as the function that on each $d \in \mathbb{Z}$ takes value $\dim_{\mathbb{K}} M_d$. This is a fundamental tool in algebraic geometric, and is still much studied. To let readers who are not acquainted with algebraic geometry get a taste of the fundamental nature of Hilbert function, let us mention that the degree and the dimension of an algebraic set can easily be get from a naturally associated Hilbert function. More precisely, let $X \subseteq \mathbb{P}(S_1)$ be the set of all points on which some system of homogenous forms in $S^\bullet$ vanishes. Then, the Hilbert function $H_X$ of $S^\bullet/I(X)$ coincides with a polynomial $p_X$ (the *Hilbert polynomial* of $X$) for all sufficiently large degrees. The degree of $p_X$ gives the dimension $n$ of $X$ and $n!$ times the leading coefficient of $p_X$ gives the degree of $X$. In the case when $X$ is a finite set of $r$ points, which by Lemma 1 is of our interest here, we get that $H_X(d) = r$ for all sufficiently large $d$. Below, we take a few lines to directly show this fact in an elementary way; readers who are interested in the general properties of Hilbert functions can find them in many basic textbooks of algebraic geometry (e.g., in [16], Chapter I, Section 7).

Let $X = \{\langle v_1 \rangle, \ldots, \langle v_r \rangle\} \subseteq \mathbb{P}(S_1)$ be a set of $r$ points and let $H_X$ be the Hilbert function of $S^\bullet/I(X)$. The degree $d$ component of that quotient is $S^d/I(X)_d$, where $I(X)_d = S^d \cap I(X)$ is the space of degree $d$ forms that vanish on $X$. From the evaluation in Equation (4), it follows that $I(X)_d = \langle v_1^d, \ldots, v_r^d \rangle^\perp$ (this fact has been already noticed in the proof of Lemma 1). It follows that

$$H_X(d) = \dim \langle v_1^d, \ldots, v_r^d \rangle .$$

Note that it is easy to find $l^1, \ldots, l^{r-1} \in S^1$ such that $l^i(v_i) = 0$ and $l^i(v_r) \neq 0$ for each $i \in \{1, \ldots, r-1\}$. It follows that when $d \geq r-1$ the hyperplane $(l^{1^{d-r+2}} l^2 \cdots l^{r-1})^\perp < S_d$ contains $v_1^d, \ldots, v_{r-1}^d$, but not $v_r^d$. In a similar way, it can be found hyperplanes that do not contain a given $v_i$ but contain all $v_j$ with $j \neq i$. This shows that $v_1^d, \ldots, v_r^d$ are linearly independent, hence

$$\dim \langle v_1^d, \ldots, v_r^d \rangle = r, \qquad \forall d \geq r-1 .$$

Therefore, $H_X(d) = r$ for all sufficiently large $d$ ($\geq r-1$, in this case).

### 4.3. What Hyperplane Sections Tell Us

Let $\overline{S}^\bullet = S^\bullet/I$ be the quotient of $S^\bullet$ by a homogeneous ideal $I$ and suppose that $\bar{l} \in \overline{S}^1$ is not a zero divisor in $\overline{S}^\bullet$. Then, the multiplication by $\bar{l}$ in $\overline{S}^\bullet$ injects each homogeneous component $\overline{S}^d$ into $\overline{S}^{d+1}$. Let $H$ and $H'$ be, respectively, the Hilbert functions of $\overline{S}^\bullet$ and of its quotient $\overline{S}^\bullet/(\bar{l})$ over the ideal generated by $\bar{l}$. Thus, we have

$$H'(d) = H(d) - H(d-1) .$$

Consequently, we also have

$$H(d) = \sum_{i=0}^{d} H'(i) .$$

In this way, relevant properties of $H$ can be deduced from properties of $H'$.

The quotient $\overline{S}^\bullet/(\bar{l})$ is naturally isomorphic to the quotient of $S^\bullet$ over the ideal $I + (l)$, with $l \in S^1$ being a representative of $\bar{l}$. When $I$ is the ideal of an algebraic set $X \subseteq \mathbb{P}(S^1)$ (that is, the set of all points where some set of homogeneous elements of $S^\bullet$ vanish), the algebraic set defined by $I + (l)$ (that is, the set of all points where all the homogeneous elements of $I + (l)$ vanish) is the intersection of $X$

with the hyperplane given by $l$. From a geometric viewpoint, the idea is that features of hyperplane sections of $X$ give relevant information on $X$. This idea is ubiquitous in algebraic geometry.

Let us see what we get in the case of our interest. When $X$ is a finite set of $r$ points, if a product of homogeneous elements $xy$ vanishes on $X$ but $y$ does not, then $x$ must vanish on some point of $X$. Conversely, if $x$ vanishes on some point of $X$, it is not difficult to find a nonzero $y$ in some $S^d$ such that $xy$ vanishes on $X$. Therefore, to find $\bar{l} \in S^{\bullet}/I(X)$ that is not a zero divisor is to find an hyperplane that does not meet $X$. In this case, since we know that $H$ takes values $r$ for $d \geq r-1$, we conclude that

$$r = \sum_{i=0}^{r-1} H'(i),$$

with $H'$ being the Hilbert function of $S^{\bullet}/(I(X) + (l))$.

### 4.4. Rank of Monomials

The result we have just discussed, in conjunction with the apolarity lemma gives a way to bound the rank of $f \in S_d$ from below. If we fix $l \in S^1 \setminus \{0\}$, for every finite set of $r$ points $X$ such that $I(X) \subseteq \operatorname{Ann} f$ we obviously have $I(X) + (l) \subseteq \operatorname{Ann} f + (l)$. Denoting by $H_f$ the Hilbert function of $S^{\bullet}/(\operatorname{Ann} f + (l))$, we have $H'(i) \geq H_f(i)$ for all $i$, hence

$$r \geq \sum_{i=0}^{d} H_f(i) =: b(l).$$

Taking into account Lemma 1, we have that every power sum decomposition $f = v_1^d + \cdots + v_r^d$ for which no $\langle v_i \rangle$ lies on the hyperplane $\langle l \rangle^{\perp}$, has at least $b(l)$ summands. On this basis, a first rough idea to find a lower bound on the rank of $f$ is to find the minimum $b(l)$, with $\langle l \rangle$ varying in an infinite subset of $\mathbb{P}(S^1)$ such that each point of $\mathbb{P}(S_1)$ lies on at most a finite number of the hyperplanes $\langle l \rangle^{\perp}$ (e.g., one may take an irreducible curve in $\mathbb{P}(S^1)$ contained in no hyperplane). However, Carlini, Catalisano and Geramita followed another interesting path.

For whatever $X$, the ideal $(I(X) : l) = \{g \in S^{\bullet} : gl \in I(X)\}$ is clearly the ideal of the set $X' := X \setminus \mathbb{P}\langle l \rangle^{\perp}$. Hence, for the Hilbert function $H'_f$ of $S^{\bullet}/((\operatorname{Ann} f : l) + (l))$,

$$\sum_{i=0}^{d} H'_f(i)$$

cannot exceed the number of points in $X'$, and consequently the number of points in $X$, for whatever $X$. This holds for whatever choice of $l \in S^1 \setminus \{0\}$, which therefore can be chosen to maximize the above sum.

Now, let us consider a positive degree monomial $f = x_1^{a_1} \cdots x_n^{a_n}$ for a given basis $(x_1, \ldots, x_n)$ of $S_1$. Let $(x^1, \ldots, x^n)$ be the dual basis in $S^1$, and note that a monomial $x^{1^{b_1}} \cdots x^{n^{b_n}}$ is apolar to $f$ if and only if $b_i > a_i$ for some $i$. Moreover, for two different (monic) monomials $m$ and $m'$ that are not apolar to $f$, we have that $\partial_m f$ and $\partial_{m'} f$ cannot be proportional. This easily implies that $\operatorname{Ann} f$ is the ideal generated by $x^{1^{a_1+1}}, \ldots, x^{n^{a_n+1}}$.

With no loss of generality, we can assume $a_1 \leq \cdots \leq a_n$, and let $a_i$ be the first nonzero exponent. It is quite easy to recognize that $(\operatorname{Ann} f : x^i)$ is generated by

$$x^1, \ldots, x^{i-1}, x^{i^{a_i}}, \ldots, x^{n^{a_n+1}}$$

and $(\operatorname{Ann} f : x^i) + (x^i)$ by

$$x^1, \ldots, x^i, x^{i+1^{a_{i+1}+1}}, \ldots, x^{n^{a_n+1}}.$$

To calculate that for the Hilbert function $H'_f$ of $S^\bullet/((\operatorname{Ann} f : x^i) + (x^i))$, one has

$$\sum_{i=0}^{d} H'_f(i) = (a_{i+1}+1)\cdots(a_n+1)$$

is not difficult (and quite easy if one is familiar with Hilbert functions).

Since we are concerned with lower bounds on rank, we could end here the subsection. However, to agree that $r := (a_{i+1}+1)\cdots(a_{i+1}+1)$ is actually the rank of $f$ is quite easy because, similar to what is smartly remarked in [19], we have that

$$\left(x^1,\ldots,x^{i-1},x^{i+1^{a_{i+1}+1}} - x^{i^{a_{i+1}+1}},\ldots,x^{n^{a_n+1}} - x^{i^{a_n+1}}\right)$$

is the ideal of a set of $r$ distinct points and is contained in $\operatorname{Ann} f$.

### 4.5. Beyond Monomials and Generic Forms

As anticipated before, once the rank of monomials has been determined, one can find ternary monomials with rank much higher than the rank of generic forms of the same degree (which is known by the Alexander–Hirschowitz theorem). When the number of indeterminates is four or greater, the rank of generic forms can not be exceeded by monomials. We give now a brief account of how Buczyński and Teitler were able to beat both ternary monomials and generic quaternary forms, with one and the same argument. A more informative description can be directly found in [11].

Let $f \in S_d$ and $l \in S^1 \setminus \{0\}$. In the calculations before, the sum of all values of the Hilbert function of quotient algebras of the type $S^\bullet/(I+(l))$ turned out to be useful. When $I = \operatorname{Ann} f$, that sum bounds from below the number of summands of a decomposition whose linear forms are outside the hyperplane $\langle l \rangle^\perp$. When $I = (\operatorname{Ann} f : l)$, that sum directly bounds from below the rank of $f$. Note that the sum under consideration is nothing but the dimension of $S^\bullet/(I+(l))$ as a $\mathbb{K}$-vector space. We can give the following useful description of this dimension, using a relation between $I+(l)$ and $(I:l)$ that one often encounters when dealing with hyperplane sections.

When $S^\bullet/I$ is finite-dimensional, we have

$$\dim_{\mathbb{K}} \frac{S^\bullet}{I+(l)} = \dim_{\mathbb{K}} \frac{S^\bullet}{I} - \dim_{\mathbb{K}} \frac{I+(l)}{I} = \dim_{\mathbb{K}} \frac{S^\bullet}{I} - \dim_{\mathbb{K}} \frac{(l)}{(l) \cap I}.$$

Since the kernel of the homomorphism

$$S^\bullet \twoheadrightarrow \frac{(l)}{(l) \cap I}, \qquad x \mapsto [xl]_{(l) \cap I},$$

is $(I:l)$, we deduce that

$$\dim_{\mathbb{K}} \frac{S^\bullet}{I+(l)} = \dim_{\mathbb{K}} \frac{S^\bullet}{I} - \dim_{\mathbb{K}} \frac{S^\bullet}{(I:l)}.$$

When $I = \operatorname{Ann} f$, $S^\bullet/I$ is called the *apolar algebra*, and its dimension the *apolar length* of $f$. They can be denoted by $A_f$ and $\operatorname{al} f$. Note that even when $I = (\operatorname{Ann} f : l)$, $S^\bullet/I$ is an apolar algebra. Indeed, for whatever $x, y \in S^\bullet$, we have

$$x \in (\operatorname{Ann} f : y) \iff xy \in \operatorname{Ann} f \iff \partial_{xy} f = 0 \iff \partial_x \partial_y f = 0 \iff x \in \operatorname{Ann} \partial_y f,$$

that is, $(\operatorname{Ann} f : y) = \operatorname{Ann} \partial_y f$. Hence, when $I = (\operatorname{Ann} f : l)$, the quotient $A$ is the apolar algebra of $\partial_l f$.

From the formula $(\operatorname{Ann} f : y) = \operatorname{Ann} \partial_y f$ and the fact that apolarity is a perfect pairing in every degree, we get another interesting fact: the apolar length of $f$ equals the dimension of the vector space of all $\partial_y f$ with $y \in S^\bullet$.

We end up with

$$\dim_{\mathbb{K}} \frac{S^{\bullet}}{\operatorname{Ann} f + (l)} = \operatorname{al} f - \operatorname{al} \partial_l f,$$

$$\dim_{\mathbb{K}} \frac{S^{\bullet}}{(\operatorname{Ann} f : l) + (l)} = \operatorname{al} \partial_l f - \operatorname{al} \partial_{l^2} f,$$

the former being a lower bound on the number of summands of a decomposition of $f$ whose linear forms are outside the hyperplane $\langle l \rangle^{\perp}$, and the latter a lower bound on the rank of $f$. Based on these remarks, a good knowledge of apolar algebras, which can be obtained for instance from [14], clearly gives precious information on lower bounds on rank.

A first obvious step is to try to maximize $\operatorname{al} \partial_l f - \operatorname{al} \partial_{l^2} f$. To use a coordinate description, let us fix dual bases $(x_1 \ldots, x_n)$ and $(x^1 \ldots, x^n)$ of $S_1$ and $S^1$. To get $\operatorname{al} \partial_{l^2} f = 0$ we choose $l = x_1$ and consider a form

$$f := x_1 g + k, \qquad g, k \in \mathbb{K}[x_2, \ldots x_n].$$

Then, $\partial_l f = g$, and we have to choose $g$ with maximum apolar length. From [14], we can find the value of that maximum and learn that it is reached by a generic $g$ (that is, for all $g$ in a suitable nonempty open set in the Zariski topology). In conclusion, there exist degree $d$ forms in $n$ indeterminates with rank not less than the maximum apolar length of degree $d - 1$ forms in $n - 1$ indeterminates. Surprisingly, that maximum equals the maximum rank of degree $d$ monomials when $n = 3$, and the rank of generic forms of degree $d$ when $n = 4$ and $d$ is odd.

When $(\operatorname{Ann} f : l) = \operatorname{Ann} \partial_l f$ is considered instead of $\operatorname{Ann} f$ to get the lower bound, one might hope that for some special $f$ some of the linear forms might be forced to lie on the hyperplane $\langle l \rangle^{\perp}$. However, for geometric reasons, we expect that forms of high rank have many decompositions, which can therefore easily escape out of the hyperplane. Note also that, when $f$ is a monomial $x_1^{a_1} \cdots x_n^{a_n}$, $\operatorname{al} f - \operatorname{al} \partial_{x^i} f = \operatorname{al} \partial_{x^i} f - \operatorname{al} \partial_{x^{i^2}} f$ whenever $a_i \neq 0$, so that we have no loss in cutting out the part on the hyperplane. This might give some indication on why in the high rank examples found in [11] the first thing considered is to raise the value of $\operatorname{al} f - \operatorname{al} \partial_{x^i} f$

For a form $f$ of the type $x_1 g + k$, one can raise $\operatorname{al} f - \operatorname{al} \partial_{x^1} f$ by one by lowering $\operatorname{al} g$ by one, but at the cost of lowering $\operatorname{al}_{x^1} f - \operatorname{al} \partial_{x^1 2} f$ too. This causes a problem for decompositions that have some linear forms on the hyperplane $\langle x^1 \rangle^{\perp}$. The idea was to show that for suitable choices of the form $k$, such decompositions must have at least *two* forms on the hyperplane. To this end, note that if a decomposition of $f$ involve exactly one linear form $v$ that lies in $\langle x_1 \rangle^{\perp}$, then $f - v^d$ has a decomposition with all linear factors outside that hyperplane. Using the fact (pointed out before) that the apolar length equals the dimension of the vector space of all derivatives, it turns out that $\operatorname{al}(f - v^d) - \operatorname{al} \partial_{x^1}(f - v^d)$ can be kept hight for whatever choice of $v$. This is incompatible with the fact that for some $v$, $f - v^d$ has a decomposition with all linear factors outside $\langle x^1 \rangle^{\perp}$.

In view of determination of maximum rank, the Buczyński–Teitler lower bound is particularly interesting because, in conjunction with the upper bound in [18], shows that the maximum rank of ternary quintics is ten. Now, at the end of [12], Introduction, a possible guess for the maximum rank for ternary forms of an arbitrarily given degree $d$ is outlined, and if it is correct then the Buczyński–Teitler lower bound is the best possible for $d$ odd (and $n = 3$). In [13], the lower bound for ternary forms given by monomials is raised by one for even degrees too, and if the guess in [12] is correct, it cannot be improved further. Basically, the lower bound in [13] follows the second advice in [11], Remark 19, but uses a more specific example, similar to that in [11], Theorem 18 (which gives the lower bound of ten for ternary quintics), and the arguments are of a purely algebraic nature (do not involve geometric dimension counts).

## 5. Upper Bounds

### 5.1. Rise and (Relative) Fall of an Upper Bound of Genuinely Geometric Nature

We know that the rank of generic forms gives a lower bound for maximum rank. A simple geometric argument shows that twice that rank gives an upper bound. This can be proven in a more general context. Indeed, the rank of a point $P$ of a projective space $\mathbb{P}$, with respect to a variety in $\mathbb{P}$, is defined as the minimum number of points that can be fixed on the variety such that $P$ lies in their projective span. For the Veronese variety we recover the Waring rank.

Let us suppose that all forms in a nonempty open subset of a projective space $\mathbb{P}$ are of a certain rank $r_{gen}$ with respect to some given variety (as it is the case when $\mathbb{P} = \mathbb{P}S_d$, and $r_{gen}$ is the rank of generic forms). When the base field is $\mathbb{C}$ (or even $\mathbb{R}$), one can consider a small ball whose points all have rank $r_{gen}$. Every point in a projective line joining two points in the ball has rank at most the sum of the ranks of the two points, hence at most $2r_{gen}$. Since all such lines cover $\mathbb{P}$, we deduce that the maximum rank is at most $2r_{gen}$. When $\mathbb{K} \neq \mathbb{C}$ (but is algebraically closed according to our standing assumption, or at least infinite), to consider the Zariski topology causes no problems, since a nonempty intersection of a line with a Zariski open set is always an infinite set.

The upper bound $2r_{gen}$, due to Blekherman and Teitler (see [7]), in the case of the Waring rank has dramatically improved the previously known upper bounds from [6,21,22]. In addition, if we recall that, for generic binary forms of degree $d$, $r_{gen} = \lfloor (d+2)/2 \rfloor$, we have that the maximum rank for degree $d$ binary forms, which is $d$, it equals $2r_{gen} - 1$ for odd degrees and $2r_{gen} - 2$ for even degrees. This might induce to hope that the Blekherman–Teitler upper bound is nearly sharp.

Somewhat symmetrically to what they had done for lower bounds, Buczyński and Teitler, jointly with Han and Mella, showed that the maximum rank is at most $2r_{gen} - 1$. In some special cases, which include binary forms of even degrees, they also lowered the upper bound down to $2r_{gen} - 2$ (see [23], Theorem 3.9 and Example 3.10). This result, which would have given a strong evidence in favor of the sharpness of such bounds, had no such effect because, when [23] was announced, it was already known that for ternary forms the upper bound $2r_{gen}$ is quite mild (because of the asymptotic estimate in [12]).

The Blekherman–Teitler bound keeps its relevance in the general context of rank with respect to varieties, which, apart from its intrinsic interest, by means of Segre varieties can be readily applied to arbitrary (not necessarily symmetric) tensors too (see [1], 4.3.4). At the same time, because of the special nature of Veronese and Segre varieties, it is reasonable to expect that the Blekherman–Teitler bound on tensor rank can be sensibly lowered, by means of appropriate algebraic techniques.

### 5.2. Linear Algebraic Tools for Upper Bounds on Waring Rank

A widely shared attitude is that hyperplane sections should provide us with a good insight on how to build short power sum decompositions. Apart from the ubiquitous nature of this tool, in this specific case we have the following easy fact. Let $f \in S_d$ and let us take a hyperplane, defined by a nonzero $l \in S^1$. A power sum decomposition of the derivative

$$\partial_l f = v_1^{d-1} + \cdots + v_r^{d-1},$$

such that $l(v_i) \neq 0$ for all $i$, leads to a "lifted" form

$$F = \frac{1}{d\,l(v_1)} v_1^d + \cdots + \frac{1}{d\,l(v_r)} v_r^d, \tag{5}$$

because, from the Leibnitz rule for $\partial_l$ and Equation (4), it readily follows that

$$\partial_l F = v_1^{d-1} + \cdots + v_r^{d-1},$$

that is, $\partial_l F = \partial_l f$. The fact that $\partial_l(f - F) = 0$ implies that $f - F$ can be regarded as a form in one indeterminate less. Indeed, note that if $(x_1, \ldots, x_n)$ and $(x^1, \ldots, x^n)$ are dual bases with $l = x^1$, then $f - F$ is a form in $x_2, \ldots, x_n$. More invariantly, $T_\bullet := \ker \partial_l$ is a graded subring of $S_\bullet$, and since the elements of the ideal $(l)$ annihilate every element in $T_\bullet$, the apolar bilinear map $S^\bullet \times S_\bullet \to S_\bullet$ induces a bilinear map $T^\bullet \times T_\bullet \to T_\bullet$, where $T^\bullet := S^\bullet/(l)$, which is an apolar bilinear map as well, and $\dim T_1 = \dim S_1 - 1$. This allows us to design an inductive procedure, which basically is the one that has given the upper bound in [6].

The main difficulty in the mentioned procedure is the condition that $l(v_i) \neq 0$ for all $i$. The improvement given in [22] to the upper bound in [6] basically consists in a slightly better handling of that condition. The fact that the bounds in [6,22] are much milder than the Blekherman–Teitler bound should not be taken as an indication to abandon their path. Indeed, the scope of [6] is wider than upper bounds on rank, and Jelisiejew opens a line of investigation in that direction (in [21] Introduction one can find some additional motivation on why that line is worthy of being pursued).

To explain what is the further idea that led to the presently known best upper bounds on maximum rank of ternary forms of given degree, let us look at binary forms, for which the actual maximum rank is known. It does not come as a surprise that in a context where duality plays an important role, the case when points and hyperplanes are the same thing, at least from a geometric viewpoint, turns out to be relatively simple. Then, a reasonable basic principle is to give an important role to both points and hyperplanes.

The idea that underlies [12], Proposition 3.3 (that supersedes the Blekherman–Teitler bound for ternary forms), and the sharp upper bounds for ternary quartics and quintics (see [17,18]), is to look for decompositions that "split along a few lines". More precisely, in [18], it is shown that every ternary quintic $f$ can be decomposed as a sum $v_1^5 + \cdots + v_{10}^5$ such that the points $\langle v_1 \rangle, \ldots, \langle v_{10} \rangle$ belong to a union of four distinct lines $\mathbb{P}\langle l^1 \rangle^\perp, \ldots, \mathbb{P}\langle l^4 \rangle^\perp$ in $\mathbb{P}S_1$. In [12] is shown that every degree $d$ ternary form has a decomposition $v_1^d + \cdots + v_r^d$, with $r$ giving the upper bound, such that the points $\langle v_1 \rangle, \ldots, \langle v_r \rangle$ belong to a union of $d$ distinct lines $\mathbb{P}\langle l^1 \rangle^\perp, \ldots, \mathbb{P}\langle l^d \rangle^\perp$ in $\mathbb{P}S_1$. Of course, even a bare induction procedure based on hyperplane sections, when followed step by step, leads to a decomposition of $f \in S_d$ that splits along $d$ hyperplanes. What is new here is that the configuration of lines is fixed in advance, and is used to drive the construction step by step of the decomposition. Let us now describe in more detail how the construction works.

> Henceforth, we assume that $\dim S_1 = 3$.

To begin with, given a nonzero $x \in S^1$, a decomposition of $f \in S_d$ gives rise to a decomposition of $\partial_x f$ where all linear forms $v \in \langle x \rangle^\perp$ disappear. Hence, to have a decomposition that splits along lines given by $l^1, \ldots, l^d \in S^1 \setminus \{0\}$, a necessary condition is that $\partial_{l^1 \ldots l^d} f = 0$. At each step, we have to lift with respect to $l^i$ a decomposition with all linear forms outside the line $\langle l^i \rangle^\perp$. Hence, we need that $\langle l^1 \rangle, \ldots, \langle l^d \rangle$ are distinct. A further mild technical need is that $\partial_{l^1 \ldots \widehat{l^i} \ldots l^d} f \neq 0$, with the hat denoting omission. It is not difficult to find such $l^1, \ldots, l^d$ (see [12], Proposition 3.1).

To start the procedure, let us consider $\partial_{l^2 \ldots l^d} f$, which of course is a linear form and has a trivial 1-power decomposition with just one summand. However, we choose a redundant decomposition with two summands lying in $\langle l^1 \rangle^\perp$, such that the summands lie on no one of the lines $\mathbb{P}\langle l^2 \rangle^\perp, \ldots \mathbb{P}\langle l^d \rangle^\perp$, and fulfill a further condition that we shall explain later. Then, we can lift that decomposition with respect to $\partial_{l^2}$ as in Equation (5) and take the difference, denoted $g'_1$, with $\partial_{l^3 \ldots l^d} f$. Since the lift is annihilated by $\partial_{l^1}$, we have $\partial_{l^1} g'_1 = \partial_{l^1 l^3 \ldots l^d} f \neq 0$, and since the lift has the same $\partial_{l^2}$-derivative as $\partial_{l^3 \ldots l^d} f$, we also have $\partial_{l^2} g'_1 = 0$. Hence, $g'_1$ can be considered as a binary form and can be decomposed along the line $\mathbb{P}\langle l^2 \rangle^\perp$.

Now, we have come to a delicate step. If $g'_1$ is a square $v^2$, with $v \in S_1$, on the one hand, we have a cheap decomposition with one summand, while, on the other hand, if $l^i(v) = 0$ for some $i \geq 3$, the procedure cannot proceed. Unfortunately, from Proposition 1, it follows that all the decompositions with two summands must involve $v$ (or, more accurately, every such decomposition must have a zero summand; a fact that also follows from elementary considerations). To overcome this problem, a crucial condition has to be imposed on $g'_1$: the least degree of a generator of Ann $g'_1$, a number which in [12] is called the *binary length* of $g'_1$, must be 2 (the highest possible binary length for degree 2 forms). This way we can find a sufficiently cheap decomposition that does not stop the procedure. A similar condition is needed in the subsequent steps.

The binary length coincides with the *border rank*, which is an important invariant of a form in the context of tensor rank, but we do not need this fact. The important fact in the proof is that the condition on $g'_1$ (and on other forms related with $l^3, \ldots, l^d$) can be assured at the previous step, when the redundant decomposition of the linear form $\partial_{l2\ldots ld} f$ is chosen. A similar care has to be taken when choosing the decomposition of $g'_1$ with two summands, and so on for all decompositions that subsequently arise by lifting with respect to $l^3, l^4$, etc. To keep control of these conditions, a technical lemma is needed: see [12], Lemma 2.7, based on [12], Lemma 2.6.

That is how the upper bound

$$\left\lceil \frac{d^2 + 6d + 1}{4} \right\rceil$$

in [12], Proposition 3.3 has been obtained. The splitting that has given the sharp upper bound for quintics in [18] was obtained in a direct, not inductive, way. However, a new and probably simpler proof can be organized in a way that is closer to that outlined above.

### 5.3. A Nontrivial Feature of Split Decompositions

A rough dimension count shows that, to say that every degree $d$ ternary form has a power sum decomposition with $r$ summands that splits along $d$ (or $d-1$) lines, imposes a nontrivial constraint on $r$. Indeed, the variety of sets of $r$ points belonging to the union of given $d$ lines is of dimension $r$. Each of those sets gives a space of decompositions of dimension at most $r-1$. Finally, the set of all unions of $d$ lines has dimension $2d$. To reach the dimension of the space of all degree $d$ ternary forms, we need that

$$2r - 1 + 2d \geq \binom{d+2}{2} - 1,$$

hence

$$r \geq \frac{d^2 - d + 2}{4}.$$

This is quite near to (and in fact a bit less than) the lower bound on ternary forms discussed in the previous section, that is,

$$\left\lceil \frac{d^2 + 2d + 5}{4} \right\rceil$$

(for $d \geq 2$). Note that, in most cases, a generic degree $d$ ternary form does not have a decomposition that splits along $d$ lines. Hence, the above dimension count indicates that this procedure could be particularly suitable for finding the maximum rank.

## 6. Summary

Let us summarize the state of the knowledge on the maximum rank $r_{max}(n, d)$ of forms of degree $d > 0$ in $n$ variables, which has been presented in this article. For ternary forms:

$$\left\lceil \frac{d^2 + 2d + 5}{4} \right\rceil \leq r_{max}(3, d) \leq \left\lceil \frac{d^2 + 6d + 1}{4} \right\rceil, \quad \forall d \geq 2.$$

For $n \geq 4$, we have the lower and upper bounds given by the rank $r_{gen}$ of generic forms and its double, which hold in general for the rank with respect to a variety. The enhancements obtained in [11,23] allow raising by one the lower bound when $d$ is odd, and lower the upper bound by one. Let us mention that, according to the list in [1], 5.4.1, the exceptional cases from the Alexander and

Hirschowitz's theorem, $(3,4), (4,4), (5,3), (5,4)$, are included in the special cases for which the upper bound $2r_{gen}$ can be lowered *by two* according to Theorem 3.9 in [23].

The other special cases are a bit more cumbersome to be detected, so we do not take them into account in the following summary:

$$\left\lceil \frac{1}{n}\binom{n+d-1}{n-1} \right\rceil + \varepsilon \leq r_{max}(n,d) \leq 2\left\lceil \frac{1}{n}\binom{n+d-1}{n-1} \right\rceil + \varepsilon' - 1, \quad \forall n \geq 4, \forall d > 0,$$

with

$$\varepsilon = \begin{cases} 1 & \text{when } n = 4 \text{ and } d \text{ is odd and} \geq 3, \text{ or } (n,d) \in \{(4,4), (5,3), (5,4)\} \\ 0 & \text{otherwise} \end{cases},$$

$$\varepsilon' = \begin{cases} 1 & \text{when } (n,d) \in \{(4,4), (5,3), (5,4)\} \\ 0 & \text{otherwise} \end{cases}.$$

To give a more concrete idea of the above values, we explicitly report some ranges in Table 2, which enrich Table 1.

**Table 2.** Ranges for $r_{max}(n,d)$ in low degree.

|   | 1 | 2 | 3 | 4 | 5 | 6 |
|---|---|---|---|---|---|---|
| 1 | 1 | 1 | 1 | 1 | 1 | 1 |
| 2 | 1 | 2 | 3 | 4 | 5 | 6 |
| 3 | 1 | 3 | 5 | 7 | 10 | 13–18 |
| 4 | 1 | 4 | 7 | 10–18 | 15–27 | 21–41 |
| 5 | 1 | 5 | 8–14 | 15–28 | 26–51 | 42–83 |
| 6 | 1 | 6 | 10–19 | 21–41 | 42–83 | 77–153 |

Let us conclude by recalling that, if the symmetric rank was well understood, it would be easy to determine $r_{max}(n,d)$. That is why the techniques invented to find $r_{max}(n,d)$ may hopefully indicate some ways to understand tensor rank, which would be a considerable achievement, because of the recently recognized high applicative interest of this topic.

**Funding:** Financial support provided by Università degli Studi di Napoli Federico II.

**Acknowledgments:** We are grateful to Zach Teitler for providing us with some updated information on the recent literature.

**Conflicts of Interest:** The author declares no conflict of interest.

## References

1. Landsberg, J. *Tensors: Geometry and Applications*; American Mathematical Society (AMS): Providence, RI, USA, 2012; pp. 1–439.
2. Geramita, A. Exposé I A: Inverse systems of fat points: Waring's problem, secant varieties of Veronese varieties and parameter spaces for Gorenstein ideals. In *The Curves Seminar at Queen's, Vol. X*; Queen's University: Kingston, ON, Canada, 1996; pp. 2–114.
3. Alexander, J.; Hirschowitz, A. Polynomial interpolation in several variables. *J. Algebr. Geom.* **1995**, *4*, 201–222.
4. Brambilla, M.C.; Ottaviani, G. On the Alexander–Hirschowitz theorem. *J. Pure Appl. Algebra* **2008**, *212*, 1229–1251, doi:10.1016/j.jpaa.2007.09.014. [CrossRef]
5. Kleppe, J. Representing a Homogeneous Polynomial as a Sum of Powers of Linear Forms. Master's Thesis, Candidatum Scientiarum, Department of Mathematics, University of Oslo, Oslo, Norway, 1999.
6. Białynicki-Birula, A.; Schinzel, A. Representations of multivariate polynomials by sums of univariate polynomials in linear forms. *Colloq. Math.* **2008**, *112*, 201–233. [CrossRef]
7. Blekherman, G.; Teitler, Z. On maximum, typical and generic ranks. *Math. Ann.* **2015**, *362*, 1021–1031. [CrossRef]
8. Ellison, W.J. A "Waring's problem" for homogeneous forms. *Math. Proc. Camb. Philos. Soc.* **1969**, *65*, 663–672. [CrossRef]

9. Segre, B. The non-singular cubic surfaces. *Bull. Amer. Math. Soc.* **1943**, *49*, 350–352.
10. Baker, H.F. *Principles of Geometry. Volume 3. Solid Geometry. Quadrics, Cubic Curves in Space, Cubic Surfaces*; Cambridge Library Collection; Cambridge University Press: Cambridge, UK, 1923.
11. Buczyński, J.; Teitler, Z. Some examples of forms of high rank. *Collect. Math.* **2016**, *67*, 431–441. [CrossRef]
12. De Paris, A. The asymptotic leading term for maximum rank of ternary forms of a given degree. *Linear Algebra Appl.* **2016**, *500*, 15–29. [CrossRef]
13. De Paris, A. High-rank ternary forms of even degree. *Arch. Math.* **2017**, *109*, 505–510. [CrossRef]
14. Iarrobino, A.; Kanev, V. *Power Sums, Gorenstein Algebras, and Determinantal Loci. With an Appendix "The Gotzmann Theorems and the Hilbert Scheme' by Anthony Iarrobino and Steven L. Kleiman*; Lecture Notes in Mathematics; Springer: Berlin, Germany, 1999; Volume 1721, doi:10.1007/BFb0093426.
15. Ehrenborg, R.; Rota, G. Apolarity and canonical forms for homogeneous polynomials. *Eur. J. Comb.* **1993**, *14*, 157–181. [CrossRef]
16. Hartshorne, R. *Algebraic Geometry*; Graduate Texts in Mathematics, No. 52; Springer: New York, NY, USA; Heidelberg, Germany, 1977.
17. De Paris, A. A proof that the maximum rank for ternary quartics is seven. *Matematiche* **2015**, *70*, 3–18. [CrossRef]
18. De Paris, A. Every ternary quintic is a sum of ten fifth powers. *Int. J. Algebra Comput.* **2015**, *25*, 607–631. [CrossRef]
19. Carlini, E.; Catalisano, M.; Geramita, A. The solution to the Waring problem for monomials and the sum of coprime monomials. *J. Algebra* **2012**, *370*, 5–14. [CrossRef]
20. Holmes, E.; Plummer, P.; Siegert, J.; Teitler, Z. Maximum Waring ranks of monomials and sums of coprime monomials. *Commun. Algebra* **1996**, *44*, 4212–4219. [CrossRef]
21. Ballico, E.; De Paris, A. Generic Power Sum Decompositions and Bounds for the Waring Rank. *Discret. Comput. Geom.* **2017**, *57*, 896–914, doi:10.1007/s00454-017-9886-7. [CrossRef]
22. Jelisiejew, J. An upper bound for the Waring rank of a form. *Arch. Math.* **2014**, *102*, 329–336. [CrossRef]
23. Buczyński, J.; Han, K.; Mella, M.; Teitler, Z. On the locus of points of high rank. *Eur. J. Math.* **2018**, *4*, 113–136. [CrossRef]

© 2018 by the author. Licensee MDPI, Basel, Switzerland. This article is an open access article distributed under the terms and conditions of the Creative Commons Attribution (CC BY) license (http://creativecommons.org/licenses/by/4.0/).

Article

# The Hitchhiker Guide to: Secant Varieties and Tensor Decomposition [†]

**Alessandra Bernardi** [1], **Enrico Carlini** [2], **Maria Virginia Catalisano** [3], **Alessandro Gimigliano** [4,*] **and Alessandro Oneto** [5]

1. Dipartimento di Matematica, Università di Trento, 38123 Trento, Italy; alessandra.bernardi@unitn.it
2. Dipartimento di Scienze Matematiche, Politecnico di Torino, 10129 Turin, Italy; enrico.carlini@polito.it
3. Dipartimento di Ingegneria Meccanica, Energetica, Gestionale e dei Trasporti, Università degli studi di Genova, 16145 Genoa, Italy; catalisano@diptem.unige.it
4. Dipartimento di Matematica, Università di Bologna, 40126 Bologna, Italy
5. Barcelona Graduate School of Mathematics, and Universitat Politècnica de Catalunya, 08034 Barcelona, Spain; alessandro.oneto@upc.edu
* Correspondence: Alessandr.Gimigliano@unibo.it
† The initial elaboration of this work had Anthony V. Geramita as one of the authors. The paper is dedicated to him.

Received: 9 October 2018; Accepted: 14 November 2018; Published: 8 December 2018

**Abstract:** We consider here the problem, which is quite classical in Algebraic geometry, of studying the secant varieties of a projective variety $X$. The case we concentrate on is when $X$ is a Veronese variety, a Grassmannian or a Segre variety. Not only these varieties are among the ones that have been most classically studied, but a strong motivation in taking them into consideration is the fact that they parameterize, respectively, symmetric, skew-symmetric and general tensors, which are decomposable, and their secant varieties give a stratification of tensors via tensor rank. We collect here most of the known results and the open problems on this fascinating subject.

**Keywords:** additive decompositions; secant varieties; Veronese varieties; Segre varieties; Segre-Veronese varieties; Grassmannians; tensor rank; Waring rank; algorithm

## 1. Introduction

### 1.1. The Classical Problem

When considering finite dimensional vector spaces over a field $\Bbbk$ (which for us, will always be algebraically closed and of characteristic zero, unless stated otherwise), there are three main functors that come to attention when doing multilinear algebra:

- the tensor product, denoted by $V_1 \otimes \cdots \otimes V_d$;
- the symmetric product, denoted by $S^d V$;
- the wedge product, denoted by $\bigwedge^d V$.

These functors are associated with three classically-studied projective varieties in algebraic geometry (see e.g., [1]):

- the Segre variety;
- the Veronese variety;
- the Grassmannian.

We will address here the problem of studying the higher secant varieties $\sigma_s(X)$, where $X$ is one of the varieties above. We have:

$$\sigma_s(X) := \overline{\bigcup_{P_1,\ldots,P_s \in X} \langle P_1, \ldots, P_s \rangle} \qquad (1)$$

i.e., $\sigma_s(X)$ is the Zariski closure of the union of the $\mathbb{P}^{s-1}$'s, which are $s$-secant to $X$.

The problem of determining the dimensions of the higher secant varieties of many classically-studied projective varieties (and also projective varieties in general) is quite classical in algebraic geometry and has a long and interesting history. By a simple count of parameters, the expected dimension of $\sigma_s(X)$, for $X \subset \mathbb{P}^N$, is $\min\{s(\dim X) + (s-1), N\}$. This is always an upper-bound of the actual dimension, and a variety $X$ is said to be defective, or $s$-defective, if there is a value $s$ for which the dimension of $\sigma_s(X)$ is strictly smaller than the expected one; the difference:

$$\delta_s(X) := \min\{s(\dim X) + (s-1), N\} - \dim \sigma_s(X)$$

is called the $s$-defectivity of $X$ (or of $\sigma_s(X)$); a variety $X$ for which some $\delta_s$ is positive is called defective.

The first interest in the secant variety $\sigma_2(X)$ of a variety $X \subset \mathbb{P}^N$ lies in the fact that if $\sigma_2(X) \neq \mathbb{P}^N$, then the projection of $X$ from a generic point of $\mathbb{P}^N$ into $\mathbb{P}^{N-1}$ is an isomorphism. This goes back to the XIX Century with the discovery of a surface $X \subset \mathbb{P}^5$, for which $\sigma_2(X)$ is a hypersurface, even though its expected dimension is five. This is the Veronese surface, which is the only surface in $\mathbb{P}^5$ with this property. The research on defective varieties has been quite a frequent subject for classical algebraic geometers, e.g., see the works of F. Palatini [2], A. Terracini [3,4] and G. Scorza [5,6].

It was then in the 1990s that two new articles marked a turning point in the study about these questions and rekindled the interest in these problems, namely the work of F. Zak and the one by J. Alexander and A. Hirshowitz.

Among many other things, like, e.g., proving Hartshorne's conjecture on linear normality, the outstanding paper of F. Zak [7] studied Severi varieties, i.e., non-linearly normal smooth $n$-dimensional subvarieties $X \subset \mathbb{P}^N$, with $\frac{2}{3}(N-1) = n$. Zak found that all Severi varieties have defective $\sigma_2(X)$, and, by using invariant theory, classified all of them as follows.

**Theorem 1.** *Over an algebraically-closed field of characteristic zero, each Severi variety is projectively equivalent to one of the following four projective varieties:*

- *Veronese surface $v_2(\mathbb{P}^2) \subset \mathbb{P}^5$;*
- *Segre variety $v_{1,1}(\mathbb{P}^2 \times \mathbb{P}^2) \subset \mathbb{P}^8$;*
- *Grassmann variety $Gr(1,5) \subset \mathbb{P}^{14}$;*
- *Cartan variety $\mathbb{E}^{16} \subset \mathbb{P}^{26}$.*

Moreover, later in the paper, also Scorza varieties are classified, which are maximal with respect to defectivity and which generalize the result on Severi varieties.

The other significant work is the one done by J. Alexander and A. Hirschowitz; see [8] and Theorem 2 below. Although not directly addressed to the study of secant varieties, they confirmed the conjecture that, apart from the quadratic Veronese varieties and a few well-known exceptions, all the Veronese varieties have higher secant varieties of the expected dimension. In a sense, this result completed a project that was underway for over 100 years (see [2,3,9]).

### 1.2. Secant Varieties and Tensor Decomposition

Tensors are multidimensional arrays of numbers and play an important role in numerous research areas including computational complexity, signal processing for telecommunications [10] and scientific data analysis [11]. As specific examples, we can quote the complexity of matrix multiplication [12], the P versus NP complexity problem [13], the study of entanglement in quantum physics [14,15], matchgates in computer science [13], the study of phylogenetic invariants [16], independent component analysis [17], blind identification in signal processing [18], branching structure in diffusion images [19] and other multilinear data analysis techniques in bioinformatics and spectroscopy [20]. Looking at this literature shows how knowledge about this subject used to be quite scattered and suffered a bit from the fact that the same type of problem can be considered in different areas using a different language.

In particular, tensor decomposition is nowadays an intensively-studied argument by many algebraic geometers and by more applied communities. Its main problem is the decomposition of a tensor with a given structure as a linear combination of decomposable tensors of the same structure called rank-one tensors. To be more precise: let $V_1, \ldots, V_d$ be $\Bbbk$-vector spaces of dimensions $n_1 + 1$, $\ldots, n_d + 1$, respectively, and let $V = V_1^* \otimes \cdots \otimes V_d^* \simeq (V_1 \otimes \ldots \otimes V_d)^*$. We call a decomposable, or rank-one, tensor an element of the type $v_1^* \otimes \cdots \otimes v_d^* \in \mathbf{V}$. If $T \in V$, one can ask:

What is the minimal length of an expression of $T$ as a sum of decomposable tensors?

We call such an expression a tensor decomposition of $T$, and the answer to this question is usually referred to as the tensor rank of $T$. Note that, since $V$ is a finite-dimensional vector space of dimension $\prod_{i=1}^{d} \dim_{\Bbbk} V_i$, which has a basis of decomposable tensors, it is quite trivial to see that every $T \in V$ can be written as the sum of finitely many decomposable tensors. Other natural questions to ask are:

What is the rank of a generic tensor in $V$? What is the dimension of the closure of the set of all tensors of tensor rank $\leq r$?

Note that it is convenient to work up to scalar multiplication, i.e., in the projective space $\mathbb{P}(V)$, and the latter questions are indeed meant to be considered in the Zariski topology of $\mathbb{P}(V)$. This is the natural topology used in algebraic geometry, and it is defined such that closed subsets are zero loci of (homogeneous) polynomials and open subsets are always dense. In this terminology, an element of a family is said to be generic in that family if it lies in a proper Zariski open subset of the family. Hence, saying that a property holds for a generic tensor in $\mathbb{P}(V)$ means that it holds on a proper Zariski subset of $\mathbb{P}(V)$.

In the case $d = 2$, tensors correspond to ordinary matrices, and the notion of tensor rank coincides with the usual one of the rank of matrices. Hence, the generic rank is the maximum one, and it is the same with respect to rows or to columns. When considering multidimensional tensors, we can check that in general, all these usual properties for tensor rank fail to hold; e.g., for $(2 \times 2 \times 2)$-tensors, the generic tensor rank is two, but the maximal one is three, and of course, it cannot be the dimension of the space of "row vectors" in whatever direction.

It is well known that studying the dimensions of the secant varieties to Segre varieties gives a first idea of the stratification of $V$, or equivalently of $\mathbb{P}(V)$, with respect to tensor rank. In fact, the Segre variety $\nu_{1,\ldots,1}(\mathbb{P}^{n_1} \times \ldots \times \mathbb{P}^{n_d})$ can be seen as the projective variety in $\mathbb{P}(\mathbf{V})$, which parametrizes rank-one tensors, and consequently, the generic point of $\sigma_s(\nu_{1,\ldots,1}(\mathbb{P}^{n_1} \times \cdots \times \mathbb{P}^{n_d}))$ parametrizes a tensor of tensor rank equal to $s$ (e.g., see [21,22]).

If $V_1 = \cdots = V_d = V$ of dimension $n + 1$, one can just consider symmetric or skew-symmetric tensors. In the first case, we study the $S^d V^*$, which corresponds to the space of homogeneous polynomials in $n + 1$ variables. Again, we have a notion of symmetric decomposable tensors, i.e, elements of the type $(v^*)^d \in S^d V^*$, which correspond to powers of linear forms. These are parametrized by the Veronese variety $\nu_d(\mathbb{P}^n) \subset \mathbb{P}(S^d V^*)$. In the skew-symmetric case, we consider $\wedge^d V^*$, whose skew-symmetric decomposable tensors are the elements of the form $v_1^* \wedge \ldots \wedge v_d^* \in \wedge^d V^*$. These are parametrized by the Grassmannian $Gr(d, n+1)$ in its Plücker embedding. Hence, we get a notion of symmetric-rank and of $\wedge$-rank for which one can ask the same questions as in the case of arbitrary tensors. Once again, these are translated into algebraic geometry problems on secant varieties of Veronese varieties and Grassmannians.

Notice that actually, Veronese varieties embedded in a projective space corresponding to $\mathbb{P}(S^d V^*)$ can be thought of as sections of the Segre variety in $\mathbb{P}((V^*)^{\otimes d})$ defined by the (linear) equations given by the symmetry relations.

Since the case of symmetric tensors is the one that has been classically considered more in depth, due to the fact that symmetric tensors correspond to homogeneous polynomials, we start from analyzing secant varieties of Veronese varieties in Section 2. Then, we pass to secant varieties of Segre varieties in Section 3. Then, Section 4 is dedicated to varieties that parametrize other types of structured tensors, such as Grassmannians, which parametrize skew-symmetric tensors, Segre–Veronese varieties, which parametrize decomposable partially-symmetric tensors, Chow varieties, which parametrize homogeneous polynomials, which factorize as product of linear forms, varieties of powers, which parametrize homogeneous polynomials, which are pure $k$-th powers in the space of degree $kd$, or varieties that parametrize homogeneous polynomials with a certain prescribed factorization structure. In Section 5, we will consider other problems related to these kinds of questions, e.g., what is known about maximal ranks, how to find the actual value of (or bounds on) the rank of a given tensor, how to determine the number of minimal decompositions of a tensor, what is known about the equations of the secant varieties that we are considering or what kind of problems we meet when treating this problem over $\mathbb{R}$, a case that is of course very interesting for applications.

## 2. Symmetric Tensors and Veronese Varieties

A symmetric tensor $T$ is an element of the space $S^d V^*$, where $V^*$ is an $(n+1)$-dimensional $\Bbbk$-vector space and $\Bbbk$ is an algebraically-closed field. It is quite immediate to see that we can associate a degree $d$ homogeneous polynomial in $\Bbbk[x_0, \ldots, x_n]$ with any symmetric tensor in $S^d V^*$.

In this section, we address the problem of symmetric tensor decomposition.

What is the smallest integer $r$ such that a given symmetric tensor $T \in S^d V^*$ can be written as a sum of $r$ symmetric decomposable tensors, i.e., as a sum of $r$ elements of the type $(v^*)^{\otimes d} \in (V^*)^{\otimes d}$?

We call the answer to the latter question the *symmetric rank* of $T$. Equivalently,

What is the smallest integer $r$ such that a given homogeneous polynomial $F \in S^d V^*$ (a $(n+1)$-ary $d$-ic, in classical language) can be written as a sum of $r$ $d$-th powers of linear forms?

We call the answer to the latter question the Waring rank, or simply rank, of $F$; denoted $R_{\text{sym}}(F)$. Whenever it will be relevant to recall the base field, it will be denoted by $R_{\text{sym}}^{\Bbbk}(F)$. Since, as we have said, the space of symmetric tensors of a given format can be naturally seen as the space of homogeneous polynomials of a certain degree, we will use both names for the rank.

The name "Waring rank" comes from an old problem in number theory regarding expressions of integers as sums of powers; we will explain it in Section 2.1.1.

The first naive remark is that there are $\binom{n+d}{d}$ coefficients $a_{i_0,\ldots,i_n}$ needed to write:

$$F = \sum a_{i_0,\ldots,i_n} x_0^{i_0} \cdots x_n^{i_n},$$

and $r(n+1)$ coefficients $b_{i,j}$ to write the same $F$ as:

$$F = \sum_{i=1}^{r} (b_{i,0} x_0 + \cdots + b_{i,n} x_n)^d.$$

Therefore, for a general polynomial, the answer to the question should be that $r$ has to be at least such that $r(n+1) \geq \binom{n+d}{d}$. Then, the minimal value for which the previous inequality holds is $\left\lceil \frac{1}{n+1} \binom{n+d}{d} \right\rceil$. For $n=2$ and $d=2$, we know that this bound does not give the correct answer because a regular quadratic form in three variables cannot be written as a sum of two squares. On the other hand, a straightforward inspection shows that for binary cubics, i.e., $d=3$ and $n=1$, the generic rank is as expected. Therefore, the answer cannot be too simple.

The most important general result on this problem has been obtained by J. Alexander and A. Hirschowitz, in 1995; see [8]. It says that the generic rank is as expected for forms of degree $d \geq 3$ in $n \geq 1$ variables except for a small number of peculiar pairs $(n,d)$; see Theorem 2.

What about non-generic forms? As in the case of binary cubics, there are special forms that require a larger $r$, and these cases are still being investigated. Other presentations of this topic from different points of view can be found in [23–26].

As anticipated in the Introduction, we introduce Veronese varieties, which parametrize homogeneous polynomials of symmetric-rank-one, i.e., powers of linear forms; see Section 2.1.2. Then, in order to study the symmetric-rank of a generic form, we will use the concept of secant varieties as defined in (1). In fact, the order of the first secant that fills the ambient space will give the symmetric-rank of a generic form. The dimensions of secant varieties to Veronese varieties were completely classified by J. Alexander and A. Hirschowitz in [8] (Theorem 2). We will briefly review their proof since it provides a very important constructive method to compute dimensions of secant varieties that can be extended also to other kinds of varieties parameterizing different structured tensors. In order to do that, we need to introduce apolarity theory (Section 2.1.4) and the so-called Horace method (Sections 2.2.1 and 2.2.2).

The second part of this section will be dedicated to a more algorithmic approach to these problems, and we will focus on the problem of computing the symmetric-rank of a given homogeneous polynomial.

In the particular case of binary forms, there is a very well-known and classical result firstly obtained by J. J. Sylvester in the XIX Century. We will show a more modern reformulation of the same algorithm presented by G. Comas and M. Seiguer in [27] and a more efficient one presented in [28]; see Section 2.3.1. In Section 2.3.2, we will tackle the more general case of the computation of the symmetric-rank of any homogeneous polynomial, and we will show the only theoretical algorithm (to our knowledge) that is able to do so, which was developed by J. Brachat, P. Comon, B. Mourrain and E. Tsigaridas in [29] with its reformulation [30,31].

The last subsection of this section is dedicated to an overview of open problems.

## 2.1. On Dimensions of Secant Varieties of Veronese Varieties

This section is entirely devoted to computing the symmetric-rank of a generic form, i.e., to the computation of the generic symmetric-rank. As anticipated, we approach the problem by computing dimensions of secant varieties of Veronese varieties. Recall that, in algebraic geometry, we say that a property holds for a generic form of degree $d$ if it holds on a Zariski open, hence dense, subset of $\mathbb{P}(S^d V^*)$.

### 2.1.1. Waring Problem for Forms

The problem that we are presenting here takes its name from an old question in number theory. In 1770, E. Waring in [9] stated (without proofs) that:

"Every natural number can be written as sum of at most 9 positive cubes, Every natural number can be written as sum of at most 19 biquadratics."

Moreover, he believed that:

"For all integers $d \geq 2$, there exists a number $g(d)$ such that each positive integer $n \in \mathbb{Z}^+$ can be written as sum of the $d$-th powers of $g(d)$ many positive integers, i.e., $n = a_1^d + \cdots + a_{g(d)}^d$ with $a_i \geq 0$."

E. Waring's belief was shown to be true by D. Hilbert in 1909, who proved that such a $g(d)$ indeed exists for every $d \geq 2$. In fact, we know from the famous four-squares Lagrange theorem (1770) that $g(2) = 4$, and more recently, it has been proven that $g(3) = 9$ and $g(4) = 19$. However, the exact number for higher powers is not yet known in general. In [32], H. Davenport proved that any sufficiently large integer can be written as a sum of 16 fourth powers. As a consequence, for any integer $d \geq 2$, a new number $G(d)$ has been defined, as the least number of $d$-th powers of positive integers to write any sufficiently large positive integer as their sum. Previously, C. F. Gauss proved that any integer congruent to seven modulo eight can be written as a sum of four squares, establishing that $G(2) = g(2) = 4$. Again, the exact value $G(d)$ for higher powers is not known in general.

This fascinating problem of number theory was then formulated for homogeneous polynomials as follows.

Let $\Bbbk$ be an algebraically-closed field of characteristic zero. We will work over the projective space $\mathbb{P}^n = \mathbb{P}V$ where $V$ is an $(n+1)$-dimensional vector space over $\Bbbk$. We consider the polynomial ring $S = \Bbbk[x_0, \ldots, x_n]$ with the graded structure $S = \bigoplus_{d \geq 0} S_d$, where $S_d = \langle x_0^d, x_0^{d-1} x_1, \ldots, x_n^d \rangle$ is the vector space of homogeneous polynomials, or forms, of degree $d$, which, as we said, can be also seen as the space $S^d V$ of symmetric tensors of order $d$ over $V$. In geometric language, those vector spaces $S_d$ are called complete linear systems of hypersurfaces of degree $d$ in $\mathbb{P}^n$. Sometimes, we will write $\mathbb{P}S_d$ in order to mean the projectivization of $S_d$, namely $\mathbb{P}S_d$ will be a $\mathbb{P}^{\binom{n+d}{d}-1}$ whose elements are classes of forms of degree $d$ modulo scalar multiplication, i.e., $[F] \in \mathbb{P}S_d$ with $F \in S_d$.

In analogy to the Waring problem for integer numbers, the so-called little Waring problem for forms is the following.

**Problem 1** (little Waring problem). *Find the minimum $s \in \mathbb{Z}$ such that all forms $F \in S_d$ can be written as the sum of at most $s$ $d$-th powers of linear forms.*

The answer to the latter question is analogous to the number $g(d)$ in the Waring problem for integers. At the same time, we can define an analogous number $G(d)$, which considers decomposition in sums of powers of all numbers, but finitely many. In particular, the big Waring problem for forms can be formulated as follows.

**Problem 2** (big Waring problem). *Find the minimum $s \in \mathbb{Z}$ such that the generic form $F \in S_d$ can be written as a sum of at most $s$ $d$-th powers of linear forms.*

In order to know which elements of $S_d$ can be written as a sum of $s$ $d$-th powers of linear forms, we study the image of the map:

$$\phi_{d,s} : \underbrace{S_1 \times \cdots \times S_1}_{s} \longrightarrow S_d, \quad \phi_{d,s}(L_1, \ldots, L_s) = L_1^d + \cdots + L_s^d. \qquad (2)$$

In terms of maps $\phi_{d,s}$, the little Waring problem (Problem 1) is to find the smallest $s$, such that $\mathrm{Im}(\phi_{d,s}) = S_d$. Analogously, to solve the big Waring problem (Problem 2), we require $\overline{\mathrm{Im}(\phi_{d,s})} = S_d$, which is equivalent to finding the minimal $s$ such that $\dim(\mathrm{Im}(\phi_{d,s})) = \dim S_d$.

The map $\phi_{d,s}$ can be viewed as a polynomial map between affine spaces:

$$\phi_{d,s} : \mathbb{A}^{s(n+1)} \longrightarrow \mathbb{A}^N, \quad \text{with } N = \binom{n+d}{n}.$$

In order to know the dimension of the image of such a map, we look at its differential at a general point $P$ of the domain:

$$d\phi_{d,s}|_P : T_P(\mathbb{A}^{s(n+1)}) \longrightarrow T_{\phi_{d,s}(P)}(\mathbb{A}^N).$$

Let $P = (L_1, \ldots, L_s) \in \mathbb{A}^{s(n+1)}$ and $v = (M_1, \ldots, M_s) \in T_P(\mathbb{A}^{s(n+1)}) \simeq \mathbb{A}^{s(n+1)}$, where $L_i, M_i \in S_1$ for $i = 1, \ldots, s$. Let us consider the following parameterizations $t \mapsto (L_1 + M_1 t, \ldots, L_s + M_s t)$ of a line $\mathcal{C}$ passing through $P$ whose tangent vector at $P$ is $M$. The image of $\mathcal{C}$ via $\phi_{d,s}$ is $\phi_{d,s}(L_1 + M_1 t, \ldots, L_s + M_s t) = \sum_{i=1}^{s} (L_i + M_i t)^d$. The tangent vector to $\phi_{d,s}(\mathcal{C})$ in $\phi_{d,s}(P)$ is:

$$\left. \frac{d}{dt} \right|_{t=0} \left( \sum_{i=1}^{s} (L_i + M_i t)^d \right) = \sum_{i=1}^{s} \left. \frac{d}{dt} \right|_{t=0} (L_i + M_i t)^d = \sum_{i=1}^{s} d L_i^{d-1} M_i. \qquad (3)$$

Now, as $v = (M_1, \ldots, M_s)$ varies in $\mathbb{A}^{s(n+1)}$, the tangent vectors that we get span $\langle L_1^{d-1} S_1, \ldots, L_s^{d-1} S_1 \rangle$. Therefore, we just proved the following.

**Proposition 1.** *Let $L_1, \ldots, L_s$ be linear forms in $S = \Bbbk[x_0, \ldots, x_n]$, where $L_i = a_{i,0} x_0 + \cdots + a_{i,n} x_n$, and consider the map:*

$$\phi_{d,s} : \underbrace{S_1 \times \cdots \times S_1}_{s} \longrightarrow S_d, \quad \phi_{d,s}(L_1, \ldots, L_s) = L_1^d + \cdots + L_s^d;$$

*then:*

$$\mathrm{rk}(d\phi_{d,s})|_{(L_1,\ldots,L_s)} = \dim_{\Bbbk} \langle L_1^{d-1} S_1, \ldots, L_s^{d-1} S_1 \rangle.$$

It is very interesting to see how the problem of determining the latter dimension has been solved, because the solution involves many algebraic and geometric tools.

2.1.2. Veronese Varieties

The first geometric objects that are related to our problem are the Veronese varieties. We recall that a *Veronese variety* can be viewed as (is projectively equivalent to) the image of the following $d$-pleembedding of $\mathbb{P}^n$, where all degree $d$ monomials in $n+1$ variables appear in lexicographic order:

$$\begin{array}{rccc} \nu_d: & \mathbb{P}^n & \hookrightarrow & \mathbb{P}^{\binom{n+d}{d}-1} \\ & [u_0 : \ldots : u_n] & \mapsto & [u_0^d : u_0^{d-1} u_1 : u_0^{d-1} u_2 : \ldots : u_n^d]. \end{array} \quad (4)$$

With a slight abuse of notation, we can describe the Veronese map as follows:

$$\begin{array}{rccc} \nu_d: & \mathbb{P}S_1 = (\mathbb{P}^n)^* & \hookrightarrow & \mathbb{P}S_d = \left(\mathbb{P}^{\binom{n+d}{d}-1}\right)^* \\ & [L] & \mapsto & [L^d] \end{array} \quad (5)$$

Let $X_{n,d} := \nu_d(\mathbb{P}^n)$ denote a Veronese variety.

Clearly, "$\nu_d$ as defined in (4)" and "$\nu_d$ as defined in (5)" are not the same map; indeed, from (5),

$$\nu_d([L]) = \nu_d([u_0 x_0 + \cdots + u_n x_n]) = [L^d] =$$
$$= \left[u_0^d : d u_0^{d-1} u_1 : \binom{d}{2} u_0^{d-1} u_2 : \ldots : u_n^d\right] \in \mathbb{P}S_d.$$

However, the two images are projectively equivalent. In order to see that, it is enough to consider the monomial basis of $S_d$ given by:

$$\left\{ \binom{d}{\alpha} x^\alpha \mid \alpha = (\alpha_0, \ldots, \alpha_n) \in \mathbb{N}^{n+1}, |\alpha| = d \right\}.$$

Given a set of variables $x_0, \ldots, x_n$, we let $x^\alpha$ denote the monomial $x_0^{\alpha_0} \cdots x_n^{\alpha_n}$, for any $\alpha \in \mathbb{N}^{n+1}$. Moreover, we write $|\alpha| = \alpha_0 + \ldots + \alpha_n$ for its degree. Furthermore, if $|\alpha| = d$, we use the standard notation $\binom{d}{\alpha}$ for the multi-nomial coefficient $\frac{d!}{\alpha_0! \cdots \alpha_n!}$.

Therefore, we can view the Veronese variety either as the variety that parametrizes $d$-th powers of linear forms or as the one parameterizing completely decomposable symmetric tensors.

**Example 1** (Twisted cubic). *Let $V = \Bbbk^2$ and $d = 3$, then:*

$$\begin{array}{rccc} \nu_3: & \mathbb{P}^1 & \hookrightarrow & \mathbb{P}^3 \\ & [a_0 : a_1] & \mapsto & [a_0^3 : a_0^2 a_1 : a_0 a_1^2 : a_1^3] \end{array}.$$

If we take $\{z_0, \ldots, z_3\}$ to be homogeneous coordinates in $\mathbb{P}^3$, then the Veronese curve in $\mathbb{P}^3$ (classically known as twisted cubic) is given by the solutions of the following system of equations:

$$\begin{cases} z_0 z_2 - z_1^2 = 0 \\ z_0 z_3 - z_1 z_2 = 0 \\ z_1 z_3 - z_2^2 = 0 \end{cases}.$$

Observe that those equations can be obtained as the vanishing of all the maximal minors of the following matrix:

$$\begin{pmatrix} z_0 & z_1 & z_2 \\ z_1 & z_2 & z_3 \end{pmatrix}. \tag{6}$$

Notice that the matrix (6) can be obtained also as the defining matrix of the linear map:

$$S^2 V^* \to S^1 V, \quad \partial_{x_i}^2 \mapsto \partial_{x_i}^2(F)$$

where $F = \sum_{i=0}^{3} \binom{d}{i}^{-1} z_i x_0^{3-i} x_1^i$ and $\partial_{x_i} := \frac{\partial}{\partial x_i}$.

Another equivalent way to obtain (6) is to use the so-called flattenings. We give here an intuitive idea about flattenings, which works only for this specific example.

Write the $2 \times 2 \times 2$ tensor by putting in position $ijk$ the variable $z_{i+j+k}$. This is an element of $V^* \otimes V^* \otimes V^*$. There is an obvious isomorphism among the space of $2 \times 2 \times 2$ tensors $V^* \otimes V^* \otimes V^*$ and the space of $4 \times 2$ matrices $(V^* \otimes V^*) \otimes V^*$. Intuitively, this can be done by slicing the $2 \times 2 \times 2$ tensor, keeping fixed the third index. This is one of the three obvious possible flattenings of a $2 \times 2 \times 2$ tensor: the other two flattenings are obtained by considering as fixed the first or the second index. Now, after having written all the possible three flattenings of the tensor, one could remove the redundant repeated columns and compute all maximal minors of the three matrices obtained by this process, and they will give the same ideal.

The phenomenon described in Example 1 is a general fact. Indeed, Veronese varieties are always defined by $2 \times 2$ minors of matrices constructed as (6), which are usually called catalecticant matrices.

**Definition 1.** *Let $F \in S_d$ be a homogeneous polynomial of degree $d$ in the polynomial ring $S = \mathbb{k}[x_0, \ldots, x_n]$. For any $i = 0, \ldots, d$, the $(i, d-i)$-th catalecticant matrix associated to $F$ is the matrix representing the following linear maps in the standard monomial basis, i.e.,*

$$\mathrm{Cat}_{i,d-i}(F): \quad S_i^* \quad \longrightarrow \quad S_{d-i},$$
$$\partial_{x^\alpha}^i \quad \mapsto \quad \partial_{x^\alpha}^i(F),$$

*where, for any $\alpha \in \mathbb{N}^{n+1}$ with $|\alpha| = d - i$, we denote $\partial_{x^\alpha}^{d-i} := \frac{\partial^{d-i}}{\partial x_0^{\alpha_0} \cdots \partial x_n^{\alpha_n}}$.*

Let $\{z_\alpha \mid \alpha \in \mathbb{N}^{n+1}, |\alpha| = d\}$ be the set of coordinates on $\mathbb{P}S^d V$, where $V$ is $(n+1)$-dimensional. The $(i, d-i)$-th catalecticant matrix of $V$ is the $\binom{n+i}{n} \times \binom{n+d-i}{n}$ matrix whose rows are labeled by $\mathcal{B}_i = \{\beta \in \mathbb{N}^{n+1} \mid |\beta| = i\}$ and columns are labeled by $\mathcal{B}_{d-i} = \{\beta \in \mathbb{N}^{n+1} \mid |\beta| = d - i\}$, given by:

$$\mathrm{Cat}_{i,d-i}(V) = \left(z_{\beta_1 + \beta_2}\right)_{\substack{\beta_1 \in \mathcal{B}_i \\ \beta_2 \in \mathcal{B}_{d-i}}}.$$

**Remark 1.** *Clearly, the catalecticant matrix representing $\mathrm{Cat}_{d-i,i}(F)$ is the transpose of $\mathrm{Cat}_{i,d-i}(F)$. Moreover, the most possible square catalecticant matrix is $\mathrm{Cat}_{\lfloor d/2 \rfloor, \lceil d/2 \rceil}(F)$ (and its transpose).*

Let us describe briefly how to compute the ideal of any Veronese variety.

**Definition 2.** *A hypermatrix $A = (a_{i_1, \ldots, i_d})_{0 \leq i_j \leq n, j=1,\ldots,d}$ is said to be symmetric, or completely symmetric, if $a_{i_1, \ldots, i_d} = a_{i_{\sigma(1)}, \ldots, i_{\sigma(d)}}$ for all $\sigma \in \mathfrak{S}_d$, where $\mathfrak{S}_d$ is the permutation group of $\{1, \ldots, d\}$.*

**Definition 3.** Let $H \subset V^{\otimes d}$ be the $\binom{n+d}{d}$-dimensional subspace of completely symmetric tensors of $V^{\otimes d}$, i.e., $H$ is isomorphic to the symmetric algebra $S^d V$ or the space of homogeneous polynomials of degree $d$ in $n+1$ variables. Let $S$ be a ring of coordinates of $\mathbb{P}^{\binom{n+d}{d}-1} = \mathbb{P}H$ obtained as the quotient $S = \tilde{S}/I$ where $\tilde{S} = \mathbb{k}[x_{i_1,\ldots,i_d}]_{0 \leq i_j \leq n, j=1,\ldots,d}$ and $I$ is the ideal generated by all:

$$x_{i_1,\ldots,i_d} - x_{i_{\sigma(1)},\ldots,i_{\sigma(d)}}, \forall \sigma \in \mathfrak{S}_d.$$

The hypermatrix $(\overline{x}_{i_1,\ldots,i_d})_{0 \leq i_j \leq n, j=1,\ldots,d}$, whose entries are the generators of $S$, is said to be a generic symmetric hypermatrix.

Let $A = (x_{i_1,\ldots,i_d})_{0 \leq i_j \leq n, j=1,\ldots,d}$ be a generic symmetric hypermatrix, then it is a known result that the ideal of any Veronese variety is generated in degree two by the $2 \times 2$ minors of a generic symmetric hypermatrix, i.e.,

$$I(\nu_d(\mathbb{P}^n)) = I_2(A) := (2 \times 2 \text{ minors of } A) \subset \tilde{S}. \tag{7}$$

See [33] for the set theoretical point of view. In [34], the author proved that the ideal of the Veronese variety is generated by the two-minors of a particular catalecticant matrix. In his PhD thesis [35], A. Parolin showed that the ideal generated by the two-minors of that catalecticant matrix is actually $I_2(A)$, where $A$ is a generic symmetric hypermatrix.

2.1.3. Secant Varieties

Now, we recall the basics on secant varieties.

**Definition 4.** Let $X \subset \mathbb{P}^N$ be a projective variety of dimension $n$. We define the $s$-th secant variety $\sigma_s(X)$ of $X$ as the closure of the union of all linear spaces spanned by $s$ points lying on $X$, i.e.,

$$\sigma_s(X) := \overline{\bigcup_{P_1,\ldots,P_s \in X} \langle P_1,\ldots,P_s \rangle} \subset \mathbb{P}^N.$$

For any $\mathcal{F} \subset \mathbb{P}^n$, $\langle \mathcal{F} \rangle$ denotes the linear span of $\mathcal{F}$, i.e., the smallest projective linear space containing $\mathcal{F}$.

**Remark 2.** The closure in the definition of secant varieties is necessary. Indeed, let $L_1, L_2 \in S_1$ be two homogeneous linear forms. The polynomial $L_1^{d-1} L_2$ is clearly in $\sigma_2(\nu_d(\mathbb{P}(V)))$ since we can write:

$$L_1^{d-1} L_2 = \lim_{t \to 0} \frac{1}{t} \left( (L_1 + t L_2)^d - L_1^d \right); \tag{8}$$

however, if $d > 2$, there are no $M_1, M_2 \in S_1$ such that $L_1^{d-1} L_2 = M_1^d + M_2^d$. This computation represents a very standard concept of basic calculus: tangent lines are the limit of secant lines. Indeed, by (3), the left-hand side of (8) is a point on the tangent line to the Veronese variety at $[L_1^d]$, while the elements inside the limit on the right-hand side of (8) are lines secant to the Veronese variety at $[L_1^d]$ and another moving point; see Figure 1.

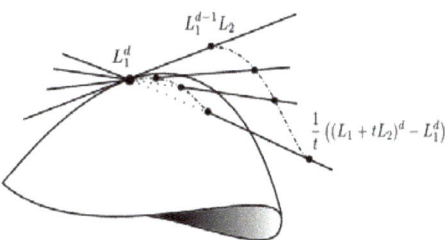

**Figure 1.** Representation of (8).

From this definition, it is evident that the generic element of $\sigma_s(X)$ is an element of some $\langle P_1, \ldots, P_s \rangle$, with $P_i \in X$; hence, it is a linear combination of $s$ elements of $X$. This is why secant varieties are used to model problems concerning additive decompositions, which motivates the following general definition.

**Definition 5.** *Let $X \subset \mathbb{P}^N$ be a projective variety. For any $P \in \mathbb{P}^N$, we define the X-rank of $P$ as*

$$R_X(P) = \min\{s \mid P \in \langle P_1, \ldots, P_s \rangle, \text{ for } P_1, \ldots, P_s \in X\},$$

*and we define the border X-rank of $P$ as*

$$\underline{R}_X(P) = \min\{s \mid P \in \sigma_s(X)\}.$$

If $X$ is a non-degenerate variety, i.e., it is not contained in a proper linear subspace of the ambient space, we obtain a chain of inclusions

$$X \subset \sigma_2(X) \subset \ldots \subset \sigma_s(X) = \mathbb{P}^N.$$

**Definition 6.** *The smallest $s \in \mathbb{Z}$ such that $\sigma_s(X) = \mathbb{P}^N$ is called the generic X-rank. This is the X-rank of the generic point of the ambient space.*

The generic X-rank of $X$ is an invariant of the embedded variety $X$.

As we described in (5), the image of the $d$-uple Veronese embedding of $\mathbb{P}^n = \mathbb{P}S_1$ can be viewed as the subvariety of $\mathbb{P}S_d$ made by all forms, which can be written as $d$-th powers of linear forms. From this point of view, the generic rank $s$ of the Veronese variety is the minimum integer such that the generic form of degree $d$ in $n+1$ variables can be written as a sum of $s$ powers of linear forms. In other words,

the answer to the Big Waring problem (Problem 2) is the generic rank with respect to the $d$-uple Veronese embedding in $\mathbb{P}S_d$.

This is the reason why we want to study the problem of determining the dimension of $s$-th secant varieties of an $n$-dimensional projective variety $X \subset \mathbb{P}^N$.

Let $X^s := \underbrace{X \times \cdots \times X}_{s}$, $X_0 \subset X$ be the open subset of regular points of $X$ and:

$$U_s(X) := \left\{ (P_1, \ldots, P_s) \in X^s \mid \begin{array}{c} \forall i, P_i \in X_0, \text{ and} \\ \text{the } P_i\text{'s are independent} \end{array} \right\}.$$

Therefore, for all $(P_1, \ldots, P_s) \in U_s(X)$, since the $P_i$'s are linearly independent, the linear span $H = \langle P_1, \ldots, P_s \rangle$ is a $\mathbb{P}^{s-1}$. Consider the following incidence variety:

$$\mathcal{I}^s(X) = \{(Q, H) \in \mathbb{P}^N \times U_s(X) \mid Q \in H\}.$$

If $s \leq N+1$, the dimension of that incidence variety is:

$$\dim(\mathcal{I}^s(X)) = n(s-1) + n + s - 1.$$

With this definition, we can consider the projection on the first factor:

$$\pi_1 : \mathcal{I}^s(X) \to \mathbb{P}^N;$$

the $s$-th secant variety of $X$ is just the closure of the image of this map, i.e.,

$$\sigma_s(X) = \overline{\text{Im}(\pi_1 : \mathcal{I}^s(X) \to \mathbb{P}^N)}.$$

Now, if $\dim(X) = n$, it is clear that, while $\dim(\mathcal{J}^s(X)) = ns + s - 1$, the dimension of $\sigma_s(X)$ can be smaller: it suffices that the generic fiber of $\pi_1$ has positive dimension to impose $\dim(\sigma_s(X)) < n(s-1) + n + s - 1$. Therefore, it is a general fact that, if $X \subset \mathbb{P}^N$ and $\dim(X) = n$, then,

$$\dim(\sigma_s(X)) \leq \min\{N, sn + s - 1\}.$$

**Definition 7.** *A projective variety $X \subset \mathbb{P}^N$ of dimension $n$ is said to be s-defective if $\dim(\sigma_s(X)) < \min\{N, sn + s - 1\}$. If so, we call s-th defect of X the difference:*

$$\delta_s(X) := \min\{N, sn + s - 1\} - \dim(\sigma_s(X)).$$

*Moreover, if X is s-defective, then $\sigma_s(X)$ is said to be defective. If $\sigma_s(X)$ is not defective, i.e., $\delta_s(X) = 0$, then it is said to be* regular *or of* expected dimension.

Alexander–Hirschowitz Theorem ([8]) tells us that the dimension of the $s$-th secant varieties to Veronese varieties is not always the expected one; moreover, they exhibit the list of all the defective cases.

**Theorem 2** (Alexander–Hirschowitz Theorem). *Let $X_{n,d} = \nu_d(\mathbb{P}^n)$, for $d \geq 2$, be a Veronese variety. Then:*

$$\dim(\sigma_s(X)) = \min\left\{\binom{n+d}{d} - 1, sn + s - 1\right\}$$

*except for the following cases:*

(1) $d = 2$, $n \geq 2$, $s \leq n$, where $\dim(\sigma_s(X)) = \min\left\{\binom{n+2}{2} - 1, 2n + 1 - \binom{s}{2}\right\}$;
(2) $d = 3$, $n = 4$, $s = 7$, where $\delta_s = 1$;
(3) $d = 4$, $n = 2$, $s = 5$, where $\delta_s = 1$;
(4) $d = 4$, $n = 3$, $s = 9$, where $\delta_s = 1$;
(5) $d = 4$, $n = 4$, $s = 14$, where $\delta_s = 1$.

Due to the importance of this theorem, we firstly give a historical review, then we will give the main steps of the idea of the proof. For this purpose, we will need to introduce many mathematical tools (apolarity in Section 2.1.4 and fat points together with the Horace method in Section 2.2) and some other excursuses on a very interesting and famous conjecture (the so-called SHGHconjecture; see Conjectures 1 and 2) related to the techniques used in the proof of this theorem.

The following historical review can also be found in [36].

The quadric cases ($d = 2$) are classical. The first non-trivial exceptional case $d = 4$ and $n = 2$ was known already by Clebsch in 1860 [37]. He thought of the quartic as a quadric of quadrics and found that $\sigma_5(\nu_4(\mathbb{P}^2)) \subsetneq \mathbb{P}^{14}$, whose dimension was not the expected one. Moreover, he found the condition that the elements of $\sigma_5(\nu_4(\mathbb{P}^2))$ have to satisfy, i.e., he found the equation of the hypersurface $\sigma_5(\nu_4(\mathbb{P}^2)) \subsetneq \mathbb{P}^{14}$: that condition was the vanishing of a $6 \times 6$ determinant of a certain catalecticant matrix.

To our knowledge, the first list of all exceptional cases was described by Richmond in [38], who showed all the defectivities, case by case, without finding any general method to describe all of them. It is remarkable that he could describe also the most difficult case of general quartics of $\mathbb{P}^4$. The same problem, but from a more geometric point of view, was at the same time studied and solved by Palatini in 1902–1903; see [39,40]. In particular, Palatini studied the general problem, proved the defectivity of the space of cubics in $\mathbb{P}^4$ and studied the case of $n = 2$. He was also able to list all the defective cases.

The first work where the problem was treated in general is due to Campbell (in 1891; therefore, his work preceded those of Palatini, but in Palatini's papers, there is no evidence of knowledge of

Campbell's work), who in [41], found almost all the defective cases (except the last one) with very interesting, but not always correct arguments (the fact that the Campbell argument was wrong for $n = 3$ was claimed also in [4] in 1915).

His approach is very close to the infinitesimal one of Terracini, who introduced in [3] a very simple and elegant argument (today known as Terracini's lemmas, the first of which will be displayed here as Lemma 1), which offered a completely new point of view in the field. Terracini showed again the case of $n = 2$ in [3]. In [42], he proved that the exceptional case of cubics in $\mathbb{P}^4$ can be solved by considering that the rational quartic through seven given points in $\mathbb{P}^4$ is the singular locus of its secant variety, which is a cubic hypersurface. In [4], Terracini finally proved the case $n = 3$ (in 2001, Roé, Zappalà and Baggio revised Terracini's argument, and they where able to present a rigorous proof for the case $n = 3$; see [43]).

In 1931, Bronowski [44] tried to tackle the problem checking if a linear system has a vanishing Jacobian by a numerical criterion, but his argument was incomplete.

In 1985, Hirschowitz ([45]) proved again the cases $n = 2, 3$, and he introduced for the first time in the study of this problem the use of zero-dimensional schemes, which is the key point towards a complete solution of the problem (this will be the idea that we will follow in these notes). Alexander used this new and powerful idea of Hirschowitz, and in [46], he proved the theorem for $d \geq 5$.

Finally, in [8,47] (1992–1995), J. Alexander and A. Hirschowitz joined forces to complete the proof of Theorem 2. After this result, simplifications of the proof followed [48,49].

After this historical excursus, we can now review the main steps of the proof of the Alexander–Hirschowitz theorem. As already mentioned, one of the main ingredients to prove is Terracini's lemma (see [3] or [50]), which gives an extremely powerful technique to compute the dimension of any secant variety.

**Lemma 1** (Terracini's lemma). *Let X be an irreducible non-degenerate variety in $\mathbb{P}^N$, and let $P_1, \ldots, P_s$ be s generic points on X. Then, the tangent space to $\sigma_s(X)$ at a generic point $Q \in \langle P_1, \ldots, P_s \rangle$ is the linear span in $\mathbb{P}^N$ of the tangent spaces $T_{P_i}(X)$ to X at $P_i$, $i = 1, \ldots, s$, i.e.,*

$$T_Q(\sigma_s(X)) = \langle T_{P_1}(X), \ldots, T_{P_s}(X) \rangle.$$

This "lemma" (we believe it is very reductive to call it a "lemma") can be proven in many ways (for example, without any assumption on the characteristic of $\Bbbk$, or following Zak's book [7]). Here, we present a proof "made by hand".

**Proof.** We give this proof in the case of $\Bbbk = \mathbb{C}$, even though it works in general for any algebraically-closed field of characteristic zero.

We have already used the notation $X^s$ for $X \times \cdots \times X$ taken $s$ times. Suppose that $\dim(X) = n$. Let us consider the following incidence variety,

$$\mathfrak{I} = \left\{ (P; P_1, \ldots, P_s) \in \mathbb{P}^N \times X^s \;\middle|\; \begin{array}{c} P \in \langle P_1, \ldots, P_s \rangle \\ P_1, \ldots, P_s \in X \end{array} \right\} \subset \mathbb{P}^N \times X^s,$$

and the two following projections,

$$\pi_1 : \mathfrak{I} \to \sigma_s(X), \quad \pi_2 : \mathfrak{I} \to X^s.$$

The dimension of $X^s$ is clearly $sn$. If $(P_1, \ldots, P_s) \in X^s$, the fiber $\pi_2^{-1}((P_1, \ldots, P_s))$ is generically a $\mathbb{P}^{s-1}$, $s < N$. Then, $\dim(\mathfrak{I}) = sn + s - 1$.

If the generic fiber of $\pi_1$ is finite, then $\sigma_s(X)$ is regular. i.e., it has the expected dimension; otherwise, it is defective with a value of the defect that is exactly the dimension of the generic fiber.

Let $(P_1, \ldots, P_s) \in X^s$ and suppose that each $P_i \in X \subset \mathbb{P}^N$ has coordinates $P_i = [a_{i,0} : \ldots : a_{i,N}]$, for $i = 1, \ldots, s$. In an affine neighborhood $U_i$ of $P_i$, for any $i$, the variety $X$ can be locally parametrized with some rational functions $f_{i,j} : \Bbbk^{n+1} \to \Bbbk$, with $j = 0, \ldots, N$, that are zero at the origin. Hence, we write:

$$X \supset U_i : \begin{cases} x_0 = a_{i,0} + f_{i,0}(u_{i,0}, \ldots, u_{i,n}) \\ \vdots \\ x_N = a_{i,N} + f_{i,N}(u_{i,0}, \ldots, u_{i,n}) \end{cases}.$$

Now, we need a parametrization $\varphi$ for $\sigma_s(X)$. Consider the subspace spanned by $s$ points of $X$, i.e.,

$$\langle (a_{1,0} + f_{1,0}, \ldots, a_{1,N} + f_{1,N}), \ldots, (a_{s,0} + f_{s,0}, \ldots, a_{s,N} + f_{s,N}) \rangle,$$

where for simplicity of notation, we omit the dependence of the $f_{i,j}$ on the variables $u_{i,j}$; thus, an element of this subspace is of the form:

$$\lambda_1 (a_{1,0} + f_{1,0}, \ldots, a_{1,N} + f_{1,N}) + \cdots + \lambda_s (a_{s,0} + f_{s,0}, \ldots, a_{s,N} + f_{s,N}),$$

for some $\lambda_1, \ldots, \lambda_s \in \Bbbk$. We can assume $\lambda_1 = 1$. Therefore, a parametrization of the $s$-th secant variety to $X$ in an affine neighborhood of the point $P_1 + \lambda_2 P_2 + \ldots + \lambda_s P_s$ is given by:

$$(a_{1,0} + f_{1,0}, \ldots, a_{1,N} + f_{1,N}) +$$
$$+ (\lambda_2 + t_2)(a_{2,1} - a_{1,0} + f_{2,1} - f_{1,0}, \ldots, a_{2,N} - a_{1,N} + f_{2,N} - f_{1,N}) +$$
$$+ \cdots +$$
$$+ (\lambda_s + t_s)(a_{s,1} - a_{1,0} + f_{s,1} - f_{1,0}, \ldots, a_{s,N} - a_{1,N} + f_{s,N} - f_{1,N}),$$

for some parameters $t_2, \ldots, t_s$. Therefore, in coordinates, the parametrization of $\sigma_s(X)$ that we are looking for is the map $\varphi : \Bbbk^{s(n+1)+s-1} \to \Bbbk^{N+1}$ given by:

$$(u_{1,0}, \ldots, u_{1,n}, u_{2,0}, \ldots, u_{2,n}, \ldots, u_{s,0}, \ldots, u_{s,n}, t_2, \ldots, t_s)$$
$$\downarrow$$
$$(\ldots, a_{1,j} + f_{1,j} + (\lambda_2 + t_2)(a_{2,j} - a_{1,j} + f_{2,j} - f_{1,j}) + \cdots + (\lambda_s + t_s)(a_{s,j} - a_{1,j} + f_{s,j} - f_{1,j}), \ldots),$$

where for simplicity, we have written only the $j$-th element of the image. Therefore, we are able to write the Jacobian of $\varphi$. We are writing it in three blocks: the first one is $(N+1) \times (n+1)$; the second one is $(N+1) \times (s-1)(n+1)$; and the third one is $(N+1) \times (s-1)$:

$$J_{\underline{0}}(\varphi) = \left( \; (1 - \lambda_2 - \cdots - \lambda_s) \frac{\partial f_{1,j}}{\partial u_{1,k}} \; \middle| \; \lambda_i \frac{\partial f_{i,j}}{\partial u_{i,k}} \; \middle| \; a_{i,j} - a_{1,j} \; \right),$$

with $i = 2, \ldots, s$; $j = 0, \ldots, N$ and $k = 0, \ldots, n$. Now, the first block is a basis of the (affine) tangent space to $X$ at $P_1$, and in the second block, we can find the bases for the tangent spaces to $X$ at $P_2, \ldots, P_s$; the rows of:

$$\begin{pmatrix} \frac{\partial f_{i,0}}{\partial u_{i,0}} & \cdots & \frac{\partial f_{i,0}}{\partial a_{i,N}} \\ \vdots & & \vdots \\ \frac{\partial f_{i,N}}{\partial a_{i,0}} & \cdots & \frac{\partial f_{i,N}}{\partial a_{i,N}} \end{pmatrix}$$

give a basis for the (affine) tangent space of $X$ at $P_i$. □

The importance of Terracini's lemma to compute the dimension of any secant variety is extremely evident. One of the main ideas of Alexander and Hirshowitz in order to tackle the specific case of Veronese variety was to take advantage of the fact that Veronese varieties are embedded in the projective space of homogeneous polynomials. They firstly moved the problem from computing the dimension of a vector space (the tangent space to a secant variety) to the computation of the dimension of its dual (see Section 2.1.4 for the precise notion of duality used in this context). Secondly, their punchline was to identify such a dual space with a certain degree part of a zero-dimensional scheme, whose Hilbert function can be computed by induction (almost always). We will be more clear on the whole technique in the sequel. Now we need to use the language of schemes.

**Remark 3.** *Schemes are locally-ringed spaces isomorphic to the spectrum of a commutative ring. Of course, this is not the right place to give a complete introduction to schemes. The reader interested in studying schemes can find the fundamental material in [51–53]. In any case, it is worth noting that we will always use only zero-dimensional schemes, i.e., "points"; therefore, for our purpose, it is sufficient to think of zero-dimensional schemes as points with a certain structure given by the vanishing of the polynomial equations appearing in the defining ideal. For example, a homogeneous ideal I contained in $\Bbbk[x,y,z]$, which is defined by the forms vanishing on a degree d plane curve C and on a tangent line to C at one of its smooth points P, represents a zero-dimensional subscheme of the plane supported at P and of length two, since the degree of intersection among the curve and the tangent line is two at P (schemes of this kind are sometimes called jets).*

**Definition 8.** *A fat point $Z \subset \mathbb{P}^n$ is a zero-dimensional scheme, whose defining ideal is of the form $\wp^m$, where $\wp$ is the ideal of a simple point and m is a positive integer. In this case, we also say that Z is a m-fat point, and we usually denote it as mP. We call the* scheme of fat points *a union of fat points $m_1 P_1 + \cdots + m_s P_s$, i.e., the zero-dimensional scheme defined by the ideal $\wp_1^{m_1} \cap \cdots \cap \wp_s^{m_s}$, where $\wp_i$ is the prime ideal defining the point $P_i$, and the $m_i$'s are positive integers.*

**Remark 4.** *In the same notation as the latter definition, it is easy to show that $F \in \wp^m$ if and only if $\partial(F)(P) = 0$, for any partial differential $\partial$ of order $\leq m-1$. In other words, the hypersurfaces "vanishing" at the m-fat point mP are the hypersurfaces that are passing through P with multiplicity m, i.e., are singular at P of order m.*

**Corollary 1.** *Let $(X, \mathcal{L})$ be an integral, polarized scheme. If $\mathcal{L}$ embeds X as a closed scheme in $\mathbb{P}^N$, then:*

$$\dim(\sigma_s(X)) = N - \dim(h^0(\mathcal{I}_{Z,X} \otimes \mathcal{L})),$$

*where Z is the union of s generic two-fat points in X.*

**Proof.** By Terracini's lemma, we have that, for generic points $P_1, \ldots, P_s \in X$, $\dim(\sigma_s(X)) = \dim(\langle T_{P_1}(X), \ldots, T_{P_s}(X) \rangle)$. Since X is embedded in $\mathbb{P}(H^0(X,\mathcal{L})^*)$ of dimension N, we can view the elements of $H^0(X,\mathcal{L})$ as hyperplanes in $\mathbb{P}^N$. The hyperplanes that contain a space $T_{P_i}(X)$ correspond to elements in $H^0(\mathcal{I}_{2P_i,X} \otimes \mathcal{L})$, since they intersect X in a subscheme containing the first infinitesimal neighborhood of $P_i$. Hence, the hyperplanes of $\mathbb{P}^N$ containing $\langle T_{P_1}(X), \ldots, T_{P_s}(X) \rangle$ are the sections of $H^0(\mathcal{I}_{Z,X} \otimes \mathcal{L})$, where Z is the scheme union of the first infinitesimal neighborhoods in X of the points $P_i$'s. □

**Remark 5.** *A hyperplane H contains the tangent space to a non-degenerate projective variety X at a smooth point P if and only if the intersection $X \cap H$ has a singular point at P. In fact, the tangent space $T_P(X)$ to X at P has the same dimension of X and $T_P(X \cap H) = H \cap T_P(X)$. Moreover, P is singular in $H \cap X$ if and only if $\dim(T_P(X \cap H)) \geq \dim(X \cap H) = \dim(X) - 1$, and this happens if and only if $H \supset T_P(X)$.*

**Example 2** (The Veronese surface of $\mathbb{P}^5$ is defective)**.** *Consider the Veronese surface $X_{2,2} = \nu_2(\mathbb{P}^2)$ in $\mathbb{P}^5$. We want to show that it is two-defective, with $\delta_2 = 1$. In other words, since the expected dimension of $\sigma_2(X_{2,2})$ is $2 \cdot 2 + 1$, i.e., we expect that $\sigma_2(X_{2,2})$ fills the ambient space, we want to prove that it is actually a hypersurface. This will imply that actually, it is not possible to write a generic ternary quadric as a sum of two squares, as expected by counting parameters, but at least three squares are necessary instead.*

Let P be a general point on the linear span $\langle R, Q \rangle$ of two general points $R, Q \in X$; hence, $P \in \sigma_2(X_{2,2})$. By Terracini's lemma, $T_P(\sigma_2(X_{2,2})) = \langle T_R(X_{2,2}), T_Q(X_{2,2}) \rangle$. The expected dimension for $\sigma_2(X_{2,2})$ is five, so $\dim(T_P(\sigma_2(X_{2,2}))) < 5$ if and only if there exists a hyperplane H containing $T_P(\sigma_2(X_{2,2}))$. The previous remark tells us that this happens if and only if there exists a hyperplane H such that $H \cap X_{2,2}$ is singular at R, Q. Now, $X_{2,2}$ is the image of $\mathbb{P}^2$ via the map defined by the complete linear system of quadrics; hence, $X_{2,2} \cap H$ is the image of a plane conic. Let $R', Q'$ be the pre-images via $\nu_2$ of R, Q respectively. Then, the double line defined by

$R', Q'$ is a conic, which is singular at $R', Q'$. Since the double line $\langle R', Q' \rangle$ is the only plane conic that is singular at $R', Q'$, we can say that $\dim(T_P(\sigma_2(X_{2,2}))) = 4 < 5$; hence, $\sigma_2(X_{2,2})$ is defective with defect equal to zero.

Since the two-Veronese surface is defined by the complete linear system of quadrics, Corollary 1 allows us to rephrase the defectivity of $\sigma_2(X_{2,2})$ in terms of the number of conditions imposed by two-fat points to forms of degree two; i.e., we say that

*two two-fat points of $\mathbb{P}^2$ do not impose independent conditions on ternary quadrics.*

As we have recalled above, imposing the vanishing at the two-fat point means to impose the annihilation of all partial derivatives of first order. In $\mathbb{P}^2$, these are three linear conditions on the space of quadrics. Since we are considering a scheme of two two-fat points, we have six linear conditions to impose on the six-dimensional linear space of ternary quadrics; in this sense, we expect to have no plane cubic passing through two two-fat points. However, since the double line is a conic passing doubly thorough the two two-fat points, we have that the six linear conditions are not independent. We will come back in the next sections on this relation between the conditions imposed by a scheme of fat points and the defectiveness of secant varieties.

Corollary 1 can be generalized to non-complete linear systems on X.

**Remark 6.** *Let D be any divisor of an irreducible projective variety X. With $|D|$, we indicate the complete linear system defined by D. Let $V \subset |D|$ be a linear system. We use the notation:*

$$V(m_1 P_1, \ldots, m_s P_s)$$

*for the subsystem of divisors of V passing through the fixed points $P_1, \ldots, P_s$ with multiplicities at least $m_1, \ldots, m_s$ respectively.*

When the multiplicities $m_i$ are equal to two, for $i = 1, \ldots, s$, since a two-fat point in $\mathbb{P}^n$ gives $n+1$ linear conditions, in general, we expect that, if $\dim(X) = n$, then:

$$\exp.\dim(V(2P_1, \ldots, 2P_s)) = \dim(V) - s(n+1).$$

Suppose that $V$ is associated with a morphism $\varphi_V : X_0 \to \mathbb{P}^r$ (if $\dim(V) = r$), which is an embedding on a dense open set $X_0 \subset X$. We will consider the variety $\overline{\varphi_V(X_0)}$.

The problem of computing $\dim(V(2P_1, \ldots, 2P_s))$ is equivalent to that one of computing the dimension of the s-th secant variety to $\overline{\varphi_V(X_0)}$.

**Proposition 2.** *Let X be an integral scheme and V be a linear system on X such that the rational function $\varphi_V : X \dashrightarrow \mathbb{P}^r$ associated with V is an embedding on a dense open subset $X_0$ of X. Then, $\sigma_s\left(\overline{\varphi_V(X_0)}\right)$ is defective if and only if for general points, we have $P_1, \ldots, P_s \in X$:*

$$\dim(V(2P_1, \ldots, 2P_s)) > \min\{-1, r - s(n+1)\}.$$

This statement can be reformulated via apolarity, as we will see in the next section.

2.1.4. Apolarity

This section is an exposition of inverse systems techniques, and it follows [54].

As already anticipated at the end of the proof of Terracini's lemma, the whole Alexander and Hirshowitz technique to compute the dimensions of secant varieties of Veronese varieties is based on the computation of the dual space to the tangent space to $\sigma_s(\nu_d(\mathbb{P}^n))$ at a generic point. Such a duality is the apolarity action that we are going to define.

**Definition 9** (Apolarity action). *Let $S = \mathbb{k}[x_0, \ldots, x_n]$ and $R = \mathbb{k}[y_0, \ldots, y_n]$ be polynomial rings and consider the action of $R_1$ on $S_1$ and of $R_1$ on $S_1$ defined by:*

$$y_i \circ x_j = \left(\frac{\partial}{\partial x_i}\right)(x_j) = \begin{cases} 0, & \text{if } i \neq j \\ 1, & \text{if } i = j \end{cases};$$

*i.e., we view the polynomials of $R_1$ as "partial derivative operators" on $S_1$.*

Now, we extend this action to the whole rings $R$ and $S$ by linearity and using properties of differentiation. Hence, we get the apolarity action:

$$\circ : R_i \times S_j \longrightarrow S_{j-i}$$

where:

$$y^\alpha \circ x^\beta = \begin{cases} \prod_{i=1}^{n} \frac{(\beta_i)!}{(\beta_i - \alpha_i)!} x^{\beta - \alpha}, & \text{if } \alpha \leq \beta; \\ 0, & \text{otherwise}. \end{cases}$$

for $\alpha, \beta \in \mathbb{N}^{n+1}$, $\alpha = (\alpha_0, \ldots, \alpha_n)$, $\beta = (\beta_0, \ldots, \beta_n)$, where we use the notation $\alpha \leq \beta$ if and only if $a_i \leq b_i$ for all $i = 0, \ldots, n$, which is equivalent to the condition that $x^\alpha$ divides $x^\beta$ in $S$.

**Remark 7.** *Here, are some basic remarks on apolarity action:*

- *the apolar action of $R$ on $S$ makes $S$ a (non-finitely generated) $R$-module (but the converse is not true);*
- *the apolar action of $R$ on $S$ lowers the degree; in particular, given $F \in S_d$, the $(i, d-i)$-th catalecticant matrix (see Definition 1) is the matrix of the following linear map induced by the apolar action*

$$\operatorname{Cat}_{i,d-i}(F) : R_i \longrightarrow S_{d-i}, \quad G \mapsto G \circ F;$$

- *the apolarity action induces a non-singular $\mathbb{k}$-bilinear pairing:*

$$R_j \times S_j \longrightarrow \mathbb{k}, \quad \forall j \in \mathbb{N}, \tag{9}$$

*that induces two bilinear maps (Let $V \times W \longrightarrow \mathbb{k}$ be a $\mathbb{k}$-bilinear parity given by $v \times w \longrightarrow v \circ w$. It induces two $\mathbb{k}$-bilinear maps: (1) $\phi : V \longrightarrow \operatorname{Hom}_{\mathbb{k}}(W, \mathbb{k})$ such that $\phi(v) := \phi_v$ and $\phi_v(w) = v \circ w$ and $\chi : W \longrightarrow \operatorname{Hom}_{\mathbb{k}}(V, \mathbb{k})$ such that $\chi(w) := \chi_w$ and $\chi_w(v) = v \circ w$; (2) $V \times W \longrightarrow \mathbb{k}$ is not singular iff for all the bases $\{w_1, \ldots, w_n\}$ of $W$, the matrix $(b_{ij} = v_i \circ w_j)$ is invertible.).*

**Definition 10.** *Let $I$ be a homogeneous ideal of $R$. The inverse system $I^{-1}$ of $I$ is the $R$-submodule of $S$ containing all the elements of $S$, which are annihilated by $I$ via the apolarity action.*

**Remark 8.** *Here are some basic remarks about inverse systems:*

- *if $I = (G_1, \ldots, G_t) \subset R$ and $F \in S$, then:*

$$F \in I^{-1} \iff G_1 \circ F = \cdots = G_t \circ F = 0,$$

*finding all such $F$'s amounts to finding all the polynomial solutions for the differential equations defined by the $G_i$'s, so one can notice that determining $I^{-1}$ is equivalent to solving (with polynomial solutions) a finite set of differential equations;*
- *$I^{-1}$ is a graded submodule of $S$, but it is not necessarily multiplicatively closed; hence in general, $I^{-1}$ is not an ideal of $S$.*

Now, we need to recall a few facts on Hilbert functions and Hilbert series.

Let $X \subset \mathbb{P}^n$ be a closed subscheme whose defining homogeneous ideal is $I := I(X) \subset S = \mathbb{k}[x_0,\ldots,x_n]$. Let $A = S/I$ be the homogeneous coordinate ring of $X$, and $A_d$ will be its degree $d$ component.

**Definition 11.** *The* Hilbert function *of the scheme X is the numeric function:*

$$\mathrm{HF}(X,\cdot) : \mathbb{N} \to \mathbb{N};$$

$$\mathrm{HF}(X,d) = \dim_\mathbb{k}(A_d) = \dim_\mathbb{k}(S_d) - \dim_\mathbb{k}(I_d).$$

*The* Hilbert series *of X is the generating power series:*

$$\mathrm{HS}(X;z) = \sum_{d \in \mathbb{N}} \mathrm{HF}(X,d) t^d \in \mathbb{k}[[z]].$$

In the following, the importance of inverse systems for a particular choice of the ideal $I$ will be given by the following result.

**Proposition 3.** *The dimension of the part of degree d of the inverse system of an ideal $I \subset R$ is the Hilbert function of $R/I$ in degree d:*

$$\dim_\mathbb{k}(I^{-1})_d = \mathrm{codim}_\mathbb{k}(I_d) = \mathrm{HF}(R/I,d). \tag{10}$$

**Remark 9.** *If $V \times W \to \mathbb{k}$ is a non-degenerate bilinear form and $U$ is a subspace of $V$, then with $U^\perp$, we denote the subspace of $W$ given by:*

$$U^\perp = \{w \in W \mid v \circ w = 0 \ \forall \ v \in U\}.$$

With this definition, we observe that:

- if we consider the bilinear map in (9) and an ideal $I \subset R$, then:

$$(I^{-1})_d \cong I_d^\perp.$$

- moreover, if $I$ is a monomial ideal, then:

$$I_d^\perp = \langle \text{monomials of } R_d \text{ that are not in } I_d \rangle;$$

- for any two ideals $I, J \subset R$: $(I \cap J)^{-1} = I^{-1} + J^{-1}$.

If $I = \wp_1^{m_1+1} \cap \cdots \cap \wp_s^{m_s+1} \subset R = \mathbb{k}[y_0,\ldots,y_n]$ is the defining ideal of the scheme of fat points $m_1 P_1 + \cdots + m_s P_s \in \mathbb{P}^n$, where $P_i = [p_{i_0} : p_{i_1} : \ldots : p_{i_n}] \in \mathbb{P}^n$, and if $L_{P_i} = p_{i_0} x_0 + p_{i_1} x_1 + \cdots + p_{i_n} x_n \in S = \mathbb{k}[x_0,\ldots,x_n]$, then:

$$(I^{-1})_d = \begin{cases} S_d, & \text{for } d \leq \max\{m_i\}, \\ L_{P_1}^{d-m_1} S_{m_1} + \cdots + L_{P_s}^{d-m_s} S_{m_s}, & \text{for } d \geq \max\{m_i + 1\}, \end{cases}$$

and also:

$$\mathrm{HF}(R/I,d) = \dim_\mathbb{k}(I^{-1})_d =$$

$$= \begin{cases} \dim_\mathbb{k} S_d, & \text{for } d \leq \max\{m_i\} \\ \dim_\mathbb{k} \langle L_{P_1}^{d-m_1} S_{m_1}, \ldots, L_{P_s}^{d-m_s} S_{m_s} \rangle, & \text{for } d \geq \max\{m_i + 1\} \end{cases} \tag{11}$$

This last result gives the following link between the Hilbert function of a set of fat points and ideals generated by sums of powers of linear forms.

**Proposition 4.** Let $I = \wp_1^{m_1+1} \cap \cdots \cap \wp_s^{m_s+1} \subset R = \Bbbk[y_0, \ldots, y_n]$, then the inverse system $(I^{-1})_d \subset S_d = \Bbbk[x_0, \ldots, x_n]_d$ is the d-th graded part of the ideal $(L_{P_1}^{d-m_1}, \ldots, L_{P_s}^{d-m_s}) \subset S$, for $d \geq \max\{m_i + 1, i = 1, \ldots, s\}$.

Finally, the link between the big Waring problem (Problem 2) and inverse systems is clear. If in (11), all the $m_i$'s are equal to one, the dimension of the vector space $\langle L_{P_1}^{d-1} S_1, \ldots, L_{P_s}^{d-1} S_1 \rangle$ is at the same time the Hilbert function of the inverse system of a scheme of $s$ double fat points and the rank of the differential of the application $\phi$ defined in (2). ▪

**Proposition 5.** Let $L_1, \ldots, L_s$ be linear forms of $S = \Bbbk[x_0, \ldots, x_n]$ such that:

$$L_i = a_{i_0} x_0 + \cdots + a_{i_n} x_n,$$

and let $P_1, \ldots, P_s \in \mathbb{P}^n$ such that $P_i = [a_{i_0}, \ldots, a_{i_n}]$. Let $\wp_i \subset R = \Bbbk[y_0, \ldots, y_n]$ be the prime ideal associated with $P_i$, for $i = 1, \ldots, s$, and let:

$$\phi_{s,d} : \underbrace{S_1 \times \cdots \times S_1}_{s} \longrightarrow S_d$$

with $\phi_{s,d}(L_1, \ldots, L_s) = L_1^d + \cdots + L_s^d$. Then,

$$R(d\phi_{s,d})|_{(L_1, \ldots, L_s)} = \dim_\Bbbk \langle L_1^{d-1} S_1, \ldots, L_s^{d-1} S_1 \rangle.$$

Moreover, by (10), we have:

$$\dim_\Bbbk(\langle L_1^{d-1} S_1, \ldots, L_s^{d-1} S_1 \rangle) = \mathrm{HF}\left(\frac{R}{\wp_1^2 \cap \cdots \cap \wp_s^2}, d\right).$$

Now, it is quite easy to see that:

$$\langle T_{P_1} X_{n,d}, \ldots, T_{P_s} X_{n,d} \rangle = \mathbb{P}\langle L_1^{d-1} S_1, \ldots, L_s^{d-1} S_1 \rangle.$$

Therefore, putting together Terracini's lemma and Proposition 5, if we assume the $L_i$'s (hence, the $P_i$'s) to be generic, we get:

$$\dim(\sigma_s(X_{n,d})) = \dim \langle T_{P_1} X_{n,d}, \ldots, T_{P_s} X_{n,d} \rangle =$$
$$= \dim_\Bbbk \langle L_1^{d-1} S_1, \ldots, L_s^{d-1} S_1 \rangle - 1 = \mathrm{HF}(R/(\wp_1^2 \cap \cdots \cap \wp_s^2), d) - 1. \quad (12)$$

**Example 3.** Let $P \in \mathbb{P}^n$, $\wp \subset S$ be its representative prime ideal and $f \in S$. Then, the order of all partial derivatives of $f$ vanishing in $P$ is almost $t$ if and only if $f \in \wp^{t+1}$, i.e., $P$ is a singular point of $V(f)$ of multiplicity greater than or equal to $t+1$. Therefore,

$$\mathrm{HF}(S/\wp^t, d) = \begin{cases} \binom{d+n}{n} & \text{if } d < t; \\ \binom{t-1+n}{n} & \text{if } d \geq t. \end{cases} \quad (13)$$

It is easy to conclude that one t-fat point of $\mathbb{P}^n$ has the same Hilbert function of $\binom{t-1+n}{n}$ generic distinct points of $\mathbb{P}^n$. Therefore, $\dim(X_{n,d}) = \mathrm{HF}(S/\wp^2, d) - 1 = (n+1) - 1$. This reflects the fact that Veronese varieties are never one-defective, or, equivalently, since $X_{n,d} = \sigma_1(X_{n,d})$, that Veronese varieties are never defective: they always have the expected dimension $1 \cdot n + 1 - 1$.

**Example 4.** Let $P_1, P_2$ be two points of $\mathbb{P}^2$, $\wp_i \subset S = \Bbbk[x_0, x_1, x_2]$ their associated prime ideals and $m_1 = m_2 = 2$, so that $I = \wp_1^2 \cap \wp_2^2$. Is the Hilbert function of $I$ equal to the Hilbert function of six points of $\mathbb{P}^2$ in general position? No; indeed, the Hilbert series of six general points of $\mathbb{P}^2$ is $1 + 3z + 6\sum_{i \geq 2} z^i$. This means that $I$ should not contain conics, but this is clearly false because the double line through $P_1$ and $P_2$ is contained in I. By (12), this implies that $\sigma_2(\nu_2(\mathbb{P}^2)) \subset \mathbb{P}^5$ is defective, i.e., it is a hypersurface, while it was expected to fill all the ambient space.

**Remark 10** (Fröberg–Iarrobino's conjecture). *Ideals generated by powers of linear forms are usually called power ideals. Besides the connection with fat points and secant varieties, they are related to several areas of algebra, geometry and combinatorics; see [55]. Of particular interest is their Hilbert function and Hilbert series. In [56], Fröberg gave a lexicographic inequality for the Hilbert series of homogeneous ideals in terms of their number of variables, number of generators and their degrees. That is, if $I = (G_1, \ldots, G_s) \subset S = \mathbb{k}[x_0, \ldots, x_n]$ with $\deg(G_i) = d_i$, for $i = 1, \ldots, s$,*

$$\mathrm{HS}(S/I; z) \succeq_{\mathrm{Lex}} \left\lceil \frac{\prod_{i=1}^{s}(1 - z^{d_i})}{(1-z)^{n+1}} \right\rceil, \tag{14}$$

*where $\lceil \cdot \rceil$ denotes the truncation of the power series at the first non-positive term. Fröberg conjectured that equality holds generically, i.e., it holds on a non-empty Zariski open subset of $\mathbb{P}S_{d_1} \times \ldots \times \mathbb{P}S_{d_s}$. By semicontinuity, fixing all the numeric parameters $(n; d_1, \ldots, d_s)$, it is enough to exhibit one ideal for which the equality holds in order to prove the conjecture for those parameters. In [57] (Main Conjecture 0.6), Iarrobino suggested to look to power ideals and asserted that, except for a list of cases, their Hilbert series coincides with the right-hand-side of (14). By (11), such a conjecture can be translated as a conjecture on the Hilbert function of schemes of fat points. This is usually referred to as the Fröberg–Iarrobino conjecture; for a detailed exposition on this geometric interpretation of Fröberg and Iarrobino's conjectures, we refer to [58]. As we will see in the next section, computing the Hilbert series of schemes of fat points is a very difficult and largely open problem.*

Back to our problem of giving the outline of the proof of Alexander and Hirshowitz Theorem (Theorem 2): Proposition 5 clearly shows that the computation of $T_Q(\sigma_s(\nu_d(\mathbb{P}^n)))$ relies on the knowledge of the Hilbert function of schemes of double fat points. Computing the Hilbert function of fat points is in general a very hard problem. In $\mathbb{P}^2$, there is an extremely interesting and still open conjecture (the SHGH conjecture). The interplay with such a conjecture with the secant varieties is strong, and we deserve to spend a few words on that conjecture and related aspects.

### 2.2. Fat Points in the Plane and SHGH Conjecture

The general problem of determining if a set of generic points $P_1, \ldots, P_s$ in the plane, each with a structure of $m_i$-fat point, has the expected Hilbert function is still an open one. There is only a conjecture due first to B. Segre in 1961 [59], then rephrased by B. Harbourne in 1986 [60], A. Gimigliano in 1987 [61], A. Hirschowitz in 1989 [62] and others. It describes how the elements of a sublinear system of a linear system $\mathcal{L}$ formed by all divisors in $\mathcal{L}$ having multiplicity at least $m_i$ at the points $P_1, \ldots, P_s$, look when the linear system does not have the expected dimension, i.e., the sublinear system depends on fewer parameters than expected. For the sake of completeness, we present the different formulations of the same conjecture, but the fact that they are all equivalent is not a trivial fact; see [63–66].

Our brief presentation is taken from [63,64], which we suggest as excellent and very instructive deepening on this topic.

Let $X$ be a smooth, irreducible, projective, complex variety of dimension $n$. Let $\mathcal{L}$ be a complete linear system of divisors on $X$. Fix $P_1, \ldots, P_s$ distinct points on $X$ and $m_1, \ldots, m_s$ positive integers. We denote by $\mathcal{L}(-\sum_{i=1}^{s} m_i P_i)$ the sublinear system of $\mathcal{L}$ formed by all divisors in $\mathcal{L}$ having multiplicity at least $m_i$ at $P_i$, $i = 1, \ldots, s$. Since a point of multiplicity $m$ imposes $\binom{m+n-1}{n}$ conditions on the divisors of $\mathcal{L}$, it makes sense to define the expected dimension of $\mathcal{L}(-\sum_{i=1}^{s} m_i P_i)$ as:

$$exp.\dim\left(\mathcal{L}\left(-\sum_{i=1}^{s} m_i P_i\right)\right) := \max\left\{\dim(\mathcal{L}) - \sum_{i=1}^{s}\binom{m_i + n - 1}{n}, -1\right\}.$$

If $\mathcal{L}(-\sum_{i=1}^{s} m_i P_i)$ is a linear system whose dimension is not the expected one, it is said to be a special linear system. Classifying special systems is equivalent to determining the Hilbert function of the zero-dimensional subscheme of $\mathbb{P}^n$ given $s$ general fat points of given multiplicities.

A first reduction of this problem is to consider particular varieties X and linear systems $\mathcal{L}$ on them. From this point of view, the first obvious choice is to take $X = \mathbb{P}^n$ and $\mathcal{L} = \mathcal{L}_{n,d} := |\mathcal{O}_{\mathbb{P}^n}(d)|$, the system of all hypersurfaces of degree $d$ in $\mathbb{P}^n$. In this language, $\mathcal{L}_{n,d}(-\sum_{i=1}^{s} m_i P_i)$ are the hypersurfaces of degree $d$ in $n+1$ variables passing through $P_1, \ldots, P_s$ with multiplicities $m_1, \ldots, m_s$, respectively.

The SHGH conjecture describes how the elements of $\mathcal{L}_{2,d}(-\sum_{i=1}^{s} m_i P_i)$ look when not having the expected dimension; here are two formulations of this.

**Conjecture 1** (Segre, 1961 [59]). *If $\mathcal{L}_{2,d}(-\sum_{i=1}^{s} m_i P_i)$ is a special linear system, then there is a fixed double component for all curves through the scheme of fat points defined by $\wp_1^{m_1} \cap \cdots \cap \wp_s^{m_s}$.*

**Conjecture 2** (Gimigliano, 1987 [61,67]). *Consider $\mathcal{L}_{2,d}(-\sum_{i=1}^{s} m_i P_i)$. Then, one has the following possibilities:*

1. *the system is non-special, and its general member is irreducible;*
2. *the system is non-special; its general member is non-reduced, reducible; its fixed components are all rational curves, except for at most one (this may occur only if the system is zero-dimensional); and the general member of its movable part is either irreducible or composed of rational curves in a pencil;*
3. *the system is non-special of dimension zero and consists of a unique multiple elliptic curve;*
4. *the system is special, and it has some multiple rational curve as a fixed component.*

This problem is related to the question of what self-intersections occur for reduced irreducible curves on the surface $X_s$ obtained by blowing up the projective plane at the $s$ points. Blowing up the points introduces rational curves (infinitely many when $s > 8$) of self-intersection $-1$. Each curve $\mathcal{C} \subset X_s$ corresponds to a curve $D_\mathcal{C} \subset \mathbb{P}^2$ of some degree $d$ vanishing to orders $m_i$ at the $s$ points:

$$\mathbb{P}^2 \dashrightarrow X_s, \quad D_\mathcal{C} \mapsto \mathcal{C},$$

and the self-intersection $\mathcal{C}^2$ is $d^2 - m_1^2 - \cdots - m_s^2$ if $D_\mathcal{C} \in \mathcal{L}_{2,d}(-\sum_{i=1}^{s} m_i P_i)$.

**Example 5.** *An example of a curve $D_\mathcal{C}$ corresponding to a curve $\mathcal{C}$ such that $\mathcal{C}^2 = -1$ on $X_s$ is the line through two of the points; in this case, $d = 1$, $m_1 = m_2 = 1$ and $m_i = 0$ for $i > 2$, so we have $d^2 - m_1^2 - \cdots - m_s^2 = -1$.*

According to the SHGH conjecture, these $(-1)$-curves should be the only reduced irreducible curves of negative self-intersection, but proving that there are no others turns out to be itself very hard and is still open.

**Definition 12.** *Let $P_1, \ldots, P_s$ be $s$ points of $\mathbb{P}^n$ in general position. The expected dimension of $\mathcal{L}(-\sum_{i=1}^{s} m_i P_i)$ is:*

$$exp.\dim\left(\mathcal{L}\left(-\sum_{i=1}^{s} m_i P_i\right)\right) := \max\left\{vir.\dim\left(\mathcal{L}\left(-\sum_{i=1}^{s} m_i P_i\right)\right), -1\right\},$$

*where:*

$$vir.\dim\left(\mathcal{L}\left(-\sum_{i=1}^{s} m_i P_i\right)\right) := \binom{n+d}{d} - 1 - \sum_{i=1}^{s}\binom{m_i + n - 1}{n},$$

*is the virtual dimension of $\mathcal{L}(-\sum_{i=1}^{s} m_i P_i)$.*

Consider the blow-up $\pi : \widetilde{\mathbb{P}}^2 \dashrightarrow \mathbb{P}^2$ of the plane at the points $P_1, \ldots, P_s$. Let $E_1, \ldots, E_s$ be the exceptional divisors corresponding to the blown-up points $P_1, \ldots, P_s$, and let $H$ be the pull-back of a general line of $\mathbb{P}^2$ via $\pi$. The strict transform of the system $\mathcal{L} := \mathcal{L}_{2,d}(\sum_{i=1}^s m_i P_i)$ is the system $\widetilde{\mathcal{L}} = |dH - \sum_{i=1}^s m_i E_i|$. Consider two linear systems $\mathcal{L} := \mathcal{L}_{2,d}(\sum_{i=1}^s m_i P_i)$ and $\mathcal{L}' := \mathcal{L}_{2,d}(\sum_{i=1}^s m'_i P_i)$. Their intersection product is defined by using the intersection product of their strict transforms on $\widetilde{\mathbb{P}}^2$, i.e.,

$$\mathcal{L} \cdot \mathcal{L}' = \widetilde{\mathcal{L}} \cdot \widetilde{\mathcal{L}'} = dd' - \sum_{i=1}^s m_i m'_i.$$

Furthermore, consider the anticanonical class $-K := -K_{\widetilde{\mathbb{P}}^2}$ of $\widetilde{\mathbb{P}}^2$ corresponding to the linear system $\mathcal{L}_{2,d}(-\sum_{i=1}^s P_i)$, which, by abusing notation, we also denote by $-K$. The adjunction formula tells us that the arithmetic genus $p_a(\widetilde{\mathcal{L}})$ of a curve in $\widetilde{\mathcal{L}}$ is:

$$p_a(\widetilde{\mathcal{L}}) = \frac{\mathcal{L} \cdot (\mathcal{L} + K)}{2} + 1 = \binom{d-1}{2} \sum_{i=1}^s \binom{m_i}{2},$$

which one defines to be the geometric genus of $\mathcal{L}$, denoted $g_\mathcal{L}$.

This is the classical Clebsch formula. Then, Riemann–Roch says that:

$$\dim(\mathcal{L}) = \dim(\widetilde{\mathcal{L}}) = \mathcal{L} \cdot (\mathcal{L} - K) + h^1(\widetilde{\mathbb{P}}^2, \widetilde{L}) - h^2(\widetilde{\mathbb{P}}^2, \widetilde{L}) =$$
$$= \mathcal{L}^2 - g_\mathcal{L} + 1 + h^1(\mathbb{P}^2, \widetilde{\mathcal{L}}) = vir.\dim(\mathcal{L}) + h^1(\mathbb{P}^2, \widetilde{\mathcal{L}})$$

because clearly, $h^2(\widetilde{\mathbb{P}}^2, \widetilde{\mathcal{L}}) = 0$. Hence,

$$\mathcal{L} \text{ is non-special if and only if } h^0(\widetilde{\mathbb{P}}^2, \widetilde{\mathcal{L}}) \cdot h^1(\widetilde{\mathbb{P}}^2, \widetilde{\mathcal{L}}) = 0.$$

Now, we can see how, in this setting, special systems can naturally arise. Let us look for an irreducible curve $\mathcal{C}$ on $\widetilde{\mathbb{P}}^2$, corresponding to a linear system $\mathcal{L}$ on $\mathbb{P}^2$, which is expected to exist, but, for example, its double is not expected to exist. It translates into the following set of inequalities:

$$\begin{cases} vir.\dim(\mathcal{L}) \geq 0; \\ g_\mathcal{L} \geq 0; \\ vir.\dim(2\mathcal{L}) \leq -1; \end{cases}$$

which is equivalent to:

$$\begin{cases} \mathcal{C}^2 - \mathcal{C} \cdot K \geq 0; \\ \mathcal{C}^2 + \mathcal{C} \cdot K \geq -2; \\ 2\mathcal{C}^2 - \mathcal{C} \cdot K \leq 0; \end{cases}$$

and it has the only solution:

$$\mathcal{C}^2 = \mathcal{C} \cdot K = -1,$$

which makes all the above inequalities equalities. Accordingly, $\mathcal{C}$ is a rational curve, i.e., a curve of genus zero, with self-intersection $-1$, i.e., a $(-1)$-curve. A famous theorem of Castelnuovo's (see [68] (p. 27)) says that these are the only curves that can be contracted to smooth points via a birational morphism of the surface on which they lie to another surface. By abusing terminology, the curve $\Gamma \subset \mathbb{P}^2$ corresponding to $\mathcal{C}$ is also called a $(-1)$-curve.

More generally, one has special linear systems in the following situation. Let $\mathcal{L}$ be a linear system on $\mathbb{P}^2$, which is not empty; let $\mathcal{C}$ be a $(-1)$-curve on $\widetilde{\mathbb{P}}^2$ corresponding to a curve $\Gamma$ on $\mathbb{P}^2$, such that $\widetilde{\mathcal{L}} \cdot \mathcal{C} = -N < 0$. Then, $\mathcal{C}$ (respectively, $\Gamma$) splits off with multiplicity $N$ as a fixed component from all curves of $\widetilde{\mathcal{L}}$ (respectively, $\mathcal{L}$), and one has:

$$\widetilde{\mathcal{L}} = N\mathcal{C} + \widetilde{\mathcal{M}}, \quad (\text{respectively}, \mathcal{L} = N\Gamma + \mathcal{M}),$$

where $\widetilde{\mathcal{M}}$ (respectively, $\mathcal{M}$) is the residual linear system. Then, one computes:

$$\dim(\mathcal{L}) = \dim(\mathcal{M}) \geq vir.\dim(\mathcal{M}) = vir.\dim(\mathcal{L}) + \binom{N}{2},$$

and therefore, if $N \geq 2$, then $\mathcal{L}$ is special.

**Example 6.** *One immediately finds examples of special systems of this type by starting from the $(-1)$-curves of the previous example. For instance, consider $\mathcal{L} := \mathcal{L}_{2,2d}(-\sum_{i=1}^{5} dP_i)$, which is not empty, consisting of the conic $\mathcal{L}_{2,2}(\sum_{i=1}^{d} P_i)$ counted $d$ times, though it has virtual dimension $-\binom{d}{2}$.*

Even more generally, consider a linear system $\mathcal{L}$ on $\mathbb{P}^2$, which is not empty, $\mathcal{C}_1,\ldots,\mathcal{C}_k$ some $(-1)$-curves on $\widetilde{\mathbb{P}}^2$ corresponding to curves $\Gamma_1,\ldots,\Gamma_k$ on $\mathbb{P}^2$, such that $\widetilde{\mathcal{L}} \cdot \mathcal{C}_i = -N_i < 0$, $i = 1,\ldots,k$. Then, for $i = 1,\ldots,k$,

$$\mathcal{L} = \sum_{i=1}^{k} N_i \Gamma_i + \mathcal{M}, \quad \widetilde{\mathcal{L}} = \sum_{i=1}^{k} N_i \mathcal{C}_i + \widetilde{\mathcal{M}}, \quad \text{and} \quad \widetilde{\mathcal{M}} \cdot \mathcal{C}_i = 0.$$

As before, $\mathcal{L}$ is special as soon as there is an $i = 1,\ldots,k$ such that $N_i \geq 2$. Furthermore, $\mathcal{C}_i \cdot \mathcal{C}_j = \delta_{i,j}$, because the union of two meeting $(-1)$-curves moves, according to the Riemann–Roch theorem, in a linear system of positive dimension on $\widetilde{\mathbb{P}}^2$, and therefore, it cannot be fixed for $\widetilde{\mathcal{L}}$. In this situation, the reducible curve $\mathcal{C} := \sum_{i=1}^{k} \mathcal{C}_i$ (respectively, $\Gamma := \sum_{i=1}^{k} N_i \Gamma_i$) is called a $(-1)$-configuration on $\widetilde{\mathbb{P}}^2$ (respectively, on $\mathbb{P}^2$).

**Example 7.** *Consider $\mathcal{L} := \mathcal{L}_{2,d}(-m_0 P_0 - \sum_{i=1}^{s} m_i P_i)$, with $m_0 + m_i = d + N_i$, $N_i \geq 1$. Let $\Gamma_i$ be the line joining $P_0$, $P_i$. It splits off $N_i$ times from $\mathcal{L}$. Hence:*

$$\mathcal{L} = \sum_{i=1}^{s} N_i \Gamma_i + \mathcal{L}_{2,d-\sum_{i=1}^{s} N_i}\left(-\left(m_0 - \sum_{i=1}^{s} N_i\right) P_0 - \sum_{i=1}^{s} (m_i - N_i) P_i\right).$$

*If we require the latter system to have non-negative virtual dimension, e.g., $d \geq \sum_{i=1}^{s} m_i$, if $m_0 = d$ and some $N_i > 1$, we have as many special systems as we want.*

**Definition 13.** *A linear system $\mathcal{L}$ on $\mathbb{P}^2$ is $(-1)$-reducible if $\widetilde{\mathcal{L}} = \sum_{i=1}^{k} N_i \mathcal{C}_i + \widetilde{\mathcal{M}}$, where $\mathcal{C} = \sum_{i=1}^{k} \mathcal{C}_i$ is a $(-1)$-configuration, $\widetilde{\mathcal{M}} \cdot \mathcal{C}_i = 0$, for all $i = 1,\ldots,k$ and $vir.\dim(\mathcal{M}) \geq 0$. The system $\mathcal{L}$ is called $(-1)$-special if, in addition, there is an $i = 1,\ldots,k$ such that $N_i > 1$.*

**Conjecture 3** (Harbourne, 1986 [60], Hirschowitz, 1989 [62]). *A linear system of plane curves $\mathcal{L}_{2,d}(-\sum_{i=1}^{s} m_i P_i)$ with general multiple base points is special if and only if it is $(-1)$-special, i.e., it contains some multiple rational curve of self-intersection $-1$ in the base locus.*

No special system has been discovered except $(-1)$-special systems.

Eventually, we signal a concise version of the conjecture (see [67] (Conjecture 3.3)), which involves only a numerical condition.

**Conjecture 4.** *A linear system of plane curves $\mathcal{L}_{2,d}(-\sum_{i=1}^{s} m_i P_i)$ with general multiple base points and such that $m_1 \geq m_2 \geq \ldots \geq m_s \geq 0$ and $d \geq m_1 + m_2 + m_3$ is always non-special.*

The idea of this conjecture comes from Conjecture 3 and by working on the surface $X = \widetilde{\mathbb{P}}^2$, which is the blow up of $\mathbb{P}^2$ at the points $P_i$; in this way, the linear system $\mathcal{L}_{2,d}(-\sum_{i=1}^{s} m_i P_i)$ corresponds to the linear system $\widetilde{\mathcal{L}} = dE_0 - m_1 E_1 - \ldots - E_s$ on $X$, where $(E_0, E_1, \ldots, E_s)$ is a basis for $\text{Pic}(X)$, and $E_0$ is the strict transform of a generic line of $\mathbb{P}^2$, while the divisors $E_1, \ldots, E_s$ are the exceptional divisors on $P_1, \ldots, P_s$. If we assume that the only special linear systems $\mathcal{L}_{2,d}(-\sum_{i=1}^s m_i P_i)$ are those that contain a fixed multiple $(-1)$-curve, this would be the same for $\tilde{\mathcal{L}}$ in $\text{Pic}(X)$, but this implies that either we have $m_s < -1$, or we can apply Cremona transforms until the fixed multiple $(-1)$-curve becomes of type $-m_i' E_i'$ in $\text{Pic}(X)$, where the $E_i'$'s are exceptional divisors in a new basis for $\text{Pic}(X)$. Our conditions in Conjecture 4 prevent these possibilities, since the $m_i$ are positive and the condition $d \geq m_1 + m_2 + m_3$ implies that, by applying a Cremona transform, the degree of a divisor with respect to the new basis cannot decrease (it goes from $d$ to $d' = 2d - m_i - m_j - m_k$, if the Cremona transform is based on $P_i$, $P_j$ and $P_k$), hence cannot become of degree zero (as $-m_i' E_i'$ would be).

One could hope to address a weaker version of this problem. Nagata, in connection with his negative solution of the fourteenth Hilbert problem, made such a conjecture.

**Conjecture 5** (Nagata, 1960 [69]). *The linear system $\mathcal{L}_{2,d}(-\sum_{i=1}^s m_i P_i)$ is empty as soon as $s \geq 10$ and $d \leq \sqrt{s}$.*

Conjecture 5 is weaker than Conjecture 3, yet still open for every non-square $n \geq 10$. Nagata's conjecture does not rule out the occurrence of curves of self-intersection less than $-1$, but it does rule out the worst of them. In particular, Nagata's conjecture asserts that $d^2 \geq sm^2$ must hold when $s \geq 10$, where $m = (m_1 + \cdots + m_s)/s$. Thus, perhaps there are curves with $d^2 - m_1^2 - \cdots - m_s^2 < 0$, such as the $(-1)$-curves mentioned above, but $d^2 - m_1^2 - \cdots - m_s^2$ is (conjecturally) only as negative as is allowed by the condition that after averaging the multiplicities $m_i$ for $n \geq 10$, one must have $d^2 - sm^2 \geq 0$.

Now, we want to find a method to study the Hilbert function of a zero-dimensional scheme. One of the most classical methods is the so-called Horace method ([8]), which has also been extended with the Horace differential technique and led J. Alexander and A. Hirschowitz to prove Theorem 2. We explain these methods in Sections 2.2.1 and 2.2.2, respectively, and we resume in Section 2.2.3 the main steps of the Alexander–Hirschowitz theorem.

2.2.1. La Méthode D'Horace

In this section, we present the so-called Horace method. It takes this name from the ancient Roman legend (and a play by Corneille: Horace, 1639) about the duel between three Roman brothers, the "Orazi", and three brothers from the enemy town of Albalonga, the "Curiazi". The winners were to have their town take over the other one. After the first clash among them, two of the Orazi died, while the third remained alive and unscathed, while the Curiazi were all wounded, the first one slightly, the second more severely and the third quite badly. There was no way that the survivor of the Orazi could beat the other three, even if they were injured, but the Roman took to his heels, and the three enemies pursued him; while running, they got separated from each other because they were differently injured and they could run at different speeds. The first to reach the Orazio was the healthiest of the Curiazi, who was easily killed. Then, came the other two who were injured, and it was easy for the Orazio to kill them one by one.

This idea of "killing" one member at a time was applied to the three elements in the exact sequence of an ideal sheaf (together with the ideals of a residual scheme and a "trace") by A. Hirschowitz in [45] (that is why now, we keep the french version "Horace" for Orazi) to compute the postulation of multiple points and count how many conditions they impose.

Even if the following definition extends to any scheme of fat points, since it is the case of our interest, we focus on the scheme of two-fat points.

**Definition 14.** *We say that a scheme $Z$ of $r$ two-fat points, defined by the ideal $I_Z$, imposes independent conditions on the space of hypersurfaces of degree $d$ in $n+1$ variable $\mathcal{O}_{\mathbb{P}^n}(d)$ if $\text{codim}_{\Bbbk}(I_Z)_d$ in $S^d V$ is $\min\left\{\binom{n+d}{d}, r(n+1)\right\}$.*

This definition, together with the considerations of the previous section and (12) allows us to reformulate the problem of finding the dimension of secant varieties to Veronese varieties in terms of independent conditions imposed by a zero-dimensional scheme of double fat points to forms of a certain degree.

**Corollary 2.** *The s-th secant variety $\sigma_s(X_{n,d})$ of a Veronese variety has the expected dimension if and only if a scheme of s generic two-fat points in $\mathbb{P}^n$ imposes independent conditions on $\mathcal{O}_{\mathbb{P}^n}(d)$.*

**Example 8.** *The linear system $\mathcal{L} := \mathcal{L}_{n,2}(-\sum_{i=1}^{s} 2P_i)$ is special if $s \leq n$. Actually, quadrics in $\mathbb{P}^n$ singular at s independent points $P_1, \ldots, P_s$ are cones with the vertex $\mathbb{P}^{s-1}$ spanned by $P_1, \ldots, P_s$. Therefore, the system is empty as soon as $s \geq n+1$, whereas, if $s \leq n$, one easily computes:*

$$\dim(\mathcal{L}) = vir.\dim(\mathcal{L}) + \binom{s}{2}.$$

*Therefore, by (12), this equality corresponds to the fact that $\sigma_s(\nu_2(\mathbb{P}^n))$ are defective for all $s \leq n$; see Theorem 2 (1).*

We can now present how Alexander and Hirschowitz used the Horace method in [8] to compute the dimensions of the secant varieties of Veronese varieties.

**Definition 15.** *Let $Z \subset \mathbb{P}^n$ be a scheme of two-fat points whose ideal sheaf is $\mathcal{I}_Z$. Let $H \subset \mathbb{P}^n$ be a hyperplane. We define the following:*

- *the trace of Z with respect to H is the scheme-theoretic intersection:*

$$\mathrm{Tr}_H(Z) := Z \cap H;$$

- *the residue of Z with respect to H is the zero-dimensional scheme defined by the ideal sheaf $\mathcal{I}_Z : \mathcal{O}_{\mathbb{P}^n}(-H)$ and denoted $\mathrm{Res}_H(Z)$.*

**Example 9.** *Let $Z = 2P_0 \subset \mathbb{P}^n$ be the two-fat point defined by $\wp^2 = (x_1, \ldots, x_n)^2$, and let H be the hyperplane $\{x_n = 0\}$. Then, the residue $\mathrm{Res}_H(Z) \subset \mathbb{P}^n$ is defined by:*

$$I_{\mathrm{Res}_H(Z)} = \wp^2 : (x_n) = (x_1, \ldots, x_n) = \wp,$$

*hence, it is a simple point of $\mathbb{P}^n$; the trace $\mathrm{Tr}_H(Z) \subset H \simeq \mathbb{P}^{n-1}$ is defined by:*

$$I_{\mathrm{Tr}_H(Z)} = (x_1, \ldots, x_n)^2 \otimes \mathbb{k}[x_0, \ldots, x_n]/(x_n) = (\bar{x}_1, \ldots, \bar{x}_{n-1})^2,$$

*where the $\bar{x}_i$'s are the coordinate of the $\mathbb{P}^{n-1} \simeq H$, i.e., $\mathrm{Tr}_H(Z)$ is a two-fat point in $\mathbb{P}^{n-1}$ with support at $P_0 \in H$.*

The idea now is that it is easier to compute the conditions imposed by the residue and by the trace rather than those imposed by the scheme $Z$; in particular, as we are going to explain in the following, this gives us an inductive argument to prove that a scheme $Z$ imposes independent conditions on hypersurfaces of certain degree. In particular, for any $d$, taking the global sections of the restriction exact sequence:

$$0 \to \mathcal{I}_{\mathrm{Res}_H(Z)}(d-1) \to \mathcal{I}_Z(d) \to \mathcal{I}_{\mathrm{Tr}_H(Z)}(d) \to 0,$$

we obtain the so-called Castelnuovo exact sequence:

$$0 \to (I_{\mathrm{Res}_H(Z)})_{d-1} \to (I_Z)_d \to (I_{\mathrm{Tr}_H(Z)})_d, \tag{15}$$

from which we get the inequality:

$$\dim_\Bbbk(I_Z)_d \leq \dim_\Bbbk(I_{\text{Res}_H(Z)})_{d-1} + \dim_\Bbbk(I_{\text{Tr}_H(Z)})_d. \tag{16}$$

Let us assume that the supports of $Z$ are $r$ points such that $t$ of them lie on the hyperplane $H$, i.e., $\text{Res}_H(Z)$ is the union of $r - t$ many two-fat points and $t$ simple points in $\mathbb{P}^n$ and $\text{Tr}_H(Z)$ is a scheme of $t$ many two-fat points in $\mathbb{P}^{n-1}$ i.e., with the notation of linear systems introduced above,

$$\dim_\Bbbk(I_{\text{Res}_H(Z)})_{d-1} = \dim\left(\mathcal{L}_{n,d-1}\left(-2\sum_{i=1}^{r-t} P_i - \sum_{i=r-t+1}^{r} P_i\right)\right) + 1;$$

$$\dim_\Bbbk(I_{\text{Tr}_H(Z)})_d = \dim\left(\mathcal{L}_{n-1,d}\left(-2\sum_{i=1}^{t} P_i\right)\right) + 1.$$

Assuming that:

1. $\text{Res}_H(Z)$ imposes independent conditions on $\mathcal{O}_{\mathbb{P}^n}(d-1)$, i.e.,

$$\dim_\Bbbk(I_{\text{Res}_H(Z)})_{d-1} = \max\left\{\binom{d-1+n}{n} - (r-t)(n+1) - t, 0\right\},$$

2. and $\text{Tr}_H(Z)$ imposes independent conditions on $\mathcal{O}_{\mathbb{P}^{n-1}}(d)$, i.e.,

$$\dim_\Bbbk(I_{\text{Tr}_H(Z)})_d = \max\left\{\binom{d+n-1}{n+1} - tn, 0\right\},$$

then, by (16) and since the expected dimension (Definition 12) is always a lower bound for the actual dimension, we conclude the following.

**Theorem 3** (Brambilla–Ottaviani [36]). *Let $Z$ be a union of $r$ many two-fat points in $\mathbb{P}^n$, and let $H \subset \mathbb{P}^n$ be a hyperplane such that $t$ of the $r$ points of $Z$ have support on $H$. Assume that $\text{Tr}_H(Z_r)$ imposes independent conditions on $\mathcal{O}_H(d)$ and that $\text{Res}_H Z_r$ imposes independent conditions on $\mathcal{O}_{\mathbb{P}^n}(d-1)$. If one of the pairs of the following inequalities occurs:*

1. $tn \leq \binom{d+n-1}{n-1}$ and $r(n+1) - tn \leq \binom{d+n-1}{n}$,
2. $tn \geq \binom{d+n-1}{n-1}$ and $r(n+1) - tn \geq \binom{d+n-1}{n}$,

*then $Z$ imposes independent conditions on the system $\mathcal{O}_{\mathbb{P}^n}(d)$.*

The technique was used by Alexander and Hirschowitz to compute the dimension of the linear system of hypersurfaces with double base points, and hence, the dimension of secant varieties of Veronese varieties is mainly the Horace method, via induction.

The regularity of secant varieties can be proven as described above by induction, but non-regularity cannot. Defective cases have to be treated case by case. We have already seen that the case of secant varieties of Veronese surfaces (Example 4) and of quadrics (Example 8) are defective, so we cannot take them as the first step of the induction.

Let us start with $\sigma_s(X_{3,3}) \subset \mathbb{P}^{19}$. The expected dimension is $4s - 1$. Therefore, we expect that $\sigma_5(X_{3,3})$ fills up the ambient space. Now, let $Z$ be a scheme of five many two-fat points in general position in $\mathbb{P}^3$ defined by the ideal $I_Z = \wp_1^2 \cap \ldots \cap \wp_5^2$. Since the points are in general position, we may assume that they are the five fundamental points of $\mathbb{P}^3$ and perform our computations for this explicit set of points. Then, it is easy to check that:

$$\text{HF}(S/\wp_1^2 \cap \ldots \cap \wp_5^2, 3) = 19 - \dim_\Bbbk(I_Z)_3 = 19 - 0.$$

Hence, $\sigma_5(X_{3,3}) = \mathbb{P}^{19}$, as expected. This implies that:

$$\dim(\sigma_s(X_{3,3})) \text{ is the expected one for all } s \leq 5. \tag{17}$$

Indeed, as a consequence of the following proposition, if the $s$-th secant variety is regular, so it is the $(s-1)$-th secant variety.

**Proposition 6.** *Assume that $X$ is $s$-defective and that $\sigma_{s+1}(X) \neq \mathbb{P}^N$. Then, $X$ is also $(s+1)$-defective.*

**Proof.** Let $\delta_s$ be the $s$-defect of $X$. By assumptions and by Terracini's lemma, if $P_1, \ldots, P_s \in X$ are general points, then the span $T_{P_1,\ldots,P_s} := \langle T_{P_1}X, \ldots, T_{P_s}X \rangle$, which is the tangent space at a general point of $\sigma_s(X)$, has projective dimension $\min(N, sn + s - 1) - \delta_s$. Hence, adding one general point $P_{s+1}$, the space $T_{P_1,\ldots,P_s,P_{s+1}}$, which is the span of $T_{P_1,\ldots,P_s}$ and $T_{P_{s+1}}X$, has dimension at most $\min\{N, sn + s - 1\} - \delta_s + n + 1$. This last number is smaller than $N$, while it is clearly smaller than $(s+1)n + s$. Therefore, $X$ is $(s+1)$-defective. □

In order to perform the induction on the dimension, we would need to study the case of $d = 4$, $s = 8$ in $\mathbb{P}^3$, i.e., $\sigma_8(X_{3,4})$. We need to compute $\mathrm{HF}_Z(4) = \mathrm{HF}(\Bbbk[x_0, \ldots, x_3]/(\wp_1^2 \cap \cdots \cap \wp_8^2), 4)$. In order to use the Horace lemma, we need to know how many points in the support of scheme $Z$ lie on a given hyperplane. The good news is upper semicontinuity, which allows us to specialize points on a hyperplane. In fact, if the specialized scheme has the expected Hilbert function, then also the general scheme has the expected Hilbert function (as before, this argument cannot be used if the specialized scheme does not have the expected Hilbert function: this is the main reason why induction can be used to prove the regularity of secant varieties, but not the defectiveness). In this case, we choose to specialize four points on $H$, i.e., $Z = 2P_1 + \ldots + 2P_8$ with $P_1, \ldots, P_4 \in H$. Therefore,

- $\mathrm{Res}_H(Z) = P_1 + \cdots + P_4 + 2P_5 + \cdots + 2P_8 \subset \mathbb{P}^3$;
- $\mathrm{Tr}_H(Z_8) = 2\widetilde{P}_1 + \cdots + 2\widetilde{P}_4 \subset H$, where $2\widetilde{P}_i$'s are two-fat points in $\mathbb{P}^2$

Consider Castelnuovo Inequality (16). Four two-fat points in $\mathbb{P}^3$ impose independent conditions to $\mathcal{O}_{\mathbb{P}^3}(3)$ by (17), then adding four simple general points imposes independent conditions; therefore, $\mathrm{Res}_H Z$ imposes the independent condition on $\mathcal{O}_{\mathbb{P}^3}(3)$. Again, assuming that the supports of $\mathrm{Tr}_H(Z)$ are the fundamental points of $\mathbb{P}^2$, we can check that it imposes the independent condition on $\mathcal{O}_{\mathbb{P}^2}(4)$. Therefore,

$$\max\left\{\binom{4+3}{3} - 8 \cdot 4, 0\right\} = 3 = exp.\dim_{\Bbbk}(I_Z)_4 \leq \dim_{\Bbbk}(I_Z)_4$$

$$\leq \dim_{\Bbbk}(I_{\mathrm{Res}_H(Z)})_3 + \dim_{\Bbbk}(I_{\mathrm{Tr}_H(Z)})_4$$

$$= \max\left\{\binom{3+3}{3} - 4 \cdot 4 - 4, 0\right\} + \max\left\{\binom{2+4}{2} - 4 \cdot 3, 0\right\}$$

$$= \max\{20 - 16 - 4, 0\} + \max\{15 - 12, 0\} = 3.$$

In conclusion, we have proven that

$$\sigma_s(X_{3,4}) \text{ has the expected dimension for any } s \leq 8.$$

Now, this argument cannot be used to study $\sigma_9(X_{3,4})$ because it is one of the defective cases, but we can still use induction on $d$.

In order to use induction on the degree $d$, we need a starting case, that is the case of cubics. We have done $\mathbb{P}^3$ already; see (17). Now, $d = 3$, $n = 4$, $s = 7$ corresponds to a defective case. Therefore, we need to start with $d = 3$ and $n = 5$. We expect that $\sigma_{10}(X_{5,3})$ fills up the ambient space. Let us try to apply the Horace method as above. The hyperplane $H$ is a $\mathbb{P}^4$; one two-fat point in $\mathbb{P}^4$ has degree five, so we can specialize up to seven points on $H$ (in $\mathbb{P}^4$, there are exactly $35 = 7 \times 5$ cubics), but seven two-fat points in $\mathbb{P}^4$ are defective in degree three; in fact, if $Z = 2P_1 + \ldots + 2P_7 \subset \mathbb{P}^4$, then $\dim_{\Bbbk}(I_Z)_3 = 1$. Therefore, if we specialize seven two-fat points on a generic hyperplane $H$, we are "not using all the room that we have at our disposal", and (16) does not give the correct upper bound. In other words, if we want to get a zero in the trace term of the Castelunovo exact sequence, we have to "add one more condition on $H$"; but, to do that, we need a more refined version of the Horace method.

2.2.2. La méthode d'Horace Differentielle

The description we are going to give follows the lines of [70].

**Definition 16.** *An ideal $I$ in the algebra of formal functions $\Bbbk[[x,y]]$, where $x = (x_1,\ldots,x_{n-1})$, is called a vertically-graded (with respect to $y$) ideal if:*

$$I = I_0 \oplus I_1 y \oplus \cdots \oplus I_{m-1} y^{m-1} \oplus (y^m)$$

*where, for $i = 0,\ldots,m-1$, $I_i \subset \Bbbk[[x]]$ is an ideal.*

**Definition 17.** *Let $Q$ be a smooth n-dimensional integral scheme, and let $D$ be a smooth irreducible divisor on $Q$. We say that $Z \subset Q$ is a vertically-graded subscheme of $Q$ with base $D$ and support $z \in D$, if $Z$ is a zero-dimensional scheme with support at the point $z$ such that there is a regular system of parameters $(x,y)$ at $z$ such that $y = 0$ is a local equation for $D$ and the ideal of $Z$ in $\widehat{\mathcal{O}}_{Q,z} \cong \Bbbk[[x,y]]$ is vertically graded.*

**Definition 18.** *Let $Z \in Q$ be a vertically-graded subscheme with base $D$, and let $p \geq 0$ be a fixed integer. We denote by $\mathrm{Res}_D^p(Z) \in Q$ and $\mathrm{Tr}_D^p(Z) \in D$ the closed subschemes defined, respectively, by the ideals sheaves:*

$$\mathcal{I}_{\mathrm{Res}_D^p(Z)} := \mathcal{I}_Z + (\mathcal{I}_Z : \mathcal{I}_D^{p+1})\mathcal{I}_D^p, \quad \text{and} \quad \mathcal{I}_{\mathrm{Tr}_D^p(Z),D} := (\mathcal{I}_Z : \mathcal{I}_D^p) \otimes \mathcal{O}_D.$$

In $\mathrm{Res}_D^p(Z)$, we remove from $Z$ the $(p+1)$-th "slice" of $Z$, while in $\mathrm{Tr}_D^p(Z)$, we consider only the $(p+1)$-th "slice". Notice that for $p = 0$, this recovers the usual trace $\mathrm{Tr}_D(Z)$ and residual schemes $\mathrm{Res}_D(Z)$.

**Example 10.** *Let $Z \subset \mathbb{P}^2$ be a three-fat point defined by $\wp^3$, with support at a point $P \in H$ lying on a plane $H \subset \mathbb{P}^3$. We may assume $\wp = (x_1, x_2)$ and $H = \{x_2 = 0\}$. Then, $3P$ is vertically graded with respect to $H$:*

$$\wp^3 = (x_1^3) \oplus (x_1^2)x_2 \oplus (x_1)x_2^2 \oplus (x_2^3),$$

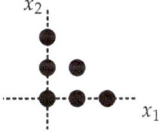

Visualization of a three-fat point in $\mathbb{P}^2$: each dot corresponds to a generator of the local ring, which is Artinian.

Now, we compute all residues (white dots) and traces (black dots) as follows:

Case $p = 0$.  $\mathrm{Res}_H^0(P) \subset \mathbb{P}^2$ is defined by:

$$\wp^3 : (x_2) = \wp^2 = (x_1^2) \oplus (x_1)x_2 \oplus (x_2^2),$$

i.e., it is a two-fat point in $\mathbb{P}^2$, while $\mathrm{Tr}_H^0(P)$ is defined by:

$$\widetilde{\wp}^3 = \wp^3 \otimes \Bbbk[[x_0, x_1, x_2]]/(x_2) = (\overline{x}_1^3),$$

where $\overline{x}_0, \overline{x}_1$ are the coordinates on $H$, i.e., it is a three-fat point in $\mathbb{P}^1$.

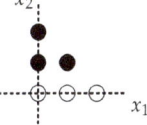

**Case $p=1$.** $\text{Res}^1_H(3P) \subset \mathbb{P}^2$ is a zero-dimensional subscheme of $\mathbb{P}^2$ of length four given by:

$$\wp^3 + (\wp^3 : (x_2^2))x_2 = (x_2^2, x_1x_2, x_1^3) =$$
$$= (x_1^3) \oplus (x_1)x_2 \oplus (x_2)^2;$$

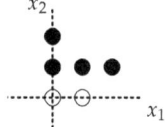

roughly speaking, we "have removed the second slice" of $3P$; while, $\text{Tr}^1_H(3P)$ is given by:

$$(\wp^3 : (x_2)) \otimes \Bbbk[[x_0, x_1, x_2]]/(x_2) = (\bar{x}_1^2).$$

**Case $p=2$.** $\text{Res}^2_H(P) \subset \mathbb{P}^2$ is a zero-dimensional scheme of length five given by:

$$\wp^3 + (\wp^3 : (x_2^3))x_2^2 = (x_2^2, x_1^2x_2, x_1^3) =$$
$$= (x_1^3) \oplus (x_1^2)x_2 \oplus (x_2)^2;$$

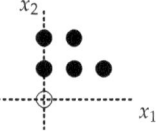

roughly speaking, we "have removed the third slice" of $3P$; while $\text{Tr}^2_H(P)$ is given by:

$$(\wp^3 : (x_2^2)) \otimes \Bbbk[[x_0, x_1, x_2]]/(x_2) = (\bar{x}_1).$$

Finally, let $Z_1, \ldots, Z_r \in Q$ be vertically-graded subschemes with base $D$ and support $z_i$; let $Z = Z_1 \cup \cdots \cup Z_r$, and set $\mathbf{p} = (p_1, \ldots, p_r) \in \mathbb{N}^r$. We write:

$$\text{Tr}^{\mathbf{p}}_D(Z) := \text{Tr}^{p_1}_D(Z_1) \cup \cdots \cup \text{Tr}^{p_r}_D(Z_r), \quad \text{Res}^{\mathbf{p}}_D(Z) := \text{Res}^{p_1}_D(Z_1) \cup \cdots \cup \text{Res}^{p_r}_D(Z_r).$$

We are now ready to formulate the Horace differential lemma.

**Proposition 7** (Horace differential lemma, [71] (Proposition 9.1))**.** *Let $H$ be a hyperplane in $\mathbb{P}^n$, and let $W \subset \mathbb{P}^n$ be a zero-dimensional closed subscheme. Let $Y_1, \ldots, Y_r, Z_1, \ldots, Z_r$ be zero-dimensional irreducible subschemes of $\mathbb{P}^n$ such that $Y_i \cong Z_i$, $i = 1, \ldots, r$, $Z_i$ has support on $H$ and is vertically graded with base $H$, and the supports of $Y = Y_1 \cup \cdots \cup Y_r$ and $Z = Z_1 \cup \cdots \cup Z_r$ are generic in their respective Hilbert schemes. Let $\mathbf{p} = (p_1, \ldots, p_r) \in \mathbb{N}^r$. Assume:*

1. $H^0(\mathcal{I}_{\text{Tr}_H W \cup \text{Tr}^{\mathbf{p}}_H(Z), H}(d)) = 0$ *and*
2. $H^0(\mathcal{I}_{\text{Res}_H W \cup \text{Res}^{\mathbf{p}}_H(Z)}(d-1)) = 0;$

*then,*
$$H^0(\mathcal{I}_{W \cup Y}(d)) = 0.$$

For two-fat points, the latter result can be rephrased as follows.

**Proposition 8** (Horace differential lemma for two-fat points)**.** *Let $H \subset \mathbb{P}^n$ be a hyperplane, $P_1, \ldots, P_r \in \mathbb{P}^n$ be generic points and $W \subset \mathbb{P}^n$ be a zero-dimensional scheme. Let $Z = 2P_1 + \cdots + 2P_r \subset \mathbb{P}^n$, and let $Z' = 2P'_1 + \ldots + 2P'_r$ such that the $P'_i$'s are generic points on $H$. Let $D_{2,H}(P'_i) = 2P'_i \cap H$ be zero-dimensional schemes in $\mathbb{P}^n$. Hence, let:*

$$\overline{Z} = \text{Res}_H(W) + D_{2,H}(P'_1) + \ldots + D_{2,H}(P'_r) \subset \mathbb{P}^n, \quad \text{and}$$
$$\overline{T} = \text{Tr}_H(W) + P'_1 + \ldots + P'_r \subset H \simeq \mathbb{P}^{n-1}.$$

Then, if the following two conditions are satisfied:

$$degue: \dim_\Bbbk(I_{\overline{Z}})_{d-1} = 0;$$
$$dime: \dim_\Bbbk(I_{\overline{T}})_d = 0,$$

then, $\dim_\Bbbk(I_{W+Z})_d = 0$.

Now, with this proposition, we can conclude the computation of $\sigma_{10}(X_{5,3})$. Before Section 2.2.2, we were left with the problem of computing the Hilbert function in degree three of a scheme $Z = 2P_1 + \cdots + 2P_{10}$ of ten two-fat points with generic support in $\mathbb{P}^5$: since a two-fat point in $\mathbb{P}^5$ imposes six conditions, the expected dimension of $(I_Z)_3$ is zero. In this case, the "standard" Horace method fails, since if we specialize seven points on a generic hyperplane, we lose one condition that we miss at the end of the game. We apply the Horace differential method to this situation. Let $P'_1, \ldots, P'_8$ be generic points on a hyperplane $H \simeq \mathbb{P}^4 \subset \mathbb{P}^5$. Consider:

$$\overline{Z} = P'_1 + \ldots + P'_7 + D_{2,H}(P_8) + 2P_9 + 2P_{10} \subset \mathbb{P}^5, \quad \text{and}$$
$$\overline{T} = 2P'_1 + \ldots + 2P'_7 + P'_8 \subset H \simeq \mathbb{P}^4.$$

Now, *dime* is satisfied because we have added on the trace exactly the one condition that we were missing. It is not difficult to prove that *degue* is also satisfied: quadrics through $\overline{Z}$ are cones with the vertex the line between $P_9$ and $P_{10}$; hence, the dimension of the corresponding linear system equals the dimension of a linear system of quadrics in $\mathbb{P}^4$ passing through a scheme of seven simple points and two two-fat points with generic support. Again, such quadrics in $\mathbb{P}^4$ are all cones with the vertex the line passing through the support of the two two-fat points: hence, the dimension of the latter linear system equals the dimension of a linear system of quadrics in $\mathbb{P}^3$ passing through a set of eight simple points with general support. This has dimension zero, since the quadrics of $\mathbb{P}^3$ are ten. In conclusion, we obtain that the Hilbert function in degree three of a scheme of ten two-fat points in $\mathbb{P}^5$ with generic support is the expected one, i.e., by (12), we conclude that $\sigma_{10}(X_{5,3})$ fills the ambient space.

### 2.2.3. Summary of the Proof of the Alexander–Hirshowitz Theorem

We finally summarize the main steps of the proof of Alexander–Hirshowitz theorem (Theorem 2):

1. The dimension of $\sigma_s(X_{n,d})$ is equal to the dimension of its tangent space at a general point $Q$;
2. By Terracini's lemma (Lemma 1), if $Q$ is general in $\langle P_1, \ldots, P_s \rangle$, with $P_1, \ldots, P_s \in X$ general points, then:
$$\dim(T_Q \sigma_s(X)) = \dim(\langle T_{P_1} X, \ldots, T_{P_s} X \rangle);$$
3. By using the apolarity action (see Definition 9), one can see that:
$$\dim(\langle T_{P_1} X_{n,d}, \ldots, T_{P_s} X_{n,d} \rangle) = \mathrm{HF}(R/(\wp_1^2 \cap \cdots \cap \wp_s^2), d) - 1,$$
where $\wp_1^2 \cap \cdots \cap \wp_s^2$ is the ideal defining the scheme of two-fat points supported by $\mathbb{P}^n$ corresponding to the $P_i$'s via the $d$-th Veronese embedding;
4. Non-regular cases, i.e., where the Hilbert function of the scheme of two-fat points is not as expected, have to be analyzed case by case; regular cases can be proven by induction:

    (a) The list of non-regular cases corresponds to defective Veronese varieties and is very classical; see Section 2.1.3, page 11 and [36] for the list of all papers where all these cases were investigated. We explained a few of them in Examples 2, 4 and 8;

    (b) The proof of the list of non-regular cases classically known is complete and can be proven by a double induction procedure on the degree $d$ and on the dimension $n$ (see Theorem 3 and Proposition 6):

- The starting step of the induction for the degree is $d = 3$ since quadrics are defective (Example 8):
    - The first case to study is therefore $X_{3,3}$: in order to prove that $\sigma_5(X_{3,3}) = \mathbb{P}^{19}$ as expected (see page 26), the Horace method (Section 2.2.1) is introduced.
- The starting step of the induction for the dimension is $n = 5$:
    - $\sigma_8(X_{3,4})$ has the expected dimension thanks to upper semicontinuity (see page 26), so also the smallest secant varieties are regular (page 26);
    - $\sigma_9(X_{3,4})$ is defective ([4,43,62]);
    - Therefore, one has to start with $\sigma_{10}(X_{3,5})$, which cannot be done with the standard Horace method (see page 27), while it can be solved (see page 29) by using the Horace differential method (Section 2.2.2).

## 2.3. Algorithms for the Symmetric-Rank of a Given Polynomial

The goal of the second part of this section is to compute the symmetric-rank of a given symmetric tensor. Here, we have decided to focus on algorithms rather than entering into the details of their proofs, since most of them, especially the more advanced ones, are very technical and even an idea of the proofs would be too dispersive. We believe that a descriptive presentation is more enlightening on the difference among them, the punchline of each one and their weaknesses, rather than a precise proof.

### 2.3.1. On Sylvester's Algorithm

In this section, we present the so-called Sylvester's algorithm (Algorithm 1). It is classically attributed to Sylvester, since he studied the problem of decomposing a homogeneous polynomial of degree $d$ into two variables as a sum of $d$-th powers of linear forms and solved it completely, obtaining that the decomposition is unique for general polynomials of odd degree. The first modern and available formulation of this algorithm is due to Comas and Seiguer; see [27].

Despite the "age" of this algorithm, there are modern scientific areas where it is used to describe very advanced tools; see [14] for the measurements of entanglement in quantum physics. The following description follows [28].

If $V$ is a two-dimensional vector space, there is a well-known isomorphism between $\bigwedge^{d-r+1}(S^d V)$ and $S^{d-r+1}(S^r V)$; see [72]. In terms of projective algebraic varieties, this isomorphism allows us to view the $(d-r+1)$-th Veronese embedding of $\mathbb{P}^r \simeq \mathbb{P} S^r V$ as the set of $(r-1)$-dimensional linear subspaces of $\mathbb{P}^d$ that are $r$-secant to the rational normal curve. The description of this result, via coordinates, was originally given by Iarrobino and Kanev; see [25]. Here, we follow the description appearing in [73] (Lemma 2.1). We use the notation $G(k, W)$ for the Grassmannian of $k$-dimensional linear spaces in a vector space $W$ and the notation $\mathbb{G}(k, n)$ for the Grassmannian of $k$-dimensional linear spaces in $\mathbb{P}^n$.

**Lemma 2.** *Consider the map $\phi_{r,d-r+1} : \mathbb{P}(S^r V) \to G(d-r+1, S^d V)$ that sends the projective class of $F \in S^r V$ to the $(d-r+1)$-dimensional subspace of $S^d V$ made by the multiples of $F$, i.e.,*

$$\phi_{r,d-r+1}([F]) = F \cdot S^{d-r} V \subset S^d V.$$

*Then, the following hold:*

1. *the image of $\phi_{r,d-r+1}$, after the Plücker embedding of $G(d-r+1, S^d V)$ inside $\mathbb{P}(\bigwedge^{d-r+1} S^d V)$, is the $(d-r+1)$-th Veronese embedding of $\mathbb{P} S^r V$;*
2. *identifying $G(d-r+1, S^d V)$ with $\mathbb{G}(r-1, \mathbb{P} S^d V^*)$, the above Veronese variety is the set of linear spaces $r$-secant to a rational normal curve $\mathcal{C}_d \subset \mathbb{P} S^d V^*$.*

For the proof, we follow the constructive lines of [28], which we keep here, even though we take the proof as it is, since it is short and we believe it is constructive and useful.

**Proof.** Let $\{x_0, x_1\}$ be the variables on $V$. Then, write $F = u_0 x_0^r + u_1 x_0^{r-1} x_1 + \cdots + u_r x_1^r \in S^r V$. A basis of the subspace of $S^d V$ of forms of the type $FH$ is given by:

$$\begin{cases} x_0^{d-r} F &= u_0 x_0^d + \cdots + u_r x_0^{d-r} x_1^r, \\ x_0^{d-r} x_1 F &= u_0 x_0^{d-1} x_1 + \cdots + u_r x_0^{d-r-1} x_1^{r+1}, \\ \vdots & \ddots \\ x_1^{d-r} F &= u_0 x_0^r x_1^{d-r} + \cdots + u_r x_1^d. \end{cases} \quad (18)$$

The coordinates of these elements with respect to the standard monomial basis $\{x_0^d, x_0^{d-1} x_1, \ldots, x_1^d\}$ of $S^d V$ are thus given by the rows of the following $(r+1) \times (d+1)$ matrix:

$$\begin{pmatrix} u_0 & u_1 & \cdots & u_r & 0 & \cdots & 0 & 0 \\ 0 & u_0 & u_1 & \cdots & u_r & 0 & \cdots & 0 \\ \vdots & \ddots & \ddots & \ddots & & \ddots & \ddots & \vdots \\ 0 & \cdots & 0 & u_0 & u_1 & \cdots & u_r & 0 \\ 0 & \cdots & 0 & 0 & u_0 & \cdots & u_{r-1} & u_r \end{pmatrix}.$$

The standard Plücker coordinates of the subspace $\phi_{r,d-r+1}([F])$ are the maximal minors of this matrix. It is known (see for example [74]) that these minors form a basis of $\Bbbk[u_0, \ldots, u_r]_{d-r+1}$, so that the image of $\phi_{r,d-r+1}([F])$ is indeed a Veronese variety, which proves (1).

To prove (2), we recall some standard facts from [74]. Consider homogeneous coordinates $z_0, \ldots, z_d$ in $\mathbb{P}(S^d V^*)$, corresponding to the dual basis of the basis $\{x_0^d, x_0^{d-1} x_1, \ldots, x_1^d\}$. Consider $\mathcal{C}_d \subset \mathbb{P}(S^d V^*)$, the standard rational normal curve with respect to these coordinates. Then, the image of $[F]$ by $\phi_{r,d-r+1}$ is precisely the $r$-secant space to $\mathcal{C}_d$ spanned by the divisor on $\mathcal{C}_d$ induced by the zeros of $F$. This completes the proof of (2). □

The rational normal curve $\mathcal{C}_d \subset \mathbb{P}^d$ is the $d$-th Veronese embedding of $\mathbb{P}V \simeq \mathbb{P}^1$ inside $\mathbb{P}S^d V \simeq \mathbb{P}^d$. Hence, a symmetric tensor $F \in S^d V$ has symmetric-rank $r$ if and only if $r$ is the minimum integer for which there exists a $\mathbb{P}^{r-1} \simeq \mathbb{P}W \subset \mathbb{P}S^d V$ such that $F \in \mathbb{P}W$ and $\mathbb{P}W$ is $r$-secant to the rational normal curve $\mathcal{C}_d \subset \mathbb{P}(S^d V)$ in $r$ distinct points. Consider the maps:

$$\mathbb{P}(S^r V) \xrightarrow{\phi_{r,d-r+1}} \mathbb{G}(d-r, \mathbb{P}S^d V) \xrightarrow{\alpha_{r,d-r+1}} \mathbb{G}(r-1, \mathbb{P}S^d V^*). \quad (19)$$

Clearly, we can identify $\mathbb{P}S^d V^*$ with $\mathbb{P}S^d V$; hence, the Grassmannian $\mathbb{G}(r-1, \mathbb{P}S^d V^*)$ can be identified with $\mathbb{G}(r-1, \mathbb{P}S^d V)$. Now, by Lemma 2, a projective subspace $\mathbb{P}W$ of $\mathbb{P}S^d V^* \simeq \mathbb{P}S^d V \simeq \mathbb{P}^d$ is $r$-secant to $\mathcal{C}_d \subset \mathbb{P}S^d V$ in $r$ distinct points if and only if it belongs to $\text{Im}(\alpha_{r,d-r+1} \circ \phi_{r,d-r+1})$ and the preimage of $\mathbb{P}W$ via $\alpha_{r,d-r+1} \circ \phi_{r,d-r+1}$ is a polynomial with $r$ distinct roots. Therefore, a symmetric tensor $F \in S^d V$ has symmetric-rank $r$ if and only if $r$ is the minimum integer for which the following two conditions hold:

1. $F$ belongs to some $\mathbb{P}W \in \text{Im}(\alpha_{r,d-r+1} \circ \phi_{r,d-r+1}) \subset \mathbb{G}(r-1, \mathbb{P}S^d V)$,
2. there exists a polynomial $F \in S^r V$ that has $r$ distinct roots and such that $\alpha_{r,d-r+1}(\phi_{r,d-r+1}([F])) = \mathbb{P}(W)$.

Now, let $\mathbb{P}U$ be a $(d-r)$-dimensional linear subspace of $\mathbb{P}S^d V$. The proof of Lemma 2 shows that $\mathbb{P}U$ belongs to the image of $\phi_{r,d-r+1}$ if and only if there exist $u_0, \ldots, u_r \in \Bbbk$ such that $U = \langle F_1, \ldots, F_{d-r+1} \rangle$, where, with respect to the standard monomial basis $\mathcal{B} = \{x_0^d, x_0^{d-1} x_1, \ldots, x_1^d\}$ of $S^d V$, we have:

$$\begin{cases} F_1 & = (u_0, u_1, \ldots, u_r, 0, \ldots, 0), \\ F_2 & = (0, u_0, u_1, \ldots, u_r, 0, \ldots, 0), \\ \vdots & \quad\vdots \\ F_{d-r+1} & = (0, \ldots, 0, u_0, u_1, \ldots, u_r). \end{cases}$$

Let $\mathcal{B}^* = \{z_0, \ldots, z_d\}$ be the dual basis of $\mathcal{B}$ with respect to the apolar pairing. Therefore, there exists a $W \subset S^d V$ such that $\mathbb{P}W = \alpha_{r,d-r+1}(\mathbb{P}U)$ if and only if $W = H_1 \cap \cdots \cap H_{d-r+1}$, and the $H_i$'s are as follows:

$$\begin{cases} H_1: & u_0 z_0 + \cdots + u_r z_r = 0; \\ H_2: & u_0 z_1 + \cdots + u_r z_{r+1} = 0; \\ \vdots & \quad\ddots \\ H_{d-r+1}: & u_0 z_{d-r} + \cdots + u_r z_d = 0. \end{cases}$$

This is sufficient to conclude that $F \in \mathbb{P}S^d V$ belongs to an $(r-1)$-dimensional projective subspace of $\mathbb{P}S^d V$ that is in the image of $\alpha_{r,d-r+1} \circ \phi_{r,d-r+1}$ defined in (19) if and only if there exist $H_1, \ldots, H_{d-r+1}$ hyperplanes in $S^d V$ as above, such that $F \in H_1 \cap \ldots \cap H_{d-r+1}$. Now, given $F \in S^d V$ with coordinates $(a_0, \ldots, a_d)$ with respect to the dual basis $\mathcal{B}^*$, we have that $F \in H_1 \cap \ldots \cap H_{d-r+1}$ if and only if the following linear system admits a non-trivial solution in the $u_i$'s

$$\begin{cases} u_0 a_0 + \cdots + u_r a_r & = 0 \\ u_0 a_1 + \cdots + u_r a_{r+1} & = 0 \\ \quad\ddots & \\ u_0 a_{d-r} + \cdots + u_r a_d & = 0. \end{cases}$$

If $d - r + 1 < r + 1$, this system admits an infinite number of solutions. If $r \leq d/2$, it admits a non-trivial solution if and only if all the maximal $(r+1)$-minors of the following catalecticant matrix (see Definition 1) vanish:

$$\begin{pmatrix} a_0 & \cdots & a_r \\ a_1 & \cdots & a_{r+1} \\ \vdots & & \vdots \\ a_{d-r} & \cdots & a_d \end{pmatrix}.$$

**Remark 11.** The dimension of $\sigma_r(\mathcal{C}_d)$ is never defective, i.e., it is the minimum between $2r - 1$ and $d$. Actually, $\sigma_r(\mathcal{C}_d) \subsetneq \mathbb{P}S^d V$ if and only if $1 \leq r < \left\lceil \frac{d+1}{2} \right\rceil$. Moreover, an element $[F] \in \mathbb{P}S^d V$ belongs to $\sigma_r(\mathcal{C}_d)$ for $1 \leq r < \left\lceil \frac{d+1}{2} \right\rceil$, i.e., $\mathrm{R}_{\mathrm{sym}}(F) = r$, if and only if $\mathrm{Cat}_{r,d-r}(F)$ does not have maximal rank. These facts are very classical; see, e.g., [1].

Therefore, if we consider the monomial basis $\left\{ \binom{d}{i}^{-1} x_0^i x_1^{d-i} \mid i = 0, \ldots, d \right\}$ of $S^d V$ and write $F = \sum_{i=0}^d \binom{d}{i}^{-1} a_i x_0^i x_1^{d-i}$, then we write the $(i, d-i)$-th catalecticant matrix of $F$ as $\mathrm{Cat}_{i,d-i}(F) = (a_{h+k})_{\substack{h=0,\ldots,i \\ k=0,\ldots,d-i}}$.

**Algorithm 1:** Sylvester's algorithm.

The algorithm works as follows.

**Require:** A binary form $F = \sum_{i=0}^{d} a_i x_0^i x_1^{d-i} \in S^d V$.
**Ensure:** A minimal Waring decomposition $F = \sum_{i=1}^{r} \lambda_i L_i(x_0, x_1)^d$.

1: initialize $r \leftarrow 0$;
2: **if** rk $Cat_{r,d-r}(F)$ is maximal **then**
3:     increment $r \leftarrow r+1$;
4: **end if**
5: compute a basis of $Cat_{r,d-r}(F)$;
6: take a random element $G \in S^r V^*$ in the kernel of $Cat_{r,d-r}(F)$;
7: compute the roots of $G$: denote them $(\alpha_i, \beta_i)$, for $i = 1, \ldots, r$;
8: **if** the roots are not distinct **then**
9:     go to Step 2;
10: **else**
11:     compute the vector $\lambda = (\lambda_1, \ldots, \lambda_r) \in \mathbb{k}^r$ such that:

$$\begin{pmatrix} \alpha_1^d & \cdots & \alpha_r^d \\ \alpha_1^{d-1}\beta_1 & \cdots & \alpha_r^{d-1}\beta_r \\ \alpha_1^{d-2}\beta_1^2 & \cdots & \alpha_r^{d-2}\beta_r^2 \\ \vdots & \vdots & \vdots \\ \beta_1^d & \cdots & \beta_r^d \end{pmatrix} \begin{pmatrix} \lambda_1 \\ \lambda_2 \\ \vdots \\ \lambda_r \end{pmatrix} = \begin{pmatrix} a_0 \\ \frac{1}{d}a_1 \\ \binom{d}{2}^{-1} a_2 \\ \vdots \\ a_d \end{pmatrix}$$

12: **end if**
13: construct the set of linear forms $\{L_i = \alpha_i x_0 + \beta_i x_1\} \subset S^1 V$;
14: **return** the expression $\sum_{i=1}^{r} \lambda_i L_i^d$.

**Example 11.** *Compute the symmetric-rank and a minimal Waring decomposition of the polynomial*

$$F = 2x_0^4 - 4x_0^3 x_1 + 30x_0^2 x_1^2 - 28x_0 x_1^3 + 17x_1^4.$$

*We follow Sylvester's algorithm. The first catalecticant matrix with rank smaller than the maximal is:*

$$Cat_{2,2}(F) = \begin{pmatrix} 2 & -1 & 5 \\ -1 & 5 & -7 \\ 5 & -7 & 17 \end{pmatrix},$$

*in fact, rk $Cat_{2,2}(F) = 2$. Now, let $\{y_0, y_1\}$ the dual basis of $V^*$. We get that $\ker Cat_{2,2}(F) = \langle 2y_0^2 - y_0 y_1 - y_1^2 \rangle$. We factorize:*

$$2y_0^2 - y_0 y_1 - y_1^2 = (-y_0 + y_1)(-2y_0 - y_1).$$

*Hence, we obtain the roots $\{(1,1), (1,-2)\}$. Then, it is direct to check that:*

$$\begin{pmatrix} 1 & 1 \\ 1 & -2 \\ 1 & 4 \\ 1 & -8 \\ 1 & 16 \end{pmatrix} \begin{pmatrix} 1 \\ 1 \end{pmatrix} = \begin{pmatrix} 2 \\ -1 \\ 5 \\ -7 \\ 17 \end{pmatrix},$$

hence, a minimal Waring decomposition is given by:

$$F = (x_0 + x_1)^4 + (x_0 - 2x_1)^4.$$

The following result was proven by Comas and Seiguer in [27]; see also [28]. It describes the structure of the stratification by symmetric-rank of symmetric tensors in $S^d V$ with dim $V = 2$. This result allows us to improve the classical Sylvester algorithm (see Algorithm 2).

**Theorem 4.** *Let $C_d = \{[F] \in \mathbb{P}S^d V \mid R_{\text{sym}}(F) = 1\} = \{[L^d] \mid L \in S^1 V\} \subset \mathbb{P}^d$ be the rational normal curve of degree d parametrizing decomposable symmetric tensors. Then,*

$$\forall r, \ 2 \leq r \leq \left\lceil \frac{d+1}{2} \right\rceil, \qquad \sigma_r(C_d) \smallsetminus \sigma_{r-1}(C_d) = \sigma_{r,r}(C_d) \cup \sigma_{r,d-r+2}(C_d),$$

*where we write:*

$$\sigma_{r,s}(C_d) := \{[F] \in \sigma_r(C_d) \mid R_{\text{sym}}(F) = s\} \subset \sigma_r(C_d).$$

---

**Algorithm 2:** Sylvester's symmetric (border) rank algorithm [28].

The latter theorem allows us to get a simplified version of the Sylvester algorithm, which computes the symmetric-rank and the symmetric-border rank of a symmetric tensor, without computing any decomposition. Notice that Sylvester's Algorithm 1 for the rank is recursive: it runs for any $r$ from one to the symmetric-rank of the given polynomial, while Theorem 4 shows that once the symmetric border rank is computed, then the symmetric-rank is either equal to the symmetric border rank or it is $d - r + 2$, and this allows us to skip all the recursive process.

**Require:** A form $F \in S^d V$, with dim $V = 2$.
**Ensure:** the symmetric-rank $R_{\text{sym}}(F)$ and the symmetric-border rank $\underline{R}_{\text{sym}}(F)$.
1: $r := \operatorname{rk} \operatorname{Cat}_{\lfloor \frac{d}{2} \rfloor, \lceil \frac{d}{2} \rceil}(F)$
2: $\underline{R}_{\text{sym}}(F) = r$;
3: choose an element $G \in \ker \operatorname{Cat}_{r,d-r}(F)$;
4: **if** $G$ has distinct roots **then**
5: $\quad R_{\text{sym}}(F) = r$
6: **else**
7: $\quad R_{\text{sym}}(F) = d - r + 2$;
8: **end if**
9: **return** $R_{\text{sym}}(F)$

---

**Example 12.** *Let $F = 5x_0^2 x_1$, and let $\{y_0, y_1\}$ be the dual basis to $\{x_0, x_1\}$. The smallest catalecticant without full rank is:*

$$\operatorname{Cat}_{2,3}(F) = \begin{pmatrix} 0 & 1 & 0 & 0 \\ 1 & 0 & 0 & 0 \\ 0 & 0 & 0 & 0 \end{pmatrix},$$

*which has rank two. Therefore $[F] \in \sigma_2(C_6)$. Now, $\ker \operatorname{Cat}_{2,3}(F) = \langle y_1^2 \rangle$, which has a double root. Hence, $[F] \in \sigma_{2,6}(C_6)$.*

**Remark 12.** *When a form $F \in \mathbb{k}[x_0, \ldots, x_n]$ can be written using less variables, i.e., $F \in \mathbb{k}[L_0, \ldots, L_m]$, for $L_j \in \mathbb{k}[x_0, \ldots, x_n]_1$, with $m < n$, we say that $F$ has $m$ essential variables (in the literature, it is also said that $F$ is m-concise). That is, $F \in S^d W$, where $W = \langle L_0, \ldots, L_m \rangle \subset V$. Then, the rank of $[F]$ with respect to $X_{n,d}$ is the same one as the one with respect to $\nu_d(\mathbb{P}W) \subset X_{n,d}$; e.g., see [75,76]. As recently clearly described in [77]*

(Proposition 10) and more classically in [25], the number of essential variables of F coincides with the rank of the first catalecticant matrix $\mathrm{Cat}_{1,d-1}(F)$. In particular, when $[F] \in \sigma_r(X_{n,d}) \subset \mathbb{P}(S^d V)$ with $\dim(V) = n+1$, then, if $r < n+1$, there is a subspace $W \subset V$ with $\dim(W) = r$ such that $[F] \in \mathbb{P}S^d W$, i.e., F can be written with respect to r variables.

Let now V be $(n+1)$-dimensional, and consider the following construction:

$$\begin{array}{ccc} \mathrm{Hilb}_r(\mathbb{P}^n) & \stackrel{\phi}{\dashrightarrow} & \mathbb{G}\left(\binom{d+n}{n} - r, S^d V\right) \\ & & \| \mathcal{R} \\ \mathbb{G}(r-1, \mathbb{P}S^d V^*) & \longleftarrow & \mathbb{G}\left(\binom{d+n}{n} - r - 1, \mathbb{P}S^d V\right) \end{array} \qquad (20)$$

where the map $\phi$ in (20) sends a zero-dimensional scheme Z with $\deg(Z) = r$ to the vector space $(I_Z)_d$ (it is defined in the open set formed by the schemes Z, which impose independent conditions to forms of degree d) and where the last arrow is the identification, which sends a linear space to its perpendicular.

As in the case $n = 1$, the final image from the latter construction gives the $(r-1)$-spaces, which are r-secant to the Veronese variety in $\mathbb{P}^N \cong \mathbb{P}(\mathbb{k}[x_0, \ldots, x_n]_d)^*$. Moreover, each such space cuts the image of Z via the Veronese embedding.

**Notation 1.** *From now on, we will always use the notation $\Pi_Z$ to indicate the projective linear subspace of dimension $r-1$ in $\mathbb{P}S^d V$, with $\dim(V) = n+1$, generated by the image of a zero-dimensional scheme $Z \subset \mathbb{P}^n$ of degree r via the Veronese embedding, i.e., $\Pi_Z = \langle \nu_d(Z) \rangle \subset \mathbb{P}S^d V$.*

**Theorem 5.** *Any $[F] \in \sigma_2(X_{n,d}) \subset \mathbb{P}S^d V$, with $\dim(V) = n+1$ can only have symmetric-rank equal to 1, 2 or d. More precisely:*

$$\sigma_2(X_{n,d}) \smallsetminus X_{n,d} = \sigma_{2,2}(X_{n,d}) \cup \sigma_{2,d}(X_{n,d});$$

*more precisely, $\sigma_{2,d}(X_{n,d}) = \tau(X_{n,d}) \smallsetminus X_{n,d}$, where $\tau(X_{n,d})$ denotes the tangential variety of $X_{n,d}$, i.e., the Zariski closure of the union of the tangent spaces to $X_{n,d}$.*

**Proof.** This is actually a quite direct consequence of Remark 12 and of Theorem 4, but let us describe the geometry in some detail, following the proof of [28]. Since $r = 2$, every $Z \in \mathrm{Hilb}_2(\mathbb{P}^n)$ is the complete intersection of a line and a quadric, so the structure of $I_Z$ is well known, i.e., $I_Z = (L_0, \ldots, L_{n-2}, Q)$, where $L_i$'s are linearly independent linear forms and Q is a quadric in $S^2 V \smallsetminus (L_0, \ldots, L_{n-2})_2$.

If $F \in \sigma_2(X_{n,d})$, then we have two possibilities: either $R_{\mathrm{sym}}(F) = 2$ or $R_{\mathrm{sym}}(T) > 2$, i.e., F lies on a tangent line $\Pi_Z$ to the Veronese, which is given by the image of a scheme $Z \subset \mathbb{P}V$ of degree 2, via the maps (20). We can view F in the projective linear space $H \cong \mathbb{P}^d$ in $\mathbb{P}(S^d V)$ generated by the rational normal curve $\mathcal{C}_d \subset X_{n,d}$, which is the image of the line $\ell$ defined by the ideal $(L_0, \ldots, L_{n-2})$ in $\mathbb{P}V$, i.e., $\ell \subset \mathbb{P}^n$ is the unique line containing Z. Hence, we can apply Theorem 4 in order to get that $R_{\mathrm{sym}}(F) \leq d$. Moreover, by Remark 12, we have $R_{\mathrm{sym}}(F) = d$. □

**Remark 13.** Let us check that $\sigma_2(X_{n,d})$ is given by the annihilation of the $(3 \times 3)$-minors of the first two catalecticant matrices, $\mathrm{Cat}_{1,d-1}(V)$ and $\mathrm{Cat}_{2,d-2}(V)$ (see Definition 1); actually, such minors are the generators of $I_{\sigma_2(\nu_d(\mathbb{P}^n))}$; see [78].

Following the construction above (20), we can notice that the coefficients of the linear spaces defined by the forms $L_i \in V^*$ in the ideal $I_Z$ are the solutions of a linear system whose matrix is given by the catalecticant matrix $\mathrm{Cat}_{1,d-1}(V)$; since the space of solutions has dimension $n-1$, we get rk $\mathrm{Cat}_{1,d-1}(V) = 2$. When we consider the quadric Q in $I_Z$, instead, the analogous construction gives that its coefficients are the solutions of a linear system defined by the catalecticant matrix $\text{Cat}_{2,d-2}(V)$, and the space of solutions give $Q$ and all the quadrics in $(L_0, \ldots, L_{n-2})_2$, which are $\binom{n}{2} + 2n - 1$, hence:

$$\text{rk Cat}_{2,d-2}(V) = \binom{n+2}{2} - \left(\binom{n}{2} + 2n\right) = 2.$$

Therefore, we can write down an algorithm (Algorithm 3) to test if an element $[F] \in \sigma_2(X_{n,d})$ has symmetric rank two or $d$.

---

**Algorithm 3:** An algorithm to compute the symmetric-rank of an element lying on $\sigma_2(X_{n,d})$.

**Require:** A from $F \in S^d V$, where $\dim V = n + 1$.
**Ensure:** If $[F] \in \sigma_2(X_{n,d})$, returns the $\text{R}_{\text{sym}}(F)$.
1: compute the number of essential variables $m = \text{rk Cat}_{1,d-1}(F)$;
2: **if** $m = 1$ **then**
3:     print $F \in X_{n,d}$;
4: **else if** $m > 2$ **then**
5:     print $F \notin \sigma_2(X_{n,d})$;
6: **else**
7:     let $W = (\ker \text{Cat}_{1,d-1}(F))^\perp$ and view $F \in S^d W$;
8: **end if**
9: **return** apply Algorithm 2 to $F$.

---

**Example 13.** *Compute the symmetric-rank of*

$$F = x_0^3 x_2 + 3x_0^2 x_1 x_2 + 3x_0 x_1^2 x_2 + x_1^3 x_2.$$

*First of all, note that $(y_0 - y_1) \circ F = 0$; in particular, $\ker \text{Cat}_{1,3}(F) = \langle y_0 - y_1 \rangle$. Hence, $F$ has two essential variables. This can also be seen by noticing that $F = (x_0 + x_1)^3 x_2$. Therefore, if we write $z_0 = x_0 + x_1$ and $z_1 = x_2$, then $F = z_0^3 z_1 \in \Bbbk[z_0, z_1]$. Hence, we can apply Algorithms 1 and 2 to compute the symmetric-rank, symmetric-border rank and a minimal decompositions of $F$. In particular, we write:*

$$\text{Cat}_{2,2}(F) = \begin{pmatrix} 0 & 1/4 & 0 \\ 1/4 & 0 & 0 \\ 0 & 0 & 0 \end{pmatrix},$$

*which has rank two, as expected. Again, as in Example 12, the kernel of $\text{Cat}_{2,2}(F)$ defines a polynomial with a double root. Hence, $\underline{R}_{\text{sym}}(F) = 2$ and $R_{\text{sym}}(F) = 4$. If we are interested in finding a minimal decomposition of $F$, we have to consider the first catalecticant whose kernel defines a polynomial with simple roots. In this case, we should get to:*

$$\text{Cat}_{0,4}(F) = \begin{pmatrix} 0 \\ 1/4 \\ 0 \\ 0 \\ 0 \end{pmatrix},$$

*whose kernel is $\langle (1,0,0,0,0), (0,0,1,0,0), (0,0,0,1,0), (0,0,0,0,1) \rangle$. If we let $\{w_0, w_1\}$ be the variables on $W^*$, we take a polynomial in this kernel, as for example $G = w_0^4 + w_0^2 w_1^2 + w_0 w_1^3 + w_1^4$. Now, if we compute the roots of $G$, we find four complex distinct roots, i.e.,*

$$(\alpha_1, \beta_1) = -\tfrac{1}{6}A - \tfrac{1}{2}\sqrt{B+C} \quad (\alpha_2, \beta_2) = -\tfrac{1}{6}A + \tfrac{1}{2}\sqrt{B+C};$$
$$(\alpha_3, \beta_3) = \tfrac{1}{6}A - \tfrac{1}{2}\sqrt{B-C} \quad (\alpha_4, \beta_4) = \tfrac{1}{6}A + \tfrac{1}{2}\sqrt{B-C};$$

where:

$$A = \sqrt{\frac{9\left(\frac{1}{18}i\sqrt{257}\sqrt{3}-\frac{43}{54}\right)^{\frac{2}{3}} - 6\left(\frac{1}{18}i\sqrt{257}\sqrt{3}-\frac{43}{54}\right)^{\frac{1}{3}} + 13}{\frac{1}{18}i\sqrt{257}\sqrt{3}-\frac{43}{54}}}^{\frac{1}{3}};$$

$$B = \sqrt{-\left(1/18i\sqrt{257}\sqrt{3}-\frac{43}{54}\right)^{1/3} - \left(\frac{\frac{13}{9}}{\frac{1}{18}i\sqrt{257}\sqrt{3}-\frac{43}{54}}\right)^{\frac{1}{3}}};$$

$$C = \frac{6}{\sqrt{\frac{9\left(\frac{1}{18}i\sqrt{257}\sqrt{3}-\frac{43}{54}\right)^{\frac{2}{3}} - 6\left(\frac{1}{18}i\sqrt{257}\sqrt{3}-\frac{43}{54}\right)^{\frac{1}{3}} + 13}{\frac{1}{18}i\sqrt{257}\sqrt{(3)}-\frac{43}{54}}}^{\frac{1}{3}}} - \frac{4}{3}$$

Hence, if we write $L_i = \alpha_i z_0 + \beta_i z_1$, for $i = 1, \ldots, 4$, we can find suitable $\lambda_i$'s to write a minimal decomposition $F = \sum_{i=1}^{4} \lambda_i L_i^4$. Observe that any hyperplane through $[F]$ that does not contain the tangent line to $C_4$ at $[z_0^4]$ intersects $C_4$ at four distinct points, so we could have chosen also another point in $\langle (1,0,0,0,0), (0,0,1,0,0), (0,0,0,1,0), (0,0,0,0,1) \rangle$, and we would have found another decomposition of $F$.

Everything that we have done in this section does not use anything more than Sylvester's algorithm for the two-variable case. In the next sections, we see what can be done if we have to deal with more variables and we cannot reduce to the binary case like in Example 13.

Sylvester's algorithm allows us to compute the symmetric-rank of any polynomial in two essential variables. It is mainly based on the fact that equations for secant varieties of rational normal curves are well known and that there are only two possibilities for the symmetric-rank of a given binary polynomial with fixed border rank (Theorem 4). Moreover, those two cases are easily recognizable by looking at the multiplicity of the roots of a generic polynomial in the kernel of the catalecticant.

The first ideas that were exploited to generalize Sylvester's result to homogeneous polynomials in more than two variables were:

- a good understanding of the inverse system (and therefore, of the scheme defined by the kernels of catalecticant matrices and possible extension of catalecticant matrices, namely Hankel matrices); we will go into the details of this idea in Section 2.3.3;
- a possible classification of the ranks of polynomials with fixed border rank; we will show the few results in this direction in Section 2.3.2.

2.3.2. Beyond Sylvester's Algorithm Using Zero-Dimensional Schemes

We keep following [28]. Let us start by considering the case of a homogeneous polynomial with three essential variables.

If $[F] \in \sigma_3(\nu_d(\mathbb{P}^n)) \smallsetminus \sigma_2(\nu_d(\mathbb{P}^n))$, then we will need more than two variables, but actually, three are always sufficient. In fact, if $[F] \in \sigma_3(\nu_d(\mathbb{P}^n))$, then there always exists a zero-dimensional scheme $\nu_d(Z)$ of length three contained in $\nu_d(\mathbb{P}^n)$, whose span contains $[F]$; the scheme $Z \subset \mathbb{P}^n$ itself spans a $\mathbb{P}^2$, which can be seen as $\mathbb{P}((L_1, L_2, L_3)_1)$ with $L_i$'s linear forms. Therefore, $F$ can be written in three variables. The following theorem computes the symmetric-rank of any polynomial in $[F] \in \sigma_3(\nu_d(\mathbb{P}^n)) \smallsetminus \sigma_2(\nu_d(\mathbb{P}^n))$, and the idea is to classify the symmetric-rank by looking at the structure of the zero-dimensional scheme of length three, whose linear span contains $[F]$.

**Theorem 6** ([28] (Theorem 37)). *Let $d \geq 3$, $X_{n,d} \subset \mathbb{P}(\Bbbk^{n+1})$. Then,*

$$\sigma_3(X_{n,3}) \smallsetminus \sigma_2(X_{n,3}) = \sigma_{3,3}(X_{n,3}) \cup \sigma_{3,4}(X_{n,3}) \cup \sigma_{3,5}(X_{n,3}),$$

while, for $d \geq 4$,

$$\sigma_3(X_{n,d}) \setminus \sigma_2(X_{n,d}) = \sigma_{3,3}(X_{n,d}) \cup \sigma_{3,d-1}(X_{n,d}) \cup \sigma_{3,d+1}(X_{n,d}) \cup \sigma_{3,2d-1}(X_{n,d}).$$

We do not give here all the details of the proof since they can be found in [28]; they are quite technical, but the main idea is the one described above. We like to stress that the relation between the zero-dimensional scheme of length three spanning $F$ and the one computing the symmetric-rank is in many cases dependent on the following Lemma 3. Probably, it is classically known, but we were not able to find a precise reference.

**Lemma 3** ([28] (Lemma 11)). *Let $Z \subset \mathbb{P}^n$, $n \geq 2$, be a zero-dimensional scheme, with $\deg(Z) \leq 2d + 1$. A necessary and sufficient condition for $Z$ to impose independent conditions on hypersurfaces of degree $d$ is that no line $\ell \subset \mathbb{P}^n$ is such that $\deg(Z \cap \ell) \geq d + 2$.*

**Remark 14.** *Notice that if $\deg(\ell \cap Z)$ is exactly $d + 1 + k$, then the dimension of the space of curves of degree $d$ through them is increased exactly by $k$ with respect to the generic case.*

It is easy to see that Lemma 3 can be improved as follows; see [79].

**Lemma 4** ([79]). *Let $Z \subset \mathbb{P}^n$, $n \geq 2$, be a zero-dimensional scheme, with $\deg(Z) \leq 2d + 1$. If $h^1(\mathbb{P}^n, \mathcal{I}_Z(d)) > 0$, there exists a unique line $\ell \subset \mathbb{P}^n$ such that $\deg(Z \cap \ell) = d + 1 + h^1(\mathbb{P}^n, \mathcal{I}_Z(d)) > 0$.*

We can go back to our problem of finding the symmetric-rank of a given tensor. The classification of symmetric-ranks of the elements in $\sigma_4(X_{n,d})$ can be treated in an analogous way as we did for $\sigma_3(X_{n,d})$, but unfortunately, it requires a very complicated analysis on the schemes of length four. This is done in [80], but because of the long procedure, we prefer to not present it here.

It is remarkable that $\sigma_4(X_{n,d})$ is the last $s$-th secant variety of Veronesean, where we can use this technique for the classification of the symmetric-rank with respect to zero-dimensional schemes of length $s$, whose span contains the given polynomial we are dealing with; for $s \geq 5$, there is a more intrinsic problem. In fact, there is a famous counterexample due to Buczyńska and Buczyśki (see [81]) that shows that, in $\sigma_5(X_{4,3})$, there is at least a polynomial for which there does not exist any zero-dimensional scheme of length five on $X_{4,3}$, whose span contains it. The example is the following.

**Example 14** (Buczyńska, Buczyński [81,82]). *One can easily check that the following polynomial:*

$$F = x_0^2 x_2 + 6 x_1^2 x_3 - 3(x_0 + x_1)^2 x_4.$$

*can be obtained as $\lim_{\epsilon \to 0} \frac{1}{3\epsilon} F_\epsilon = F$ where:*

$$F_\epsilon = (x_0 + \epsilon x_2)^3 + 6(x_1 + \epsilon x_3)^3 - 3(x_0 + x_1 + \epsilon x_4)^3 + 3(x_0 + 2x_1)^3 - (x_0 + 3x_1)^3$$

*has symmetric-rank five for $\epsilon > 0$. Therefore, $[F] \in \sigma_5(\nu_3(\mathbb{P}^4))$.*

*An explicit computation of $F^\perp$ yields the Hilbert series for $\mathrm{HS}_{R/F^\perp}(z) = 1 + 5z + 5z^2 + z^3$. Let us prove, by contradiction, that there is no saturated ideal $I \subset F^\perp$ defining a zero-dimensional scheme of length $\leq 5$. Suppose on the contrary that $I$ is such an ideal. Then, $\mathrm{HF}_{R/I}(i) \geq \mathrm{HF}_{R/F^\perp}(i)$ for all $i \in \mathbb{N}$. As $\mathrm{HF}_{R/I}(i)$ is an increasing function of $i \in \mathbb{N}$ with $\mathrm{HF}_{R/F^\perp}(i) \leq \mathrm{HF}_{R/I}(i) \leq 5$, we deduce that $\mathrm{HS}_{R/I}(t) = 1 + 5 \sum_{i=1}^\infty z^i$. This shows that $I_1 = \{0\}$ and that $I_2 = (F^\perp)_2$. As $I$ is saturated, $I_2 : (x_0, \ldots, x_4) = I_1 = \{0\}$, since $\mathrm{HF}_{R/F^\perp}(1) = 5$. However, an explicit computation of $(F^\perp)_2 : (x_0, \ldots, x_4)$ gives $\langle x_2, x_3, x_4 \rangle$. In this way, we obtain a contradiction, so that there is no saturated ideal of degree $\leq 5$ such that $I \subset F^\perp$. Consequently, the minimal zero-dimensional scheme contained in $X_{4,3}$ whose linear span contains $[F]$ has degree six.*

In the best of our knowledge, the two main results that are nowadays available to treat these "wild" cases are the following.

**Proposition 9** ([28]). *Let $X \subset \mathbb{P}^N$ be a non-degenerate smooth variety. Let $H_r$ be the irreducible component of the Hilbert scheme of zero-dimensional schemes of degree $r$ of $X$ containing $r$ distinct points, and assume that for each $y \in H_r$, the corresponding subscheme $Y$ of $X$ imposes independent conditions on linear forms. Then, for each $P \in \sigma_r(X) \setminus \sigma_r^0(X)$, there exists a zero-dimensional scheme $Z \subset X$ of degree $r$ such that $P \in \langle Z \rangle \cong \mathbb{P}^{r-1}$. Conversely, if there exists $Z \in H_r$ such that $P \in \langle Z \rangle$, then $P \in \sigma_r(X)$.*

Obviously, five points on a line do not impose independent conditions on cubics in any $\mathbb{P}^n$ for $n \geq 5$; therefore, this could be one reason why the counterexample given in Example 14 is possible. Another reason is the following.

**Proposition 10** ([81]). *Suppose there exist points $P_1, \ldots, P_r \in X$ that are linearly degenerate, that is $\dim \langle P_1, \ldots, P_r \rangle < r - 1$. Then, the join of the $r$ tangent stars (see [83] (Section 1.4) for a definition) at these points is contained in $\sigma_r(X)$. In the case that $X$ is smooth at $P_1, \ldots P_r$, then $\langle T_{P_1} X, \ldots, T_{P_r} X \rangle \subset \sigma_r(X)$.*

2.3.3. Beyond Sylvester's Algorithm via Apolarity

We have already defined in Section 2.1.4 the apolarity action of $S^\bullet V^* \simeq \Bbbk[y_0, \ldots, y_n]$ on $S^\bullet V \simeq \Bbbk[x_0, \ldots, x_n]$ and inverse systems. Now, we introduce the main algebraic tool from apolarity theory to study ranks and minimal Waring decompositions: that is the apolarity lemma; see [24,25]. First, we introduce the apolar ideal of a polynomial.

**Definition 19.** *Let $F \in S^d V$ be a homogeneous polynomial. Then, the apolar ideal of $F$ is:*

$$F^\perp = \{G \in S^\bullet V^* \mid G \circ F = 0\}.$$

**Remark 15.** *The apolar ideal is a homogeneous ideal. Clearly, $F_i^\perp = S^i V^*$, for any $i > d$, namely $A_F = S^\bullet V^* / F^\perp$ is an Artinian algebra with socle degree equal to $d$. Since $\dim_\Bbbk (A_F)_d = 1$, then it is also a Gorenstein algebra. Actually, Macaulay proved that there exists a one-to-one correspondence between graded Artinian Gorenstein algebras with socle degree $d$ and homogeneous polynomials of degree $d$; for details, see [24] (Theorem 8.7).*

**Remark 16.** *Note that, directly by the definitions, the non-zero homogeneous parts of the apolar ideal of a homogeneous polynomial $F$ coincide with the kernel of its catalecticant matrices, i.e., for $i = 0, \ldots, d$,*

$$F_i^\perp = \ker(Cat_{i,d-i}(F)).$$

The apolarity lemma tells us that Waring decompositions of a given polynomial correspond to sets of reduced points whose defining ideal is contained in the apolar ideal of the polynomial.

**Lemma 5** (Apolarity lemma). *Let $Z = \{[L_1], \ldots, [L_r]\} \subset \mathbb{P}(S^1 V)$, then the following are equivalent:*

1. $F = \sum_{i=1}^r \lambda_i L_i^d$, for some $\lambda_1, \ldots, \lambda_r \in \Bbbk$;
2. $I(Z) \subseteq F^\perp$.

*If these conditions hold, we say that $Z$ is a set of points apolar to $F$.*

**Proof.** The fact that (1) implies (2) follows from the easy fact that, for any $G \in S^d V^*$, we have that $G \circ L^d$ is equal to $d$ times the evaluation of $G$ at the point $[L] \in \mathbb{P}V$. Conversely, if $I(Z) \subset F^\perp$, then we have that $F \in I(Z)_d^\perp = \langle L_1^d, \ldots, L_r^d \rangle$; see Remark 9 and Proposition 4. □

**Remark 17** (Yet again: Sylvester's algorithm). *With this lemma, we can rephrase Sylvester's algorithm. Consider the binary form $F = \sum_{i=0}^{d} c_i \binom{d}{i} x_0^{d-i} x_1^i$. Such an F can be decomposed as the sum of r distinct powers of linear forms if and only if there exists $Q = q_0 y_0^r + q_1 y_0^{r-1} y_1 + \cdots + q_r y_1^r$ such that:*

$$\begin{pmatrix} c_0 & c_1 & \cdots & c_r \\ c_1 & & \cdots & c_{r+1} \\ \vdots & & & \vdots \\ c_{d-r} & & \cdots & c_d \end{pmatrix} \begin{pmatrix} q_0 \\ q_1 \\ \vdots \\ q_r \end{pmatrix} = 0, \qquad (21)$$

*and $Q = \mu \prod_{k=1}^{r} (\beta_k y_0 - \alpha_k y_1)$, for a suitable scalar $\mu \in \mathbb{k}$, where $[\alpha_i : \beta_i]$'s are different points in $\mathbb{P}^1$. In this case, there exists a choice of $\lambda_1, \ldots, \lambda_r$ such that $F = \sum_{k=1}^{r} \lambda_k (\alpha_k x_0 + \beta_k x_1)^d$. This is possible because of the following remarks:*

- *Gorenstein algebras of codimension two are always complete intersections, i.e.,*

$$I(Z) \subset (F^\perp) = (G_1, G_2);$$

- *Artinian Gorenstein rings have a symmetric Hilbert function, hence:*

$$\deg(G_1) + \deg(G_2) = \deg(F) + 2, \quad \text{say } \deg(G_1) \leq \deg(G_2);$$

- *If $G_1$ is square-free, i.e., has only distinct roots, we take $Q = G_1$ and $R_{\text{sym}}(F) = \deg(G_1)$; otherwise, the first degree where we get something square-free has to be the degree of $G_2$; in particular, we can take $Q$ to be a generic element in $F_{\deg(G_2)}^\perp$ and $R_{\text{sym}}(F) = \deg(G_2)$.*

By using Apolarity Theory, we can describe the following algorithm (Algorithm 4).

---
**Algorithm 4:** Iarrobino and Kanev [25].

We attribute the following generalization of Sylvester's algorithm to any number of variables to Iarrobino and Kanev: despite that they do not explicitly write the algorithm, the main idea is presented in [25]. Sometimes, this algorithm is referred to as the catalecticant method.

**Require:** $F \in S^d V$, where $\dim V = n + 1$.
**Ensure:** a minimal Waring decomposition.
1: construct the most square catalecticant of $F$, i.e., $\text{Cat}_{m,d-m}(F)$ for $m = \lceil d/2 \rceil$;
2: compute $\ker \text{Cat}_{m,d-m}(F)$;
3: if the zero-set $Z$ of the polynomials in $\ker \text{Cat}_{m,d-m}(F)$ is a reduced set of points, say $\{[L_1], \ldots, [L_r]\}$, then continue, otherwise the algorithm fails;
4: solve the linear system defined by $F = \sum_{i=1}^{s} \lambda_i L_i^d$ in the unknowns $\lambda_i$.

---

**Example 15.** *Compute a Waring decomposition of:*

$$F = 3x^4 + 12x^2 y^2 + 2y^4 - 12x^2 yz + 12xy^2 z - 4y^3 z + 12x^2 z^2 - 12xyz^2 + 6y^2 z^2 - 4yz^3 + 2z^4.$$

*The most square catalecticant matrix is:*

$$\text{Cat}_{2,2}(F) = \begin{pmatrix} 3 & 0 & 0 & 2 & -1 & 2 \\ 0 & 2 & -1 & 0 & 1 & -1 \\ 0 & -1 & 2 & 1 & -1 & 0 \\ 2 & 0 & 1 & 2 & -1 & 1 \\ -1 & 1 & -1 & -1 & 1 & -1 \\ 2 & -1 & 0 & 1 & -1 & 2 \end{pmatrix}.$$

Now, compute that the rank of $Cat_{2,2}(F)$ is three, and its kernel is:

$$\ker(Cat_{2,2}(F)) = \langle (1,0,0,-1,-1,-1),\ (0,1,0,-1,-2,0),\ (0,0,1,0,2,1)\rangle =$$
$$\langle y_0^2 - y_1^2 - y_1 y_2 - y_2^2,\ y_0 y_1 - y_1^2 - 2 y_1 y_2,\ y_0 y_2 + 2 y_1 y_2 + y_2^2\rangle \subset S^2 V^*.$$

It is not difficult to see that these three quadrics define a set of reduced points $\{[1:1:0], [1:0:-1], [1:-1:1]\} \subset \mathbb{P}V$. Hence, we take $L_1 = x_0 + x_1$, $L_2 = x_0 - x_2$ and $L_3 = x_0 - x_1 + x_2$, and, by the apolarity lemma, the polynomial $F$ is a linear combinations of those forms, in particular,

$$F = (x_0 + x_1)^4 + (x_0 - x_2)^4 + (x_0 - x_1 + x_2)^4.$$

Clearly, this method works only if $R_{\mathrm{sym}}(F) = \mathrm{rank}\ Cat_{m,d-m}(F)$, for $m = \left\lceil \frac{d}{2} \right\rceil$. Unfortunately, in many cases, this condition is not always satisfied.

Algorithm 4 has been for a long time the only available method to handle the computation of the decomposition of polynomials with more than two variables. In 2013, there was an interesting contribution due to Oeding and Ottaviani (see [84]), where the authors used vector bundle techniques introduced in [85] to find non-classical equations of certain secant varieties. In particular, the very interesting part of the paper [84] is the use of representation theory, which sheds light on the geometric aspects of this algorithm and relates these techniques to more classical results like the Sylvester pentahedral theorem (the decomposition of cubic polynomial in three variables as the sum of five cubes). For the heaviness of the representation theory background needed to understand that algorithm, we have chosen to not present it here. Moreover, we have to point out that ([84] Algorithm 4) fails whenever the symmetric-rank of the polynomial is too large compared to the rank of a certain matrix constructed with the techniques introduced in [84], similarly as happens for the catalecticant method.

Nowadays, one of the best ideas to generalize the method of catalecticant matrices is due to Brachat, Comon, Mourrain and Tsidgaridas, who in [29] developed an algorithm (Algorithm 5) that gets rid of the restrictions imposed by the usage of catalecticant matrices. The idea developed in [29] is to use the so-called Hankel matrix that in a way encodes all the information of all the catalecticant matrices. The algorithm presented in [29] to compute a Waring decomposition of a form $F \in S^d V$ passes through the computation of an affine Waring decomposition of the dehomogenization $f$ of the given form with respect to a suitable variable. Let $S = \mathbb{k}[x_1, \ldots, x_n]$ be the polynomial ring in $n$ variables over the field $\mathbb{k}$ corresponding to such dehomogenization.

We first need to introduce the definition of Hankel operator associated with any $\Lambda \in S^*$. To do so, we need to use the structure of $S^*$ as the $S$-module, given by:

$$a * \Lambda : S \to \mathbb{k},\quad b \mapsto \Lambda(ab),\quad \text{for } a \in S, \Lambda \in S^*.$$

Then, the Hankel operator associated with $\Lambda \in S^*$ is the matrix associated with the linear map:

$$H_\Lambda : S \to S^*,\ \text{such that } a \mapsto a * \Lambda.$$

Here are some useful facts about Hankel operators.

**Proposition 11.** $\ker(H_\Lambda)$ *is an ideal.*

Let $I_\Lambda = \ker(H_\Lambda)$ and $A_\Gamma = S/I_\Lambda$.

**Proposition 12.** *If* $\mathrm{rank}(H_\Lambda) = r < \infty$, *then the algebra* $A_\Lambda$ *is a* $\mathbb{k}$*-vector space of dimension* $r$, *and there exist polynomials* $l_1, \ldots l_k$ *of degree one and* $g_1, \ldots, g_k$ *of degree* $d_1, \ldots, d_k$, *respectively, in* $\mathbb{k}[\partial_1, \ldots, \partial_n]$ *such that:*

$$\Lambda = \sum_{i=1}^{k} l_i^{d-d_i} g_i.$$

Moreover, $I_\Lambda$ defines the union of affine schemes $Z_1, \ldots, Z_k$ with support on the points $l_1^*, \ldots, l_k^* \in \mathbb{k}^n$, respectively, and with multiplicity equal to the dimension of the vector space spanned by the inverse system generated by $l_i^{d-d_i} g_i$.

The original proof of this proposition can be found in [29]; for a more detailed and expanded presentation, see [30,31].

**Theorem 7** (Brachat, Comon, Mourrain, Tsigaridas [29]). *An element $\Lambda \in S^*$ can be decomposed as $\Lambda = \sum_{i=1}^r \lambda_i l_i^d$ if and only if $\mathrm{rank} H_\Lambda = r$, and $I_\Lambda$ is a radical ideal.*

Now, we consider the multiplication operators in $A_\Lambda$. Given $a \in A_\Lambda$:

$$M_a : A_\Lambda \to A_\Lambda,$$

$$b \mapsto a \cdot b,$$

and,

$$M_a^t : A_\Lambda^* \to A_\Lambda^*,$$

$$\gamma \mapsto a * \gamma.$$

Now,

$$H_{a*\Lambda} := M_a^t \cdot H_\Lambda. \tag{22}$$

**Theorem 8.** *If $\dim A_\Lambda < \infty$, then, $\Lambda = \sum_{i=1}^k l_i^{d-d_i} g_i$ and:*

- *the eigenvalues of the operators $M_a$ and $M_a^t$ are given by $\{a(l_1^*), \ldots, a(l_r^*)\}$;*
- *the common eigenvectors of the operators $(M_{x_i}^t)_{1 \leq i \leq n}$ are, up to scalar, the $l_i$'s.*

Therefore, one can recover the $l_i$'s, i.e., the points $l_i^*$'s, by eigenvector computations: take $B$ as a basis of $A_f$, i.e., say $B = \{b_1, \ldots, b_r\}$ with $r = \mathrm{rank} H_\Lambda$, and let $H_{a*\Lambda}^B = M_a^t H_\Lambda^B = H_\Lambda^B M_a$ ($M_a$ is the matrix of the multiplication by $a$ in the basis $B$). The common solutions of the generalized eigenvalue problem:

$$(H_{a*\Lambda} - \lambda H_\Lambda) v = 0,$$

for all $a \in S$ yield the common eigenvectors $H_\Lambda^B v$ of $M_a^t$, that is the evaluations at the points $l_i^*$'s. Therefore, these common eigenvectors $H_\Lambda^B v$ are up to scalar the vectors $[b_i(l_i^*), \ldots, b_r(l_i^*)]$, for $i = 1, \ldots, r$.

If $f = \sum_{i=1}^r \lambda_i l_i^d$, then the $Z_i$'s in Proposition 12 are simple, and one eigenvector computation is enough: in particular, for any $a \in S$, $M_a$ is diagonalizable, and the generalized eigenvectors $H_\Lambda^B v$ are, up to scalar, the evaluations at the points $l_i^*$'s.

Now, in order to apply this algebraic tool to our problem of finding a Waring decomposition of a homogeneous polynomial $F \in \mathbb{k}[x_0, \ldots, x_n]$, we need to consider its dehomogenization $f = F(1, x_1, \ldots, x_n)$ with respect to the variable $x_0$ (with no loss of generality, we may assume that the coefficients with respect to $x_0$ are all non-zero). Then, we associate a truncated Hankel matrices as follows.

**Definition 20.** *Let B be a subset of monomials in S. We say that B is connected to one if $\forall m \in B$ either $m = 1$ or there exists $i \in \{1, \ldots, n\}$ and $m' \in B$ such that $m = x_i m'$.*

Let $B, B' \subset S_{\leq d}$ be sets of monomials of degree $\leq d$, connected to one. For any $f = \sum_{\substack{\alpha \in \mathbb{N}^n \\ |\alpha| \leq d}} c_\alpha \binom{d}{d-|\alpha|, \alpha_1, \dots, \alpha_n} x^\alpha \in S_d$, we consider the *Hankel matrix*:

$$H_f^{B,B'} = (h_{\alpha+\beta})_{\alpha \in B, \beta \in B'},$$

where $h_\alpha = c_\alpha$ if $|\alpha| \leq d$, and otherwise, $h_\alpha$ is an unknown. The set of all these new variables is denoted $h$. Note that, by this definition, the known parts correspond to the catalecticant matrices of $F$. For simplicity, we write $H_f^B = H_f^{B,B}$. This matrix is also called *quasi-Hankel* [86].

**Example 16.** Consider $F = -4x_0x_1 + 2x_0x_2 + 2x_1x_2 + x_2^2 \in \mathbb{k}[x_0, x_1, x_2]$. Then, we look at the dehomogenization with respect to $x_0$ given by $f = -4x_1 + 2x_2 + 2x_1x_2 + x_2^2 \in \mathbb{k}[x_1, x_2]$. Then, if we consider the standard monomial basis of $S_{\leq 2}$ given by $B = \{1, x_1, x_2, x_1^2, x_1x_2, x_2^2\}$, then we get:

$$H_f^B = \begin{pmatrix} 0 & -2 & 1 & 0 & 1 & 1 \\ -2 & 0 & 1 & h_{(3,0)} & h_{(2,1)} & h_{(1,2)} \\ 1 & 1 & 1 & h_{(2,1)} & h_{(1,2)} & h_{(0,3)} \\ 0 & h_{(3,0)} & h_{(2,1)} & h_{(4,0)} & h_{(3,1)} & h_{(2,2)} \\ 1 & h_{(2,1)} & h_{(1,2)} & h_{(3,1)} & h_{(2,2)} & h_{(1,3)} \\ 1 & h_{(1,2)} & h_{(0,3)} & h_{(2,2)} & h_{(1,3)} & h_{(0,4)} \end{pmatrix},$$

where the $h$'s are unknowns.

Now, the idea of the algorithm is to find a suitable polynomial $\overline{f}$ whose Hankel matrix extends the one of $f$, has rank equal to the Waring rank of $f$ and the kernel gives a radical ideal. This is done by finding suitable values for the unknown part of the Hankel matrix of $f$. Those $\overline{f}$ are elements whose homogenization is in the following set:

$$\mathcal{E}_r^{d,0} := \left\{ [F] \in \mathbb{P}(S^d V) \;\middle|\; \begin{array}{l} \exists L \in S^1 V \smallsetminus \{0\}, \exists F' \in Y_r^{m,m'} \text{ s.t. } L^{m+m'-d}F' = F \\ \text{with } m = \max\{r, \lceil d/2 \rceil\}, m' = \max\{r-1, \lfloor d/2 \rfloor\} \end{array} \right\}$$

where $Y_r^{i,d-i} = \{[F] \in \mathbb{P}(S^d V) \mid \text{rankCat}_{i,d-i}(F) \leq r\}$. If $[F] \in \mathcal{E}_r^{d,0}$, we say that $f$ is the generalized affine decomposition of size $r$.

Suppose that $H_f^{B,B'}$ is invertible in $\mathbb{k}(h)$, then we define the formal multiplication operators:

$$M_i^{B,B'}(h) := (H_f^{B,B'})^{-1} H_{x_i f}^{B,B'}.$$

**Notation 2.** *If $B$ is a subset of monomials, then we write $B^+ = B \cup x_1 B \cup \dots \cup x_n B$. Note that, if $B$ is connected to one, then also $B^+$ is connected to one.*

The key result for the algorithm is the following.

**Theorem 9** (Brachat, Comon, Mourrain, Tsigaridas [29]). *If $B$ and $B'$ are sets of monomials connected to one, the coefficients of $f$ are known on $B^+ \times B'^+$, and if $H_f^{B,B'}$ is invertible, then $f$ extends uniquely to $S$ if and only if:*

$$M_i^{B,B'} \cdot M_j^{B,B'} = M_j^{B,B'} \cdot M_i^{B,B'}, \quad \text{for any } 1 \leq i < j \leq n.$$

**Algorithm 5:** Brachat, Comon, Mourrain, Tsigaridas [29–31].

Here is the idea of algorithm presented in [29]. In [30,31], a faster and more accurate version can be found.

**Require:** Any polynomial $f \in S$.
**Ensure:** An affine Waring decomposition of $f$.

1: $r \leftarrow 1$;
2: Compute a set $B$ of monomials of degree $\leq d$ connected to one and with $|B| = r$;
3: Find parameters $h$ such that $\det(H_f^B) \neq 0$ and the operators $M_i^B = (H_f^B)^{-1} H_{x_i f}^B$ commute;
4: **if** there is no solution **then**
5:    go back to 2 with $r \leftarrow r + 1$;
6: **else**
7:    compute the $n \cdot r$ eigenvalues $z_{i,j}$ and the eigenvectors $v_j$ such that $M_j v_j = z_{i,j} v_j$, $i = 1, \ldots, n, j = 1, \ldots, r$, until one finds $r$ different common eigenvectors;
8: **end if**
9: Solve the linear system $f = \sum_{j=1}^{r} \lambda_j z_j^d$ in the $\lambda_i$'s, where the $z_j$'s are the eigenvectors found above.

For simplicity, we give the example chosen by the authors of [29].

**Example 17.** *We look for a decomposition of:*

$$F = -1549440\, x_0 x_1 x_2^3 + 2417040\, x_0 x_1^2 x_2^2 + 166320\, x_0^2 x_1 x_2^2 - 829440\, x_0 x_1^3 x_2$$
$$- 5760\, x_0^3 x_1 x_2 - 222480\, x_0^2 x_1^2 x_2 + 38\, x_0^5 - 497664\, x_1^5 - 1107804\, x_2^5$$
$$- 120\, x_0^4 x_1 + 180\, x_0^4 x_2 + 12720\, x_0^3 x_1^2 + 8220\, x_0^3 x_2^2 - 34560\, x_0^2 x_1^3$$
$$- 59160\, x_0^2 x_2^3 + 831840\, x_0 x_1^4 + 442590\, x_0 x_2^4 - 5591520\, x_1^4 x_2$$
$$+ 7983360\, x_1^3 x_2^2 - 9653040\, x_1^2 x_2^3 + 5116680\, x_1 x_2^4.$$

1. We form a $\binom{n+d-1}{d} \times \binom{n+d-1}{d}$ matrix, the rows and the columns of which correspond to the coefficients of the polynomial with respect to the expression $f = F(1, x_1, \ldots, x_n) = \sum_{\substack{\alpha \in \mathbb{N}^n \\ |\alpha| \leq d}} c_\alpha \binom{d}{d-|\alpha|, \alpha_1, \ldots, \alpha_n} x^\alpha$.

The whole $21 \times 21$ matrix is the following.

$$\begin{pmatrix}
 & 1 & x_1 & x_2 & x_1^2 & x_1 x_2 & x_2^2 & x_1^3 & x_1^2 x_2 & x_1 x_2^2 & x_2^3 \\
1 & 38 & -24 & 36 & 1272 & -288 & 822 & -3456 & -7416 & 5544 & -5916 \\
x_1 & -24 & 1272 & -288 & -3456 & -7416 & 5544 & 166{,}368 & -41{,}472 & 80{,}568 & -77{,}472 \\
x_2 & 36 & -288 & 822 & -7416 & 5544 & -5916 & -41{,}472 & 80{,}568 & -77{,}472 & 88{,}518 \\
x_1^2 & 1272 & -3456 & -7416 & 166{,}368 & -41{,}472 & 80{,}568 & -497{,}664 & -1{,}118{,}304 & 798{,}336 & -965{,}304 \\
x_1 x_2 & -288 & -7416 & 5544 & -41{,}472 & 80{,}568 & -77{,}472 & -1{,}118{,}304 & 798{,}336 & -965{,}304 & 1{,}023{,}336 \\
x_2^2 & 822 & 5544 & -5916 & 80{,}568 & -77{,}472 & 88{,}518 & 798{,}336 & -965{,}304 & 1{,}023{,}336 & -1{,}107{,}804 \\
x_1^3 & -3456 & 166{,}368 & -41{,}472 & -497{,}664 & -1{,}118{,}304 & 798{,}336 & h_{6,0,0} & h_{5,1,0} & h_{4,2,0} & h_{3,3,0} \\
x_1^2 x_2 & -7416 & -41{,}472 & 80{,}568 & -1{,}118{,}304 & 798{,}336 & -965{,}304 & h_{5,1,0} & h_{4,2,0} & h_{3,3,0} & h_{2,4,0} \\
x_1 x_2^2 & 5544 & 80{,}568 & -77{,}472 & 798{,}336 & -965{,}304 & 1{,}023{,}336 & h_{4,2,0} & h_{3,3,0} & h_{2,4,0} & h_{1,5,0} \\
x_2^3 & -5916 & -77{,}472 & 88{,}518 & -965{,}304 & 1{,}023{,}336 & -1{,}107{,}804 & h_{3,3,0} & h_{2,4,0} & h_{1,5,0} & h_{0,6,0}
\end{pmatrix}$$

Notice that we do not know the elements in some positions of the matrix. In this case, we do not know the elements that correspond to monomials with (total) degree higher than five.

2. We extract a principal minor of full rank.

We should re-arrange the rows and the columns of the matrix so that there is a principal minor of full rank. We call this minor $\Delta_0$. In order to do that, we try to put the matrix in row echelon form, using elementary row and column operations.

In our example, the $4 \times 4$ principal minor is of full rank, so there is no need for re-arranging the matrix. The matrix $\Delta_0$ is:

$$\Delta_0 = \begin{pmatrix} 38 & -24 & 36 & 1272 \\ -24 & 1272 & -288 & -3456 \\ 36 & -288 & 822 & -7416 \\ 1272 & -3456 & -7416 & 166{,}368 \end{pmatrix}$$

Notice that the columns of the matrix correspond to the set of monomials $\{1, x_1, x_2, x_1^2\}$.

3. We compute the "shifted" matrix $\Delta_1 = x_1 \Delta_0$.

The columns of $\Delta_0$ correspond to the set of some monomials, say $\{\mathbf{x}^\alpha\}$, where $\alpha \subset \mathbb{N}^n$. The columns of $\Delta_1$ correspond to the set of monomials $\{x_1 \mathbf{x}^\alpha\}$.

The shifted matrix $\Delta_1$ is:

$$\Delta_1 = \begin{pmatrix} -24 & 1272 & -288 & -3456 \\ 1272 & -3456 & -7416 & 166{,}368 \\ -288 & -7416 & 5544 & -41{,}472 \\ -3456 & 166{,}368 & -41{,}472 & -497{,}664 \end{pmatrix}.$$

Notice that the columns correspond to the monomials $\{x_1, x_1^2, x_1 x_2, x_1^3\}$, which are just the corresponding monomials of the columns of $\Delta_0$, i.e., $\{1, x_1, x_2, x_1^2\}$, multiplied by $x_1$.

In this specific case, all the elements of the matrices $\Delta_0$ and $\Delta_1$ are known. If this is not the case, then we can compute the unknown entries of the matrix, using either necessary or sufficient conditions of the quotient algebra, e.g., it holds that the $M_{x_i} M_{x_j} - M_{x_j} M_{x_i} = 0$, for any $i, j \in \{1, \ldots, n\}$.

4. We solve the equation $(\Delta_1 - \lambda \Delta_0) X = 0$.

We solve the generalized eigenvalue/eigenvector problem [87]. We normalize the elements of the eigenvectors so that the first element is one, and we read the solutions from the coordinates of the normalized eigenvectors.

The normalized eigenvectors of the generalized eigenvalue problem are:

$$\begin{pmatrix} 1 \\ -12 \\ -3 \\ 144 \end{pmatrix}, \begin{pmatrix} 1 \\ 12 \\ -13 \\ 144 \end{pmatrix}, \begin{pmatrix} 1 \\ -2 \\ 3 \\ 4 \end{pmatrix}, \begin{pmatrix} 1 \\ 2 \\ 3 \\ 4 \end{pmatrix}.$$

The coordinates of the eigenvectors correspond to the elements of the monomial basis $\{1, x_1, x_2, x_1^2\}$. Thus, we can recover the coefficients of $x_1$ and $x_2$ in the decomposition from the coordinates of the eigenvectors.

Recall that the coefficients of $x_0$ are considered to be one because of the dehomogenization process. Thus, our polynomial admits a decomposition:

$$F = \lambda_1 (x_0 - 12x_1 - 3x_2)^5 + \lambda_2 (x_0 + 12x_1 - 13x_2)^5 + \\ \lambda_3 (x_0 - 2x_1 + 3x_2)^5 + \lambda_4 (x_0 + 2x_1 + 3x_2)^5.$$

It remains to compute $\lambda_i$'s. We can do this easily by solving an over-determined linear system, which we know that always has a solution, since the decomposition exists. Doing that, we deduce that $\lambda_1 = 3$, $\lambda_2 = 15$, $\lambda_3 = 15$ and $\lambda_4 = 5$.

## 3. Tensor Product and Segre Varieties

### 3.1. Introduction: First Approaches

As we saw in the Introduction, if we consider the space parametrizing $(n_1 + 1) \times \cdots \times (n_t + 1)$-tensors (up to multiplication by scalars), i.e., the space $\mathbb{P}^N$, with $N = \prod_{i=1}^{t}(n_i + 1) - 1$, then additive decomposition problems lead us to study secant varieties of the Segre varieties $X_n \subset \mathbb{P}^N$, $\mathbf{n} = (n_1, \ldots, n_t)$, which are the image of the Segre embedding of the multiprojective spaces $\mathbb{P}^{n_1} \times \cdots \times \mathbb{P}^{n_t}$, defined by the map:

$$\nu_{1,\ldots,1} : \mathbb{P}^{n_1} \times \cdots \times \mathbb{P}^{n_t} \to \mathbb{P}^N,$$

$$\nu_{1,\ldots,1}(P) = (a_{1,0}a_{2,0}\cdots a_{t,0}, \ldots, a_{1,n_1}\cdots a_{t,n_t}),$$

where $P = ((a_{1,0}, \ldots, a_{1,n_1}), \ldots, (a_{t,0}, \ldots, a_{t,n_t})) \in \mathbb{P}^{n_1} \times \cdots \times \mathbb{P}^{n_t}$, and the products are taken in lexicographical order. For example, if $P = ((a_0, a_1), (b_0, b_1, b_2)) \in \mathbb{P}^1 \times \mathbb{P}^2$, then we have $\nu_{1,1}(P) = (a_0b_0, a_0b_1, a_0b_2, a_1b_0, a_1b_1, a_1b_2) \in \mathbb{P}^5$.

Note that, if $\{x_{i,0}, \ldots, x_{i,n_i}\}$ are homogeneous coordinates in $\mathbb{P}^{n_i}$ and $z_{j_1,\ldots,j_t}$, $j_i \in \{0, \ldots n_i\}$ are homogeneous coordinates in $\mathbb{P}^N$, we have that $X_n$ is the variety whose parametric equations are:

$$z_{j_1,\ldots,j_t} = x_{j_1,1}\cdots x_{j_t,t}; \quad j_i \in \{1, \ldots n_i\}.$$

Since the use of tensors is ubiquitous in so many applications and to know a decomposition for a given tensor allows one to ease the computational complexity when trying to manipulate or study it, this problem has many connections with questions raised by computer scientists in complexity theory [88] and by biologists and statisticians (e.g., see [16,89,90]).

As it is to be expected with a problem with so much interest in such varied disciplines, the approaches have been varied; see, e.g., [88,91] for the computational complexity approach, [16,90] for the biological statistical approach, [22,92–94] for the classical algebraic geometry approach, [95,96] for the representation theory approach, [97] for a tropical approach and [98] for a multilinear algebra approach. Since the $t = 2$ case is easy (it corresponds to ordinary matrices), we only consider $t \geq 3$.

The first fundamental question about these secant varieties, as we have seen, is about their dimensions. Despite all the progresses made on this question, it still remains open; only several partial results are known.

Notice that the case $t = 3$, since it corresponds to the simplest tensors, which are not matrices, had been widely studied, and many previous results from several authors are collected in [22].

We start by mentioning the following result on non-degeneracy; see [22] (Propositions 2.3 and 3.7).

**Theorem 10.** *Let* $n_1 \leq n_2 \leq \cdots \leq n_t$, $t \geq 3$. *Then, the dimension of the s-th secant variety of the Segre variety* $X_\mathbf{n}$ *is as expected, i.e.,*

$$\dim \sigma_s(X_n) = s(n_1 + n_2 + \cdots + n_t + 1) - 1$$

*if either:*

- $s \leq n_1 + 1$; or
- $\max\{n_t + 1, s\} \leq \left[\frac{n_1 + n_2 + \cdots + n_t + 1}{2}\right]$.

In the paper mentioned above, these two results are obtained in two ways. The first is via combinatorics on monomial ideals in the multihomogeneous coordinate ring of $\mathbb{P}^{n_1} \times \cdots \times \mathbb{P}^{n_t}$: curiously enough, this corresponds to understanding possible arrangements of a set of rooks on an $t$-dimensional chessboard (corresponding to the array representing the tensor). There is also a reinterpretation of these problems in terms of code theory and Hamming distance (the so-called Hamming codes furnish nice examples of non-defective secants varieties to Segre's of type $\mathbb{P}^n \times \cdots \times \mathbb{P}^n$).

Combinatorics turns out to be a nice, but limited tool for those questions. The second part of Theorem 10 (and many other results that we are going to report) are obtained by the use of inverse systems and the multigraded version of apolarity theory (recall Section 2.1.4 for the standard case, and we refer to [99–102] for definitions of multigraded apolarity) or via Terracini's lemma (see Lemma 1).

The idea behind these methods is to translate the problem of determining the dimension of $\sigma_s(X)$, into the problem of determining the multihomogeneous Hilbert function of a scheme $Z \subset \mathbb{P}^{n_1} \times \cdots \times \mathbb{P}^{n_t}$ of $s$ generic two-fat points in multi-degree $\mathbf{1} = (1,\ldots,1)$. We have that the coordinate ring of the multi-projective space $\mathbb{P}^{n_1} \times \cdots \times \mathbb{P}^{n_t}$ is the polynomial ring $S = \Bbbk[x_{1,0},\ldots,x_{1,n_1},\ldots,x_{t,0},\ldots,x_{t,n_t}]$, equipped with the multi-degree given by $\deg(x_{i,j}) = \mathbf{e}_i = (0,\ldots,\underset{i}{1},\ldots,0)$. Then, the scheme $Z$ is defined by a multi-homogeneous ideal $I = I(Z)$, which inherits the multi-graded structure. Hence, recalling the standard definition of Hilbert function (Definition 11), we say that the multi-graded Hilbert function of $Z$ in multi-degree $\mathbf{d} = (d_1,\ldots,d_t)$ is:

$$\mathrm{HF}(Z;\mathbf{d}) = \dim_{\Bbbk}(S/I)_{\mathbf{d}} = \dim_{\Bbbk} S_{\mathbf{d}} - \dim_{\Bbbk} I_{\mathbf{d}}.$$

### 3.2. The Multiprojective Affine Projective Method

We describe here a way to study the dimension of $\sigma_s(X_\mathbf{n})$ by studying the multi-graded Hilbert function of a scheme of fat points in multiprojective space via a very natural reduction to the Hilbert function of fat points in the standard projective space (of equal dimension).

We start recalling a direct consequence of Terracini's lemma for any variety.

Let $Y \subset \mathbb{P}^N$ be a positive dimensional smooth variety, and let $Z \subset Y$ be a scheme of $s$ generic two-fat points, i.e., a scheme defined by the ideal sheaf $\mathcal{I}_Z = \mathcal{I}_{P_1,Y}^2 \cap \cdots \cap \mathcal{I}_{P_s,Y}^2 \subset \mathcal{O}_Y$, where the $P_i$'s are $s$ generic points of $Y$ defined by the ideal sheaves $\mathcal{I}_{P_i,Y} \subset \mathcal{O}_Y$, respectively. Since there is a bijection between hyperplanes of the space $\mathbb{P}^N$ containing the linear space $\langle T_{P_1}(Y),\ldots,T_{P_s}(Y)\rangle$ and the elements of $H^0(Y,\mathcal{I}_Z(1))$, we have the following consequence of the Terracini lemma.

**Theorem 11.** *Let $Y$ be a positive dimensional smooth variety; let $P_1,\ldots,P_s$ be generic points on $Y$; and let $Z \subset Y$ be the scheme defined by $\mathcal{I}_{P_1,Y}^2 \cap \ldots \cap \mathcal{I}_{P_s,Y}^2$. Then,*

$$\dim \sigma_s(Y) = \dim\langle T_{P_1}(Y),\ldots,T_{P_s}(Y)\rangle = N - \dim_{\Bbbk} H^0(Y,\mathcal{I}_Z(1)).$$

Now, we apply this result to the case of Segre varieties; we give, e.g., [103] as the main reference. Consider $\mathbb{P}^\mathbf{n} := \mathbb{P}^{n_1} \times \cdots \times \mathbb{P}^{n_t}$, and let $X_\mathbf{n} \subset \mathbb{P}^N$ be its Segre embedding given by $\mathcal{O}_{\mathbb{P}^{n_1}}(1) \otimes \ldots \otimes \mathcal{O}_{\mathbb{P}^{n_t}}(1)$. By applying Theorem 11 and since the scheme $Z \subset X_\mathbf{n}$ corresponds to a scheme of $s$ generic two-fat points in $X$, which, by a little abuse of notation, we call again $Z$, we get:

$$\dim \sigma_s(X_\mathbf{n}) = \mathrm{HF}(Z,\mathbf{1}) - 1.$$

Now, let $n = n_1 + \cdots + n_t$, and consider the birational map:

$$\mathbb{P}^\mathbf{n} \dashrightarrow \mathbb{A}^n,$$

where:

$$([x_{1,0}:\cdots:x_{1,n_1}],\ldots,[x_{t,0},\ldots,x_{t,n_t}])$$
$$\downarrow$$
$$\left(\frac{x_{1,1}}{x_{1,0}},\frac{x_{1,2}}{x_{1,0}},\ldots,\frac{x_{1,n_1}}{x_{1,0}};\frac{x_{2,1}}{x_{2,0}},\frac{x_{2,n_2}}{x_{2,0}};\ldots;\frac{x_{t,1}}{x_{t,0}},\ldots,\frac{x_{t,n_t}}{x_{1,0}}\right).$$

This map is defined in the open subset of $\mathbb{P}^\mathbf{n}$ given by $\{x_{1,0}x_{2,0}\cdots x_{t,0} \neq 0\}$.

Now, let $\Bbbk[z_0, z_{1,1}, \ldots, z_{1,n_1}, z_{2,1}, \ldots, z_{2,n_2}, \ldots, z_{t,1}, \ldots, z_{t,n_t}]$ be the coordinate ring of $\mathbb{P}^n$, and consider the embedding of $\mathbb{A}^n \to \mathbb{P}^n$, whose image is the affine chart $\{z_0 \neq 0\}$. By composing the two maps above, we get:
$$\varphi : \mathbb{P}^n \dashrightarrow \mathbb{P}^n,$$
with:
$$([x_{1,0} : \cdots : x_{1,n_1}], \ldots, [x_{t,0}, \ldots, x_{t,n_t}])$$
$$\downarrow$$
$$\left[1 : \frac{x_{1,1}}{x_{1,0}} : \frac{x_{1,2}}{x_{1,0}} : \cdots : \frac{x_{1,n_1}}{x_{1,0}} : \frac{x_{2,1}}{x_{2,0}} : \cdots : \frac{x_{2,n_2}}{x_{2,0}} : \cdots : \frac{x_{t,1}}{x_{t,0}} : \cdots : \frac{x_{t,n_t}}{x_{t,0}}\right].$$

Let $Z \subset X$ be a zero-dimensional scheme, which is contained in the affine chart $\{x_{0,1} x_{0,2} \cdots x_{0,t} \neq 0\}$, and let $Z' = \varphi(Z)$. We want to construct a scheme $W \subset \mathbb{P}^n$ such that $\mathrm{HF}(W; d) = \mathrm{HF}(Z; (d_1, \ldots, d_t))$, where $d = d_1 + \ldots + d_t$.

Let:
$$Q_0, Q_{1,1}, \ldots, Q_{1,n_1}, Q_{2,1}, \ldots, Q_{2,n_2}, Q_{t,1}, \ldots, Q_{t,n_t}$$

be the coordinate points of $\mathbb{P}^n$. Consider the linear space $\Pi_i \cong \mathbb{P}^{n_i - 1} \subset \mathbb{P}^n$, where $\Pi_i = \langle Q_{i,1}, \ldots, Q_{i,n_i} \rangle$. The defining ideal of $\Pi_i$ is:
$$I(\Pi_i) = (z_0, z_{1,1}, \ldots, z_{1,n_1}; \ldots; \hat{z}_{i,1}, \ldots, \hat{z}_{i,n_i}; \ldots; z_{t,1}, \ldots, z_{t,n_t}).$$

Let $W_i$ be the subscheme of $\mathbb{P}^n$ denoted by $(d_i - 1)\Pi_i$, i.e., the scheme defined by the ideal $I(\Pi_i)^{d_i - 1}$. Since $I(\Pi_i)$ is a prime ideal generated by a regular sequence, the ideal $I(\Pi_i)^{d_i - 1}$ is saturated (and even primary for $I(\Pi_i)$). Notice that $W_i \cap W_j = \emptyset$, for $i \neq j$. With this construction, we have the following key result.

**Theorem 12.** *Let $Z$, $Z'$, $W_1, \ldots, W_t$ be as above, and let $W = Z' + W_1 + \cdots + W_t \subset \mathbb{P}^n$. Let $I(W) \subset S$ and $I(Z) \subset R$ be the ideals of $W$ and $Z$, respectively. Then, we have, for all $(d_1, \ldots, d_t) \in \mathbb{N}^t$:*
$$\dim_{\Bbbk} I(W)_d = \dim_{\Bbbk} I(Z)_{(d_1, \ldots, d_t)},$$
*where $d = d_1 + \cdots + d_t$.*

Note that when studying Segre varieties, we are only interested to the case $(d_1, \ldots, d_t) = (1, \ldots, 1)$; but, in the more general case of Segre–Veronese varieties, we will have to look at Theorem 12 for any multidegree $(d_1, \ldots, d_t)$; see Section 4.2.

Note that the scheme $W$ in $\mathbb{P}^n$ that we have constructed has two parts: the part $W_1 + \cdots + W_t$ (which we shall call the part at infinity and we denote as $W_\infty$) and the part $Z'$, which is isomorphic to our original zero-dimensional scheme $Z \subset X$. Thus, if $Z = \emptyset$ (and hence, $Z' = \emptyset$), we obtain from the theorem that:
$$\dim_{\Bbbk} I(W_\infty)_d = \dim_{\Bbbk} S_{(d_1, \ldots, d_t)}, \quad d = d_1 + \cdots + d_t.$$

It follows that:
$$\mathrm{HF}(W_\infty; d) = \binom{d_1 + \cdots + d_t + n}{n} - \binom{d_1 + n_1}{n_1} \cdots \binom{d_t + n_t}{n_t}.$$

With this observation made, the following corollary is immediate.

**Corollary 3.** *Let $Z$ and $Z'$ be as above, and write $W = Z' + W_\infty$. Then,*
$$\mathrm{HF}(W; d) = \mathrm{HF}(Z; (d_1, \ldots, d_t)) + \mathrm{HF}(W_\infty; d).$$

Eventually, when $Z$ is given by $s$ generic two-fat points in multi-projective space, we get the following.

**Theorem 13.** *Let $Z \subset X$ be a generic set of $s$ two-fat points, and let $W \subset \mathbb{P}^n$ be as in the Theorem 12. Then, we have:*
$$\dim \sigma_s(X_\mathbf{n}) = \mathrm{HF}(Z, (1,\ldots,1)) - 1 = N - \dim I(W)_t,$$
*where $N = \Pi_{i=1}^t(1+n_i) - 1$.*

Therefore, eventually, we can study a projective scheme $W \subset \mathbb{P}^n$, which is made of a union of generic two-fat points and of fat linear spaces. Note that, when $n_1 = \ldots = n_t = 1$, then also $W$ is a scheme of fat points.

### 3.3. The Balanced Case

One could try to attack the problem starting with a case, which is in a sense more "regular", i.e., the "balanced" case of $n_1 = \cdots = n_t$, $t \geq 3$. Several partial results are known, and they lead Abo, Ottaviani and Peterson to propose, in their lovely paper [92], a conjecture, which states that there are only a finite number of defective Segre varieties of the form $\mathbb{P}^n \times \cdots \times \mathbb{P}^n$, and their guess is that $\sigma_4(\mathbb{P}^2 \times \mathbb{P}^2 \times \mathbb{P}^2)$ and $\sigma_3(\mathbb{P}^1 \times \mathbb{P}^1 \times \mathbb{P}^1 \times \mathbb{P}^1)$ are actually the only defective cases (as we will see later, this is just part of an even more hazardous conjecture; see Conjecture 6).

In the particular case of $n_1 = \cdots = n_t = 1$, the question has been completely solved in [104], supporting the above conjecture.

**Theorem 14** ([104])**.** *Let $t, s \in \mathbb{N}$, $t \geq 3$. Let $X_1$ be the Segre embedding of $\mathbb{P}^1 \times \cdots \times \mathbb{P}^1$, ($t$-times). The dimension of $\sigma_s(X_1) \subset \mathbb{P}^N$, with $N = 2^t - 1$, is always as expected, i.e.,*
$$\dim \sigma_s(X) = \min\{N, s(t+1) - 1\},$$
*except for $t = 4$, $s = 3$. In this last case, $\dim \sigma_3(X) = 13$, instead of $14$.*

The method that has been used to compute the multi-graded Hilbert function for schemes of two-fat points with generic support in multi-projective spaces is based first on the procedure described on the multiprojective affine projective method explained in the previous section, which brings to the study of the standard Hilbert function of the schemes $W \subset \mathbb{P}^t$. Secondly, the problem of determining the dimension of $I(W)_t$ can be attacked by induction, via the powerful tool constituted by the differential Horace method, created by Alexander and Hirschowitz; see Section 2.2.2. This is used in [104] together with other "tricks", which allow one to "move" on a hyperplane some of the conditions imposed by the fat points, analogously as we have described in the examples in Section 2.2.2. These were the key ingredients to prove Theorem 14.

The only defective secant variety in the theorem above is made by the second secant variety of $\mathbb{P}^1 \times \mathbb{P}^1 \times \mathbb{P}^1 \times \mathbb{P}^1 \subset \mathbb{P}^{15}$, which, instead of forming a hypersurface in $\mathbb{P}^{15}$, has codimension two. Via Theorem 13 above, this is geometrically related to a configuration of seven fat points; more precisely, in this case, the scheme $W$ of Theorem 13 is union of three two-fat points and four three-fat ones (see also Theorem 12 for a detailed description of $W$). These always lie on a rational normal curve in $\mathbb{P}^4$ (see, e.g., Theorem 1.18 of [1]) and do not have the expected Hilbert function in degree four, by the result in [105].

For the general "balanced" case $\mathbb{P}^n \times \cdots \times \mathbb{P}^n$, the following partial result is proven in [92].

**Theorem 15.** *Let $X_\mathbf{n}$ be the Segre embedding of $\mathbb{P}^n \times \cdots \times \mathbb{P}^n$ ($t$ times), $t \geq 3$. Let $s_t$ and $e_t$ be defined by:*
$$s_t = \left\lfloor \frac{(n+1)^t}{nt+1} \right\rfloor \quad \text{and} \quad e_t \equiv s_t \bmod (n+1) \quad \text{with} \quad e_t \in \{0,\ldots,n\}.$$

*Then:*
- *if $s \leq s_t - e_t$, then $\sigma_s(X)$ has the expected dimension;*

- if $s \geq s_t - e_t + n + 1$, then $\sigma_s(X)$ fills the ambient space.

In other words, if $s_t = q(n+1) + r$, with $1 \leq r \leq n$, then $\sigma_s(X_n)$ has the expected dimension both for $s \geq (q+1)(n+1)$ and for $s \leq q(n+1)$, but if $n+1$ divides $s_t$, then $\sigma_s(X)$ has the expected dimension for any $s$.

Other known results in the "balanced" case are the following:

- $\sigma_4(\mathbb{P}^2 \times \mathbb{P}^2 \times \mathbb{P}^2)$ is defective with defect $\delta_4 = 1$; see [21].
- $\mathbb{P}^n \times \mathbb{P}^n \times \mathbb{P}^n$ is never defective for $n \geq 3$; see [106];
- $\mathbb{P}^n \times \mathbb{P}^n \times \mathbb{P}^n \times \mathbb{P}^n$ for $2 \leq n \leq 10$ is never defective except at most for $s = 100$ and $n = 8$ or for $s = 357$ and $n = 10$; see [107].

### 3.4. The General Case

When we drop the balanced dimensions request, not many defective cases are actually known. For example:

- $\mathbb{P}^2 \times \mathbb{P}^3 \times \mathbb{P}^3$ is defective only for $s = 5$, with defect $\delta_4 = 1$; see [92];
- $\mathbb{P}^2 \times \mathbb{P}^n \times \mathbb{P}^n$ with $n$ even is defective only for $s = \frac{3n}{2} + 1$ with defect $\delta_s = 1$; see [22,94].
- $\mathbb{P}^1 \times \mathbb{P}^1 \times \mathbb{P}^n \times \mathbb{P}^n$ is defective only for $s = 2n + 1$ with defect $\delta_{2n+1} = 1$; see [92].

However, when taking into consideration cases where the $n_i$'s are far from being equal, we run into another "defectivity phenomenon", known as "the unbalanced case"; see [22,108].

**Theorem 16.** *Let $X_n$ be the Segre embedding of $\mathbb{P}^{n_1} \times \cdots \times \mathbb{P}^{n_t} \times \mathbb{P}^n \subset \mathbb{P}^M$, with $M = (n+1)\prod_{i=1}^{t}(n_i + 1) - 1$. Let $N = \prod_{i=1}^{t}(n_i + 1) - 1$, and assume $n > N - \sum_{i=1}^{t} n_i + 1$. Then, $\sigma_s(X)$ is defective for:*

$$N - \sum_{i=1}^{t} n_i + 1 < s \leq \min\{n; N\},$$

*with defect equal to $\delta_s(X) = s^2 - s(N - \sum_{i=1}^{t} n_i + 1)$.*

The examples described above are the few ones for which defectivities of Segre varieties are known. Therefore, the following conjecture has been stated in [92], where, for $n_1 \leq n_2 \leq \ldots \leq n_t$, it is proven for $s \leq 6$.

**Conjecture 6.** *The Segre embeddings of $\mathbb{P}^{n_1} \times \cdots \times \mathbb{P}^{n_t}$, $t \geq 3$, are never defective, except for:*

- $\mathbb{P}^1 \times \mathbb{P}^1 \times \mathbb{P}^1 \times \mathbb{P}^1$, for $s = 3$, with $\delta_3 = 1$;
- $\mathbb{P}^2 \times \mathbb{P}^2 \times \mathbb{P}^2$, for $s = 4$, with $\delta_4 = 1$;
- the "unbalanced case";
- $\mathbb{P}^2 \times \mathbb{P}^3 \times \mathbb{P}^3$, for $s = 5$, with $\delta_5 = 1$;
- $\mathbb{P}^2 \times \mathbb{P}^n \times \mathbb{P}^n$, with $n$ even, for $s = \frac{3n}{2} + 1$, with $\delta_s = 1$);
- $\mathbb{P}^1 \times \mathbb{P}^1 \times \mathbb{P}^n \times \mathbb{P}^n$, for $s = 2n + 1$, with $\delta_{2n+1} = 1$).

## 4. Other Structured Tensors

There are other varieties of interest, parametrizing other "structured tensors", i.e., tensors that have determined properties. In all these cases, there exists an additive decomposition problem, which can be geometrically studied similarly as we did in the previous sections. In this section, we want to present some of these cases.

In particular, we consider the following structured tensors:

1. skew-symmetric tensors, i.e., $v_1 \wedge \ldots \wedge v_k \in \bigwedge^k V$;

2. decomposable partially-symmetric tensors, i.e., $L_1^{d_1} \otimes \ldots \otimes L_t^{d_t} \in S^{d_1}V_1^* \otimes \cdots \otimes S^{d_t}V_t^*$;
3. $d$-powers of linear forms, i.e., homogeneous polynomials $L_1^{d_1} \cdots L_t^{d_t} \in S^d V^*$, for any partition $d = (d_1, \ldots, d_t) \vdash d$;
4. reducible forms, i.e., $F_1 \cdots F_t \in S^d V^*$, where $\deg(F_i) = d_i$, for any partition $\mathbf{d} = (d_1, \ldots, d_t) \vdash d$;
5. powers of homogeneous polynomials, i.e., $G^k \in S^d V^*$, for any $k | d$;

## 4.1. Exterior Powers and Grassmannians

Denote by $\mathbb{G}(k,n)$ the Grassmannian of $k$-dimensional linear subspaces of $\mathbb{P}^n \cong \mathbb{P}V$, for a fixed $n+1$ dimensional vector space $V$. We consider it with the embedding given by its Plücker coordinates as embedded in $\mathbb{P}^{N_k}$, where $N_k = \binom{n+1}{k} - 1$.

The dimensions of the higher secant varieties to the Grassmannians of lines, i.e., for $k = 1$, are well known; e.g., see [7], [109] or [110]. The secant variety $\sigma_s(Gr(1,n))$ parametrizes all $(n+1) \times (n+1)$ skew-symmetric matrices of rank at most $2s$.

**Theorem 17.** *We have that $\sigma_s(\mathbb{G}(1,n))$ is defective for $s < \lfloor \frac{n+1}{2} \rfloor$ with defect equal to $\delta_s = 2s(s-1)$.*

For $k \geq 2$, not many results can be found in the classical or contemporary literature about this problem; e.g., see [109–111]. However, they are sufficient to have a picture of the whole situation. Namely, there are only four other cases that are known to be defective (e.g., see [110]), and it is conjectured in [112] that these are the only ones. This is summarized in the following conjecture:

**Conjecture 7** (Baur–Draisma–de Graaf, [112]). *Let $k \geq 2$. Then, the secant variety $\sigma_s(\mathbb{G}(k,n))$ has the expected dimension except for the following cases:*

|  | actual codimension | expected codimension |
|---|---|---|
| $\sigma_3(\mathbb{G}(2,6))$ | 1 | 0 |
| $\sigma_3(\mathbb{G}(3,7))$ | 20 | 19 |
| $\sigma_4(\mathbb{G}(3,7))$ | 6 | 2 |
| $\sigma_4(\mathbb{G}(2,8))$ | 10 | 8 |

In [112], they proved the conjecture for $n \leq 15$ (the case $n \leq 14$ can be found in [113]). The conjecture has been proven to hold for $s \leq 6$ (see [111]) and later for $s \leq 12$ in [114].

A few more results on non-defectivity are proven in [110,111]. We summarize them in the following.

**Theorem 18.** *The secant variety $\sigma_s(\mathbb{G}(k,n))$ has the expected dimension when:*

- $k \geq 3$ and $ks \leq n+1$,
- $k = 2$, $n \geq 9$ and $s \leq s_1(n)$ or $s \geq s_2(n)$,

*where:*
$$s_1(n) = \left\lfloor \frac{n^2}{18} - \frac{20n}{27} + \frac{287}{81} \right\rfloor + \left\lfloor \frac{6n-13}{9} \right\rfloor \text{ and } s_2(n) = \left\lceil \frac{n^2}{18} - \frac{11n}{27} + \frac{44}{81} \right\rceil + \left\lceil \frac{6n-13}{9} \right\rceil;$$
*in the second case, $\sigma_s(\mathbb{G}(2,n))$ fills the ambient space.*

Other partial results can be found in [114], while in [115], the following theorem can be found.

**Definition 21.** *Given an integer $m \geq 2$, we define a function $h_m : \mathbb{N} \to \mathbb{N}$ as follows:*

- $h_m(0) = 0$;
- *for any $k \geq 1$, write $k+1 = 2^{\lambda_1} + 2^{\lambda_2} + \ldots + 2^{\lambda_\ell} + \epsilon$, for a suitable choice of $\lambda_1 > \lambda_2 > \ldots > \lambda_\ell \geq 1$, $\epsilon \in \{0,1\}$, and define:*
$$h_m(k) := m^{\lambda_1 - 1} + m^{\lambda_2 - 1} + \ldots + m^{\lambda_\ell - 1}.$$

In particular, we get $h_m(2k) = h_m(2k-1)$, and $h_2(k) = \lfloor \frac{k+1}{2} \rfloor$.

**Theorem 19** (Theorem 3.5.1 of [115]). *Assume that $k \geq 2$, and set:*

$$\alpha := \left\lfloor \frac{n+1}{k+1} \right\rfloor.$$

*If either:*

- $n \geq k^2 + 3k + 1$ and $h \leq \alpha h_\alpha(k-1)$;

*or:*

- $n < k^2 + 3k + 1$, $k$ is even, and $h \leq (\alpha - 1)h_\alpha(k-1) + h_\alpha(n-2-\alpha r)$;

*or:*

- $n < k^2 + 3k + 1$, $r$ is odd, and:

$$h \leq (\alpha-1)h_\alpha(k-2) + h_\alpha(\min\{n-3-\alpha(k-1), r-2\}),$$

*then, $\mathbb{G}(k,n)$ is not $(h+1)$-defective.*

This results strictly improves the results in [111] for $k \geq 4$, whenever $(k,n) \neq (4,10), (5,11)$; see [115].

Notice that, if we let $R_\wedge(k,n)$ be the generic rank with respect to $\mathbb{G}(k,n)$, i.e., the minimum $s$ such that $\sigma_s(\mathbb{G}(k,n)) = \mathbb{P}^{N_k}$, we have that the results above give that, asymptotically:

$$R_\wedge(2,n) \sim \frac{n^2}{18}.$$

A better asymptotic result can be found in [115].

Finally, we give some words concerning the methods involved. The approach in [110] uses Terracini's lemma and an exterior algebra version of apolarity. The main idea there is to consider the analog of the perfect pairing induced by the apolarity action that we have seen for the symmetric case in the skew-symmetric situation; see Section 2.1.4. In fact, the pairing considered here is:

$$\bigwedge^k V \times \bigwedge^{n-k} V \to \bigwedge^n V \simeq \Bbbk,$$

induced by the multiplication in $\bigwedge V$, and it defines the apolarity of a subspace $Y \subset V$ of dimension $k$ to be $Y^\perp := \{w \in \wedge^{n-k} V \mid w \wedge v = 0 \, \forall v \in Y\}$. Now, one can proceed in the same way as the symmetric case, namely by considering a generic element of the Grassmannian $\mathbb{G}(k,n)$ and by computing the tangent space at that point. Then, its orthogonal, via the above perfect pairing, turns out to be, as in the symmetric case, the degree $n-k$ part of an ideal, which is a double fat point. Hence, in [110], the authors apply Terracini's lemma to this situation in order to study all the known defective cases and in other various cases. Notice that the above definition of skew-symmetric apolarity works well for computing the dimension of secant varieties to Grassmannians since it defines the apolar of a subspace that is exactly what is needed for Terracini's lemma, but if one would like to have an analogous definition of apolarity for skew symmetric tensors, then there are a few things that have to be done. Firstly, one needs to extend by linearity the above definition to all the elements of $\wedge^k V$. Secondly, in order to get the equivalent notion of the apolar ideal in the skew symmetric setting, one has to define the skew-symmetric apolarity in any degree $\leq d$. This is done in [116], where also the skew-symmetric version of the apolarity lemma is given. Moreover, in [116], one can find the classification of all the skew-symmetric-ranks of any skew-symmetric tensor in $\wedge^3 \mathbb{C}^n$ for $n \leq n$ (the same classification can actually be found also in [117,118]), together with algorithms to get the skew-symmetric-rank and the skew-symmetric decompositions for any for those tensors (as far as we know, this is new).

Back to the results on dimensions of secant varieties of Grassmannians: in [112], a tropical geometry approach is involved. In [111], as was done by Alexander and Hirshowitz for the symmetric case, the authors needed to introduce a specialization technique, by placing a certain number of points on sub-Grassmannians and by using induction. In this way, they could prove several non-defective cases. Moreover, in the same work, invariant theory was used to describe the equation of $\sigma_3(\mathbb{G}(2,6))$, confirming the work of Schouten [119], who firstly proved that it was defective by showing that it is a hypersurface (note that by parameter count, it is expected to fill the ambient space). Lascoux [120] proved that the degree of Schouten's hypersurface is seven. In [92], with a very clever idea, an explicit description of this degree seven invariant was found by relating its cube to the determinant of a $21 \times 21$ symmetric matrix.

Eventually, in [115], the author employed a new method for studying the defectivity of varieties based on the study of osculating spaces.

### 4.2. Segre–Veronese Varieties

Now, we consider a generalization of the apolarity action that we have seen in both Section 2 and Section 3 to the multi-homogeneous setting; see [99–102]. More precisely, fixing a set of vector spaces $V_1, \ldots, V_t$ of dimensions $n_1 + 1, \ldots, n_t + 1$, respectively, and positive integers $d_1, \ldots, d_t$, we consider the space of partially-symmetric tensors:

$$S^{d_1} V_1^* \otimes \ldots \otimes S^{d_t} V_t^*.$$

The Segre–Veronese variety parametrizes decomposable partially-symmetric tensors, i.e., it is the image of the embedding:

$$\begin{aligned} \nu_{\mathbf{n},\mathbf{d}} : \mathbb{P}(S^1 V_1^*) \times \ldots \times \mathbb{P}(S^1 V_t^*) &\longrightarrow \mathbb{P}(S^{d_1} V_1^* \otimes \ldots \otimes S^{d_t} V_t^*) \\ ([L_1], \ldots, [L_t]) &\mapsto \left[ L_1^{d_1} \otimes \ldots \otimes L_t^{d_t} \right], \end{aligned}$$

where, for short, we denote $\mathbf{n} = (n_1, \ldots, n_t)$, $\mathbf{d} = (d_1, \ldots, d_t)$.

More geometrically, the Segre–Veronese variety is the image of the Segre–Veronese embedding:

$$\nu_{\mathbf{n},\mathbf{d}} : \mathbb{P}^{\mathbf{n}} := \mathbb{P}^{n_1} \times \ldots \times \mathbb{P}^{n_t} \longrightarrow \mathbb{P}^N, \quad \text{where } N = \prod_{i=1}^t \binom{d_i + n_i}{n_i} - 1,$$

given by $\mathcal{O}_{\mathbb{P}^{\mathbf{n}}}(\mathbf{d}) := \mathcal{O}_{\mathbb{P}^{n_1}}(d_1) \otimes \ldots \otimes \mathcal{O}_{\mathbb{P}^{n_t}}(d_t)$, that is via the forms of multidegree $(d_1, \ldots, d_t)$ of the multigraded homogeneous coordinate ring:

$$R = \mathbb{k}[x_{1,0}, \ldots, x_{1,n_1}; x_{2,0}, \ldots, x_{2,n_2}; \ldots; x_{t,0}, \ldots, x_{t,n_t}].$$

For instance, if $\mathbf{n} = (2,1)$, $\mathbf{d} = (1,2)$ and $P = ([a_0 : a_1 : a_2], [b_0 : b_1]) \in \mathbb{P}^2 \times \mathbb{P}^1$, we have $\nu_{\mathbf{n},\mathbf{d}}(P) = [a_0 b_0^2 : a_0 b_0 b_1 : a_0 b_1^2 : a_1 b_0^2 : a_1 b_0 b_1 : a_1 b_1^2 : a_2 b_0^2 : a_2 b_0 b_1 : a_2 b_1^2]$, where the products are taken in lexicographical order. We denote the embedded variety $\nu_{\mathbf{n},\mathbf{d}}(\mathbb{P}^{\mathbf{n}})$ by $X_{\mathbf{n},\mathbf{d}}$. Clearly, for $t = 1$, we recover Veronese varieties, while for $(d_1, \ldots, d_t) = (1, \ldots, 1)$, we get the Segre varieties.

The corresponding additive decomposition problem is as follows.

**Problem 3.** *Given a partially-symmetric tensor* $T \in S^{d_1} V_1^* \times \ldots \times \mathbb{P} S^{d_t} V_t^*$ *or, equivalently, a multihomogeneous polynomial* $T \in R$ *of multidegree* $\mathbf{d}$, *find the smallest possible length r of an expression* $T = \sum_{i=1}^r L_{i,1}^{d_1} \otimes \ldots \otimes L_{i,t}^{d_t}$.

As regards the generic tensor, a possible approach to this problem mimics what has been done for Segre and Veronese varieties. One can use Terracini's lemma (Lemma 1 and Theorem 11), as in [22,103], to translate the problem of determining the dimensions of the higher secant varieties of $X_{n,d}$ into that of calculating the value, at $\mathbf{d} = (d_1, \ldots, d_t)$, of the Hilbert function of generic sets of two-fat points in $\mathbb{P}^n$. Then, by using the multiprojective affine projective method introduced in Section 3.2, i.e., by passing to an affine chart in $\mathbb{P}^n$ and then homogenizing in order to pass to $\mathbb{P}^n$, with $n = n_1 + \cdots + n_t$, this last calculation amounts to computing the Hilbert function in degree $d = d_1 + \cdots + d_n$ for the subscheme $W \subset \mathbb{P}^n$; see Theorem 12.

There are many scattered results on the dimension of $\sigma_s(X_{n,d})$, by many authors, and very few general results. One is the following, which generalizes the "unbalanced" case considered for Segre varieties; see [92,108].

**Theorem 20.** *Let $X = X_{(n_1,\ldots,n_t,n),(d_1,\ldots,d_t,1)}$ be the Segre–Veronese embedding:*

$$\mathbb{P}^{n_1} \times \cdots \times \mathbb{P}^{n_t} \times \mathbb{P}^n \xrightarrow{(d_1,\ldots,d_t,1)} X \subset \mathbb{P}^M, \quad \text{with } M = (n+1)\left(\prod_{i=1}^t \binom{n_i+d_i}{d_i}\right) - 1.$$

*Let $N = \prod_{i=1}^t \binom{n_i+d_i}{d_i} - 1$, then, for $N - \sum_{i=1}^t n_i + 1 < s \leq \min\{n, N\}$, the secant variety $\sigma_s(X)$ is defective with $\delta_s(X) = s^2 - s(N - \sum_{i=1}^t n_i + 1)$.*

When it comes to Segre–Veronese varieties with only two factors, there are many results by many authors, which allow us to get a quite complete picture, described by the following conjectures, as stated in [121].

**Conjecture 8.** *Let $X = X_{(m,n),(a,b)}$, then $X$ is never defective, except for:*

- $b = 1$, $m \geq 2$, and it is unbalanced (as in the theorem above);
- $(m,n) = (1,n)$, $(a,b) = (2d,2)$;
- $(m,n) = (3,4)$, $(a,b) = (2,1)$;
- $(m,n) = (2,n)$, $(a,b) = (2,2)$;
- $(m,n) = (2,2k+1)$, $k \geq 1$, $(a,b) = (1,2)$;
- $(m,n) = (1,2)$, $(a,b) = (1,3)$;
- $(m,n) = (2,2)$, $(a,b) = (2,2)$;
- $(m,n) = (3,3)$, $(a,b) = (2,2)$;
- $(m,n) = (3,4)$, $(a,b) = (2,2)$.

**Conjecture 9.** *Let $X = X_{(m,n),(a,b)}$, then for $(a,b) \geq (3,3)$, $X$ is never defective.*

The above conjectures are based on results that can be found in [93,99,112,121–134].

For the Segre embeddings of many copies of $\mathbb{P}^1$, we have a complete result. First, in [99] and in [130], the cases of two and three copies of $\mathbb{P}^1$, respectively, were completely solved.

**Theorem 21.** *Let $X = X_{(1,1),(a,b)}$, $a \leq b$. Then, $X$ is never defective, except for $(a,b) = (2,2d)$; in this case, $\sigma_{2d+1}(X)$ is defective with $\delta_{2d+1} = 1$.*

**Theorem 22.** *Let $X = X_{(1,1,1),(a,b,c)}$, $a \leq b \leq c$; then $X$ is never defective, except for:*

- $(a,b,c) = (1,1,2d)$; in this case, $\sigma_{2d+1}(X)$ is defective with $\delta_{2d+1} = 1$;
- $(a,b,c) = (2,2,2)$; here, $\sigma_7(X)$ is defective, and $\delta_7 = 1$.

In [135] the authors, by using an induction approach, whose basic step was Theorem 14 about the Segre varieties $X_1$ [104], concluded that there are no other defective cases except for the previously-known ones.

**Theorem 23.** *Let $X = X_{(1,\ldots,1),(a_1,\ldots,a_r)}$. Then, $X$ is never defective, except for:*

- $X_{(1,1),(2,2d)}$ *(Theorem 21);*
- $X_{(1,1,1),(1,1,2d)}$ *and $X_{(1,1,1),(2,2,2)}$ (Theorem 22);*
- $X_{(1,1,1,1),(1,1,1,1)}$ *(Theorem 14).*

For several other partial results on the defectivity of certain Segre–Veronese varieties, see, e.g., [99,115,133,136], and for an asymptotic result about non-defective Segre–Veronese varieties, see [115,137].

### 4.3. Tangential and Osculating Varieties to Veronese Varieties

Another way of generalizing what we saw in Section 2 for secants of Veronese Varieties $X_{n,d}$ is to work with their tangential and osculating varieties.

**Definition 22.** *Let $X_{n,d} \subset \mathbb{P}^N$ be a Veronese variety. We denote by $\tau(X_{n,d})$ the tangential variety of $X_{n,d}$, i.e., the closure in $\mathbb{P}^N$ of the union of all tangent spaces:*

$$\tau(X_{n,d}) = \overline{\bigcup_{P \in X_{n,d}} T_P(X_{n,d})} \subset \mathbb{P}^N.$$

*More in general, we denote by $O^k(X_{n,d})$ the k-th osculating variety of $X_{n,d}$, i.e., the closure in $\mathbb{P}^N$ of the union of all k-th osculating spaces:*

$$O^k(X_{n,d}) = \overline{\bigcup_{P \in X_{n,d}} O_P^k(X_{n,d})} \subset \mathbb{P}^N.$$

*Hence, $\tau(X_{n,d}) = O^1(X_{n,d})$.*

These varieties are of interest also because the space $O^k(X_{n,d})$ parametrizes a particular kind of form. Indeed, if the point $P = [L^d] \in X_{n,d} \subset \mathbb{P}S_d$, then the k-th osculating space $O_P^k(X_{n,d})$ correspond to linear space $\{[L^{d-k}G] \mid G \in S_k\}$. Therefore, the corresponding additive decomposition problem asks the following.

**Problem 4.** *Given a homogeneous polynomial $F \in \mathbb{k}[x_0, \ldots, x_n]$, find the smallest length of an expression $F = \sum_{i=1}^{r} L_i^{d-k} G_i$, where the $L_i$'s are linear forms and the $G_i$'s are forms of degree k.*

The type of decompositions mentioned in the latter problem have been called generalized additive decomposition in [25] and in [29]. In the special case of $k = 1$, they are a particular case of the so-called Chow–Waring decompositions that we treat in full generality in Section 4.4. In this case, the answer to Problem 4 is called $(d-1,1)$-rank, and we denote it by $R_{(d-1,1)}(F)$.

**Remark 18.** *Given a family of homogeneous polynomials $\mathcal{F} = \{F_1, \ldots, F_m\}$, we define the simultaneous rank of $\mathcal{F}$ the smallest number of linear forms that can be used to write a Waring decomposition of all polynomials of $\mathcal{F}$.*

Now, a homogeneous polynomial $F \in S^d V$ can be seen as a partially-symmetric tensor in $S^1 V \otimes S^{d-1} V$ via the equality:

$$F = \frac{1}{d} \sum_{i=0}^{n} x_i \otimes \frac{\partial F}{\partial x_i}.$$

From this expression, it is clear that a list of linear forms that decompose simultaneously all partial derivatives of F also decompose F, i.e., the simultaneous rank of the first partial derivatives is an upper bound of the symmetric-rank of F. Actually, it is possible to prove that for every homogeneous polynomial, this is an equality (e.g., see [138] (Section 1.3) or [139] (Lemma 2.4)). For more details on relations between simultaneous ranks of higher order partial derivatives and partially-symmetric-ranks, we refer to [139].

Once again, in order to answer the latter question in the case of the generic polynomial, we study the secant varieties to the $k$-th osculating variety of $X_{n,d}$. In [21], the dimension of $\sigma_s(\tau(X_{n,d}))$ is studied ($k = 1$ case). Via apolarity and inverse systems, with an analog of Theorem 11, the problem is again reduced to the computation of the Hilbert function of some particular zero-dimensional subschemes of $\mathbb{P}^n$; namely,

$$\dim \sigma_s(\tau(X_{n,d})) = \dim_{\Bbbk}(L_1^{d-1}, \ldots, L_s^{d-1}, L_1^{d-2}M_1, \ldots, L_s^{d-2}M_s)_d - 1 =$$
$$= \mathrm{HF}(Z, d) - 1,$$

where $L_1, \ldots, L_s, M_1, \ldots, M_s$ are $2s$ generic linear forms in $\Bbbk[x_0, \ldots, x_n]$, while $\mathrm{HF}(Z, d)$ is the Hilbert function of a scheme $Z$, which is the union of $s$ generic $(2,3)$-points in $\mathbb{P}^n$, which are defined as follows.

**Definition 23.** *A $(2,3)$-point is a zero-dimensional scheme in $\mathbb{P}^n$ with support at a point $P$ and whose ideal is of type $I(P)^3 + I(\ell)^2$, where $I(P)$ is the homogeneous ideal of $P$ and $\ell \subset \mathbb{P}^n$ is a line through $P$ defining ideal $I(\ell)$.*

Note that when we say that $Z$ is a scheme of $s$ generic $(2,3)$-points in $\mathbb{P}^n$, we mean that $I(Z) = I(Q_1) \cap \ldots \cap I(Q_s)$, where the $Q_i$'s are $(2,3)$-points, i.e., $I(Q_i) = I(P_i)^3 + I(\ell_i)^2$, such that $P_1, \ldots P_s$ are generic points in $\mathbb{P}^n$, while $\ell_1, \ldots, \ell_s$ are generic lines passing though $P_1, \ldots, P_s$, respectively.

By using the above fact, in [21], several cases where $\sigma_s(\tau(X_{n,d}))$ is defective were found, and it was conjectured that these exceptions were the only ones. The conjecture has been proven in a few cases in [22] ($s \leq 5$ and $n \geq s+1$) and in [140] ($n = 2, 3$). In [70], it was proven for $n \leq 9$, and moreover, it was proven that if the conjecture holds for $d = 3$, then it holds in every case. Finally, by using this latter fact, Abo and Vannieuwenhoven completed the proof of the following theorem [141].

**Theorem 24.** *The $s$-th secant variety $\sigma_s(\tau(X_{n,d}))$ of the tangential variety to the Veronese variety has dimension as expected, except in the following cases:*
1. *$d = 2$ and $2 \leq 2s < n$;*
2. *$d = 3$ and $n = 2, 3, 4$.*

As a direct corollary of the latter result, we obtain the following answer to Problem 4 in the case of generic forms.

**Corollary 4.** *Let $F \in S_d$ be a generic form. Then,*

$$R_{(d-1,1)}(F) = \left\lceil \frac{\binom{n+d}{n}}{2n+1} \right\rceil,$$

*except for:*

1. *$d = 2$, where $R_{(d-1,1)}(F) = \lfloor \frac{n}{2} \rfloor + 1$;*
2. *$d = 3$ and $n = 2, 3, 4$, where $R_{(d-1,1)}(F) = \left\lceil \frac{\binom{n+d}{n}}{2n+1} \right\rceil + 1.$*

The general case of $\sigma_s(O^k(X_{n,d}))$ is studied in [70,142–144]. Working in analogy with the case $k = 1$, the dimension of $\sigma_s(O^k(X_{n,d}))$ is related to the Hilbert function of a certain zero-dimensional scheme $Z = Z_1 \cup \cdots \cup Z_s$, whose support is a generic union of points $P_1, \ldots, P_s \in \mathbb{P}^n$, respectively, and such that, for each $i = 1, \ldots, s$, we have that $(k+1)P_i \subset Z_i \subset (k+2)P_i$.

As one of the manifestations of the ubiquity of fat points, the following conjecture describes the conditions for the defect of this secant variety in terms of the Hilbert function of fat points:

**Conjecture 10** ([70] (Conjecture 2a)). *The secant variety $\sigma_s(O^k(X_{n,d}))$ is defective if and only if either:*

1. $h^1(\mathcal{I}_X(d)) > \max\{0, \deg Z - \binom{d+n}{n}\}$, *or*
2. $h^0(\mathcal{I}_T(d)) > \max\{0, \binom{d+n}{n} - \deg Z\}$,

*where $X$ is a generic union of $s$, $(k+1)$-fat points and $T$ is a generic union of $(k+2)$-fat points.*

In [142,144], the conjecture is proven for $n = 2, s \leq 9$, and in [70] for $n = 2$ and any $s$.

### 4.4. Chow–Veronese Varieties

Let $\mathbf{d} = (d_1, \ldots, d_t)$ be a partition of a positive integer $d$, i.e., $d_1 \geq \ldots \geq d_t$ are positive integers, which sum to $d$. Then, we consider the following problem.

Let $S = \bigoplus_{d \in \mathbb{N}} S_d$ be a polynomial ring in $n+1$ variables.

**Problem 5.** *Given a homogeneous polynomial $F \in S_d$, find the smallest length of an expression $F = \sum_{i=1}^{r} L_{i,1}^{d_1} \cdots L_{i,t}^{d_t}$, where $L_{i,j}$'s are linear forms.*

The decompositions considered in the latter question are often referred to as Chow–Waring decompositions. We call the answer to Problem 5 as the **d**-rank of $F$, and we denote it by $R_\mathbf{d}(F)$.

In this case, the summands are parametrized by the so-called Chow–Veronese variety, which is given by the image of the embedding:

$$\nu_\mathbf{d}: \mathbb{P}S_1 \times \ldots \times \mathbb{P}S_1 \longrightarrow \mathbb{P}S_d,$$
$$([L_1], \ldots, [L_t]) \mapsto \left[L_1^{d_1} \cdots L_t^{d_t}\right].$$

We denote by $X_\mathbf{d}$ the image $\nu_\mathbf{d}(\mathbb{P}^n)$. Notice that this map can be seen as a linear projection of the Segre–Veronese variety $X_{\mathbf{n},\mathbf{d}} \subset \mathbb{P}(S_{d_1} \otimes \ldots \otimes S_{d_t})$, for $\mathbf{n} = (n, \ldots, n)$, under the map induced by the linear projection of the space of partially-symmetric tensors $S_{d_1} \otimes \ldots \otimes S_{d_t}$ on the totally symmetric component $S_d$. Once again, we focus on the question posed in Problem 5 in the case of a generic polynomial, for which we study dimensions of secant varieties to $X_\mathbf{d}$.

In the case of $\mathbf{d} = (d-1, 1)$, we have that $X_\mathbf{d}$ coincides with the tangential variety of the Veronese variety $\tau(X_{n,d})$, for which the problem has been completely solved, as we have seen in the previous section (Theorem 24).

The other special case is given by $\mathbf{d} = (1, \ldots, 1)$, for $d \geq 3$. In this case, $X_d$ has been also referred to as the Chow variety or as the variety of split forms or completely decomposable forms. After the first work by Arrondo and Bernardi [73], Shin found the dimension of the second secant variety in the ternary case ($n = 2$) [145], and Abo determined the dimensions of higher secant varieties [146]. All these cases are non-defective. It is conjectured that varieties of split forms of degree $d \geq 3$ are never defective. New cases have been recently proven in [147,148].

The problem for any arbitrary partition **d** has been considered in [149]. Dimensions of all $s$-th secant varieties for any partition haves been computed in the case of binary forms ($n = 1$). In a higher number of variables, the dimensions of secant line varieties ($s = 2$) and of higher secant varieties with $s \leq 2\lfloor \frac{n}{3} \rfloor$ have been computed. This was done by using the classical Terracini's lemma (Lemma 1) in order to obtain a nice description of the generic tangent space of the $s$-th secant variety. In the following example, we explain how the binary case could be treated.

**Example 18.** If $P = [L_1^{d_1} \cdots L_t^{d_t}] \in X_{\mathbf{d}}$, then it is not difficult to prove (see Proposition 2.2 in [149]) that:

$$T_P X_{\mathbf{d}} = \mathbb{P}\left((I_P)_d\right), \quad \text{with } d = d_1 + \ldots + d_t$$

where $I_P = (L_1^{d_1-1} L_2^{d_2} \cdots L_t^{d_t}, L_1^{d_1} L_2^{d_2-1} \cdots L_t^{d_t}, \ldots, L_1^{d_1} L_2^{d_2} \cdots L_t^{d_t-1})$.

In the particular case of binary forms, some more computations show that actually, $T_P X_{\mathbf{d}} = \mathbb{P}\left((I'_P)_d\right)$, where $I'_P$ is the principal ideal $(L_1^{d_1-1} \cdots L_t^{d_t-1})$. In this way, by using Terracini's lemma, we obtain that, if $Q$ is a generic point on the linear span of $s$ generic points on $X_{\mathbf{d}}$, then:

$$T_Q \sigma_s(X_{\mathbf{d}}) = \mathbb{P}\left((L_{1,1}^{d_1-1} \cdots L_{1,t}^{d_t-1}, \ldots, L_{s,1}^{d_1-1} \cdots L_{s,t}^{d_t-1})_d\right),$$

where $L_{i,j}$'s are generic linear forms. Now, in order to compute the dimension of this tangent space, we can study the Hilbert function of the ideal on the right-hand side. By semicontinuity, we may specialize to the case $L_{i,1} = \ldots = L_{i,t}$, for any $i = 1, \ldots, s$. In this way, we obtain a power ideal, i.e., an ideal generated by powers of linear forms, whose Hilbert function is prescribed by Fröberg–Iarrobino's conjecture; see Remark 10. Now, since in [150], the authors proved that the latter conjecture holds in the case of binary forms, i.e., the Hilbert function of a generic power ideal in two variables is equal to the right-hand side of (14), we can conclude our computation of the dimension of the secant variety of $X_{\mathbf{d}}$ in the binary case. This is the way Theorem 3.1 in [149] was proven.

In the following table, we resume the current state-of-the-art regarding secant varieties and Chow–Veronese varieties.

| **d** | s | d | n | References | $\dim \sigma_s(X_{\mathbf{d}})$ |
|---|---|---|---|---|---|
| $(d-1,1)$ | any | any | any | [141] | non-defective, except for (1) $d = 2$ and $2 \leq 2s < n$; (2) $d = 3$ and $n = 2, 3, 4$. |
| $(1,\ldots,1)$ | any | $d > 2$ any some numerical constraints | $3(s-1) < n$ 2 | [73] [146] [146,148] | non-defective non-defective, except for cases above |
| any | any 2 $\leq 2 \lfloor \frac{n}{3} \rfloor$ | any any any | 1 any any | [149] | non-defective, except for cases above |

## 4.5. Varieties of Reducible Forms

In 1954, Mammana [151] considered the variety of reducible plane curves and tried to generalize previous works by, among many others, C. Segre, Spampinato and Bordiga. More recently, in [147], the authors considered the varieties of reducible forms in full generality.

Let $\mathbf{d} = (d_1, \ldots, d_t) \vdash d$ be a partition of a positive integer $d$, i.e., $d_1 \geq \ldots \geq d_t$ are positive integers, which sum up to $d$ and $t \geq 2$. Inside the space of homogeneous polynomials of degree $d$, we define the variety of **d**-reducible forms as:

$$Y_{\mathbf{d}} = \{[F] \in \mathbb{P}S_d \mid F = G_1 \cdots G_t, \text{ where } \deg(G_i) = d_i\},$$

i.e., the image of the embedding:

$$\psi_{\mathbf{d}} : \mathbb{P}S_{d_1} \times \ldots \times \mathbb{P}S_{d_t} \longrightarrow \mathbb{P}S_d,$$
$$(G_1, \ldots, G_t) \mapsto G_1^{d_1} \cdots G_t^{d_t}.$$

Clearly, if $d = 2$, then $\mathbf{d} = (1,1)$, and $Y_{(1,1)}$ is just the Chow variety $X_{(1,1)}$. In general, we may see $Y_{\mathbf{d}}$ as the linear projection of the Segre variety $X_m$ inside $\mathbb{P}(S_{d_1} \otimes \cdots \otimes S_{d_t})$, where $m = (m_1, \ldots, m_t)$ with $m_i = \binom{d_i+n}{n} - 1$. Note that, if $\mathbf{d}, \mathbf{d}'$ are two partitions of $d$ such that $\mathbf{d}$ can be recovered from $\mathbf{d}'$ by grouping and summing some of entries, then we have the obvious inclusion $Y_{\mathbf{d}'} \subset Y_{\mathbf{d}}$. Therefore, if we define the variety of reducible forms as the union over all possible partitions $\mathbf{d} \vdash d$ of the varieties $Y_{\mathbf{d}}$, we can actually write:

$$Y = \bigcup_{k=1}^{\lfloor \frac{d}{2} \rfloor} Y_{(d-k,k)} \subset \mathbb{P}S_d.$$

In terms of additive decompositions, the study of varieties of reduced forms and their secant varieties is related to the notion of the strength of a polynomial, which was recently introduced by T. Ananyan and M. Hochster [152] and then generalized to any tensor in [153].

**Problem 6.** *Given a homogeneous polynomial* $F \in S_d$, *find the smallest length of an expression* $F = \sum_{i=1}^{r} G_{i,1} G_{i,2}$, *where* $1 \leq \deg(G_{i,j}) \leq d - 1$.

The answer to Problem 6 is called the strength of $F$, and we denote it by $S(F)$.

In [147], the authors gave a conjectural formula for the dimensions of all secant varieties $\sigma_s(Y_{\mathbf{d}})$ of the variety of $\mathbf{d}$-reducible forms for any partition $\mathbf{d}$ (see Conjecture 1.1 in [147]), and they proved it under certain numerical conditions (see Theorem 1.2 in [147]). These computations have been made by using the classical Terracini's lemma and relating the dimensions of these secants to the famous Fröberg's conjecture on the Hilbert series of generic forms.

The variety of reducible forms is not irreducible and the irreducible component with biggest dimension is the one corresponding to the partition $(d-1,1)$, i.e., $\dim Y = \dim Y_{(d-1,1)}$. Higher secant varieties of the variety of reducible forms are still reduced, but understanding which is the irreducible component with the biggest dimension is not an easy task. In Theorem 1.5 of [147], the authors proved that, if $2s \leq n - 1$, then the biggest irreducible component of $\sigma_s(Y)$ is $\sigma_s(Y_{(d-1,1)})$, i.e., $\dim \sigma_s(Y) = \dim \sigma_s(Y_{(d-1,1)})$, and together with the aforementioned Theorem 1.2 of [147], this allows us to compute the dimensions of secant varieties of varieties of reducible forms and answer Problem 6 under certain numerical restrictions (see Theorem 7.4 [147]).

In conclusion, we have that Problem 6 is answered in the following cases:

1. any binary form ($n = 2$), where $S(F) = 1$,

    since every binary form is a product of linear forms;
2. generic quadric ($d = 2$), where $S(F) = \lfloor \frac{n}{2} \rfloor + 1$,

    since it forces $\mathbf{d} = (1,1)$, which is solved by Corollary 4 (2);
3. generic ternary cubic ($n = 2$, $d = 3$), where $S(F) = 2$,

    since $Y_{(2,1)}$ is seven-dimensional and non-degenerate inside $\mathbb{P}^9 = \mathbb{P}S_3$, then $\sigma_2(Y_{(2,1)})$ cannot be eight-dimensional; otherwise, we get a contradiction by one of the classical Palatini's lemmas, which states that if $\dim \sigma_{s+1}(X) = \dim \sigma_s(X) + 1$, then $\sigma_{s+1}(X)$ must be a linear space [2].

### 4.6. Varieties of Powers

Another possible generalization of the classical Waring problem for forms is given by the following.

**Problem 7.** *Given a homogeneous polynomial* $F \in S_d$ *and a positive divisor* $k > 1$ *of* $d$, *find the smallest length of an expression* $F = \sum_{i=1}^{r} G_i^k$.

The answer to Problem 7 is called the $k$-th Waring rank, or simply $k$-th rank, of $F$, and we denote it $R_d^k(F)$. In this case, we need to consider the variety of $k$-th powers, i.e.,

$$V_{k,d} = \{[G^k] \in \mathbb{P}S_d \mid G \in S_{d/k}\}.$$

That is, the variety obtained by considering the composition:

$$\pi \circ \nu_k : \mathbb{P}S_{d/k} \to \mathbb{P}S^k(S_{d/k}) \dashrightarrow \mathbb{P}S_d, \tag{23}$$

where:

1. if $W = S_{d/k}$, then $\nu_k$ is the $k$-th Veronese embedding of $\mathbb{P}W$ in $\mathbb{P}S^kW$;
2. if we consider the standard monomial basis $w_\alpha = \mathbf{x}^\alpha$ of $W$, i.e., $|\alpha| = d/k$, then $\pi$ is the linear projection from $\mathbb{P}S^kW$ to $\mathbb{P}S_d$ induced by the substitution $w_\alpha \mapsto \mathbf{x}^\alpha$. In particular, we have that the center of the projection $\pi$ is given by the homogeneous part of degree $k$ of the ideal of the Veronese variety $\nu_d(\mathbb{P}^n)$.

Problem 7 was considered by Fröberg, Shapiro and Ottaviani [154]. Their main result was that, if $F$ is generic, then:

$$R_d^k(F) \leq k^n, \tag{24}$$

i.e., the $k^n$-th secant variety of $V_{k,d}$ fills the ambient space. This was proven by Terracini's lemma. Indeed, for any $G, H \in S_{d/k}$, we have that:

$$\left.\frac{d}{dt}\right|_{t=0} (G + tH)^k = kG^{k-1}H;$$

therefore, we obtain that:

$$T_{[G^k]} V_{k,d} = \mathbb{P}\left(\langle [G^{k-1}H] \mid H \in S_{d/k} \rangle\right),$$

and, by Terracini's lemma (Lemma 1), if $Q$ is a generic point on $\langle [G_1^k], \ldots, [G_s^k] \rangle$, where the $G_i$'s are generic forms of degree $d/k$, then:

$$T_Q \sigma_s V_{k,d} = \mathbb{P}\left((G_1^{k-1}, \ldots, G_s^{k-1})_d\right). \tag{25}$$

In [154] (Theorem 9), the authors showed that the family:

$$G_{i_1,\ldots,i_n} = (x_0 + \zeta^{i_1} x_1 + \ldots + \zeta^{i_n} x_n)^{d/k} \in S_{d/k}, \quad \text{for } i_1, \ldots, i_n \in \{0, \ldots, k-1\},$$

where $\zeta^k = 1$, is such that:

$$(G_1^{k-1}, \ldots, G_s^{k-1})_d = S_d.$$

In this way, they showed that $\sigma_{k^n}(V_{k,d})$ fills the ambient space. A remarkable fact with the upperbound (24) is that it is independent of the degree of the polynomial, but it only depends on the power $k$. Now, the naive lower bound due to parameter counting is $\left\lceil \frac{\dim S_d}{\dim S_{d/k}} \right\rceil = \left\lceil \frac{\binom{n+d}{n}}{\binom{n+d/k}{n}} \right\rceil$, which tends to $k^n$ when $d$ runs to infinity.

In conclusion, we obtain that the main result of [154] is resumed as follows.

**Theorem 25** ([154] (Theorem 4)). *Let $F$ be a generic form of degree $d$ in $n+1$ variables. Then,*

$$R_d^k(F) \leq k^n.$$

*If $d \gg 0$, then the latter bound is sharp.*

This result gives an asymptotic answer to Problem 7, but, in general, it is not known for which degree $d$ the generic $k$-th Waring rank starts to be equal to $k^n$, and it is not known what happens in lower degrees.

We have explained in (23) how to explicitly see the variety of powers $V_{k,d}$ as a linear projection of a Veronese variety $X_{N,k} = \nu_k(\mathbb{P}^N)$, where $N = \binom{n+d/k}{n} - 1$. It is possible to prove that $\sigma_2(X_{N,k})$

does not intersect the base of linear projection, and therefore, $V_{k,d}$ is actually isomorphic to $X_{N,k}$. Unfortunately, higher secant varieties intersect non-trivially the base of the projection, and therefore, their images, i.e., the secant varieties of the varieties of powers, are more difficult to understand. However, computer experiments suggest that the dimensions are preserved by the linear projection; see [155] (Section 4) for more details about these computations (a Macaulay2 script with some examples is available in the ancillary files of the arXiv version of [155]). In other words, it seems that we can use the Alexander–Hirschowitz theorem to compute the dimensions of secant varieties of varieties of powers and provide an answer to Problem 7. More on this conjecture is explained in [155].

**Conjecture 11** ([155] (Conjecture 1.2)). *Let F be a generic form of degree d in n + 1 variables. Then,*

$$R_d^k(F) = \begin{cases} \min\{s \geq 1 \mid s\binom{n+d/2}{n} - \binom{s}{2} \geq \binom{n+d}{n}\} & \text{for } k = 2; \\ \min\{s \geq 1 \mid s\binom{n+d/k}{n} \geq \binom{n+d}{n}\} & \text{for } k \geq 3. \end{cases}$$

**Remark 19.** *The latter conjecture claims that for $k \geq 3$, the correct answer is given by the direct parameter count. For $k = 2$, we have that secant varieties are always defective. This is analogous to the fact that secant varieties to the two-fold Veronese embeddings are defective. Geometrically, this is motivated by Terracini's lemma and by the fact that:*

$$T_{[G^2]}V_{2,d} \cap T_{[H^2]}V_{2,d} = [GH],$$

*and not empty, as expected.*

**Example 19.** *Here, we explain how the binary case can be treated; see [155] (Theorem 2.3). By (25), the computation of the dimension of secant varieties of varieties of powers reduces to the computation of dimensions of homogeneous parts of particular ideals, i.e., their Hilbert functions. This relates Problem 7 to some variation of Fröberg's conjecture, which claims that the ideal $(G_1^k, \ldots, G_s^k)$, where the $G_i$'s are generic forms of degrees at least two, has Hilbert series equal to the right-hand side of (14); see [156]. In the case of binary forms, by semicontinuity, we may specialize the $G_i$'s to be powers of linear forms. In this way, we may employ the result of [150], which claims that power ideals in two variables satisfy Fröberg–Iarrobino's conjecture, i.e., (14) is actually an equality, and we conclude the proof of Conjecture 11 in the case of binary forms.*

By using an algebraic study on the Hilbert series of ideals generated by powers of forms, we have a complete answer to Problem 7 in the following cases (see [155]):

1. binary forms ($n = 1$), where:

$$R_d^k(F) = \left\lceil \frac{d+1}{d/k+1} \right\rceil;$$

2. ternary forms as sums of squares ($n = 2, k = 2$), where:

$$R_d^2(F) = \left\lceil \frac{\binom{d+2}{2}}{\binom{d/2+2}{2}} \right\rceil,$$

except for $d = 1, 3, 4$, where $R_d^2(F) = \left\lceil \frac{\binom{d+2}{2}}{\binom{d/2+2}{2}} \right\rceil + 1.$;

3. quaternary forms as sums of squares ($n = 3, k = 2$), where:

$$R_d^2(F) = \left\lceil \frac{\binom{d+3}{3}}{\binom{d/2+3}{3}} \right\rceil,$$

except for $d = 1, 2$, where $R_d^2(F) = \left\lceil \frac{\binom{d+3}{3}}{\binom{d/2+3}{3}} \right\rceil + 1.$

## 5. Beyond Dimensions

We want to present here, as a natural final part of this work, a list of problems about secant varieties and decomposition of tensors, which are different from merely trying to determine the dimensions of the varieties $\sigma_s(X)$ for the various $X$ we have considered before. We will consider problems such as determining maximal possible ranks, finding bounds or exact values on ranks of given tensor, understanding the set of all possible minimal decompositions of a given tensor, finding equations for the secant varieties or studying what happens when working over $\mathbb{R}$. The reader should be aware of the fact that there are many very difficult open problems around these questions.

### 5.1. Maximum Rank

A very difficult and still open problem is the one that in the Introduction we have called the "little Waring problem". We recall it here.

Which is the minimum integer $r$ such that any form can be written as a sum of $r$ pure powers of linear forms?

This corresponds to finding the maximum rank of a form of certain degree $d$ in a certain number $n+1$ of variables.

To our knowledge, the best general achievement on this problem is due to Landsberg and Teitler, who in [76] (Proposition 5.1) proved that the rank of a degree $d$ form in $n+1$ variables is smaller than or equal to $\binom{n+d}{d} - n$. Unfortunately, this bound is sharp only for $n = 1$ if $d \geq 2$ (binary forms); in fact, for example, if $n = 2$ and $d = 3,4$, then the maximum ranks are known to be $5 < \binom{2+3}{2} - 3 = 7$ and $7 < \binom{2+4}{2} - 4 = 11$, respectively; see [28] (Theorem 40 and Theorem 44). Another general bound has been obtained by Jelisiejew [157], who proved that, for $F \in S^d\mathbb{k}^{n+1}$, we have $R_{\text{sym}}(F) \leq \binom{n+d-1}{d-1} - \binom{n+d-5}{d-3}$. Again, this bound is not sharp for $n \geq 2$. Another remarkable result is the one due to Blekherman and Teitler, who proved in [158] (Theorem 1) that the maximum rank is always smaller than or equal to twice the generic rank.

**Remark 20.** *The latter inequality, which has a very short and elegant proof, holds also between maximal and generic X-ranks with respect to any projective variety X.*

In a few cases in small numbers of variables and small degrees, exact values of maximal ranks have been given. We resume them in the following table.

|  | $d$ | $n$ | Maximal Rank | Ref. |
|---|---|---|---|---|
| binary forms | any | 1 | $d$ | classical, [159] |
| quadrics | 2 | any | $n+1$ | classical |
| plane cubics | 3 | 2 | 5 | [76,160] |
| plane quartics | 4 | 2 | 7 | [161,162] |
| plane quintics | 5 | 2 | 10 | [163,164] |
| cubic surfaces | 3 | 3 | 7 | [160] |
| cubic hypersurfaces | 3 | any | $\leq \lfloor \frac{d^2+6d+1}{4} \rfloor$ | [165] |

We want to underline the fact that it is very difficult to find examples of forms having high rank, in the sense higher than the generic rank. Thanks to the complete result on monomials in [166] (see Theorem 31), we can easily see that in the case of binary and ternary forms, we can find monomials having rank higher than the generic one. However, for higher numbers of variables, monomials do not provide examples of forms of high rank. Some examples are given in [164], and the spaces of forms of high rank are studied from a geometric point of view in [167].

### 5.2. Bounds on the Rank

In the previous subsection, we discussed the problem of finding the maximal rank of a given family of tensors. However, for a given specific tensor $T$, it is more interesting, and relevant, to find explicit bounds on the rank of $T$ itself. For example, by finding good lower and upper bounds on the rank of $T$, one can try to compute actually the rank of $T$ itself, but usually, the maximal rank is going to be too large to be useful in this direction.

One typical approach to find upper bounds is very explicit: by finding a decomposition of $T$. In the case of symmetric tensors, that is in the case of homogeneous polynomials, the apolarity lemma (Lemma 5) is an effective tool to approach algebraically the study of upper bounds: by finding the ideal of a reduced set of points $\mathbb{X}$ inside $F^\perp$, we bound the rank of $F$ from above by the cardinality of $\mathbb{X}$.

**Example 20.** For $F = x_0 x_1 x_2$, in standard notation, we have $F^\perp = (y_0^2, y_1^2, y_2^2)$, and we can consider the complete intersection set of four reduced points $\mathbb{X}$ whose defining ideal is $(y_1^2 - y_0^2, y_2^2 - y_0^2)$, and thus, the rank of $F$ is at most four. Analogously, if $F = x_0 x_1^2 x_2^3$, we have $F^\perp = (y_0^2, y_1^3, y_3^4)$, and we can consider the complete intersection of 12 points defined by $(y_1^3 - y_0^3, y_2^4 - y_0^4)$.

Other upper bounds have been given by using different notions of rank.

**Definition 24.** We say that a scheme $Z \subset \mathbb{P}^N$ is curvilinear if it is a finite union of schemes of the form $\mathcal{O}_{\mathcal{C}_i, P_i} / \mathfrak{m}_{P_i}^{e_i}$, for smooth points $P_i$ on reduced curves $\mathcal{C}_i \subset \mathbb{P}^N$. Equivalently, the tangent space at each connected component of $Z$ supported at the $P_i$'s has Zariski dimension $\leq 1$. The curvilinear rank $R_{\text{curv}}(F)$ of a degree $d$ form $F$ in $n+1$ variables is:

$$R_{\text{curv}}(F) := \min \{ \deg(Z) \mid Z \subset X_{n,d}, Z \text{ curvilinear}, [F] \in \langle Z \rangle \}.$$

With this definition, in [168] (Theorem 1), it is proven that the rank of an $F \in S^d \Bbbk^{n+1}$ is bounded by $(R_{\text{curv}}(F) - 1)d + 2 - R_{\text{curv}}(F)$. This result is sharp if $R_{\text{curv}}(F) = 2, 3$; see ([28] Theorems 32 and 37). Another very related notion of rank is the following; see [82,169].

**Definition 25.** We define the smoothable rank of a form $F \in S^d \Bbbk^{n+1}$ as:

$$R_{\text{smooth}}(F) := \min \left\{ \deg(Z) \;\middle|\; \begin{array}{l} Z \text{ is a limit of smooth schemes } Z_i \text{ such that} \\ Z, Z_i \subset X_{n,d}, \text{ are zero-dim schemes with } \deg(Z_i) = \deg(Z), \\ \text{and } [F] \in \langle Z \rangle \end{array} \right\}.$$

In ([168] Section 2), it is proven that if $F$ is a ternary form of degree $d$, then $R_{\text{sym}}(F) \leq (R_{\text{smooth}}(F) - 1)d$. We refer to [82] for a complete analysis on the relations between different notions of ranks.

The use of the apolarity lemma (Lemma 5) to obtain lower bounds to the symmetric-rank of a homogeneous polynomial was first given in [170].

**Theorem 26** ([170] (Proposition 1)). *If the ideal $F^\perp$ is generated in degree $t$ and $\mathbb{X}$ is a finite scheme apolar to $F$, that is $I_\mathbb{X} \subset F^\perp$, then:*

$$\frac{1}{t} \deg F^\perp \leq \deg \mathbb{X}.$$

This result is enough to compute the rank of the product of variables.

**Example 21.** For $F = x_0 x_1 x_2$, Theorem 26 yields:

$$\frac{1}{2} 8 \leq \deg \mathbb{X}.$$

If we assume $\mathbb{X}$ to be reduced, i.e., $\deg \mathbb{X} = |\mathbb{X}|$, by the apolarity lemma, we get $R_{\text{sym}}(F) \geq 4$, and thus, by Example 20, the rank of $F$ is equal to four. However, for the monomial $x_0 x_1^2 x_2^3$, we get the lower bound of six, which does not allow us to conclude the computation of the rank, since Example 20 gives us 12 as the upper bound.

To solve the latter case, we need a more effective use of the apolarity lemma in order to produce a better lower bound for the rank; see [166,171].

**Theorem 27** ([171] (Corollary 3.4)). *Let F be a degree d form, and let $e > 0$ be an integer. Let I be any ideal generated in degree e, and let G be a general form in I. For $s \gg 0$, we have:*

$$R_{sym}(F) \geq \frac{1}{e}\sum_{i=0}^{s} \mathrm{HF}\left(R/(F^{\perp} : I + (G)), i\right).$$

A form for which there exists a positive integer $e$ such that the latter lower bound is actually sharp is called *e-computable*; see [171]. Theorem 27 was first presented in [166] in the special case of $e = 1$: this was the key point to prove Theorem 31 on the rank of monomials, by showing that monomials are one-computable. In order to give an idea of the method, we give two examples: in the first one, we compute the rank of $x_0 x_1^2 x_2^3$ by using one-computability, while in the second one, we give an example in which it is necessary to use two-computability; see [166,171].

**Example 22.** *Consider again $F = x_0 x_1^2 x_2^3$. We use Theorem 27 with $e = 1$, $G = y_0$ and $I = (y_0)$. Note that:*

$$F^{\perp} : I + (G) = (y_0^2, y_1^3, y_2^4) : (y_0) + (y_0) = (y_0, y_1^3, y_2^4).$$

*This yields to:*

$$R_{sym}(F) \geq \sum_{i=0}^{s} \mathrm{HF}\left(R/(y_0, y_1^3, y_2^4), i\right) = 12,$$

*since $\mathrm{HS}(R/(y_0, y_1^3, y_2^4), z) = 1 + 2z + 3z^2 + 3z^3 + 2z^4 + z^5$. Hence, by using Example 20, we conclude that the rank of F is actually 12.*

**Example 23.** *Consider the polynomial:*

$$F = x_0^{11} - 22x_0^9 x_1^2 + 33 x_0^7 x_1^4 - 22 x_0^9 x_2^2 + 396 x_0^7 x_1^2 x_2^2 - 462 x_0^5 x_1^4 x_2^2 + \\ 33 x_0^7 x_2^4 - 462 x_0^5 x_1^2 x_2^4 + 385 x_0^3 x_1^4 x_2^4,$$

*we show that F is two-computable and $R_{sym}(F) = 25$. By direct computation, we get:*

$$F^{\perp} = ((y_0^2 + y_1^2 + y_2^2)^2, G_1, G_2),$$

*where $G_1 = y_1^5 + y_2(y_0^2 + y_1^2 + y_2^2)^2$ and $G_2 = y_2^5 + y_0(y_0^2 + y_1^2 + y_2^2)^2$.*
*Hence, by (27), we get:*

$$R_{sym}(F) \geq \frac{1}{2}\sum_{i=0}^{\infty} \mathrm{HF}(T/(F^{\perp} : (y_0^2 + y_1^2 + y_2^2) + (y_0^2 + y_1^2 + y_2^2)), i) = 25.$$

*Moreover, the ideal $(G_1, G_2) \subset F^{\perp}$ is the ideal of 25 distinct points, and thus, the conclusion follows. It can be shown that F is not one-computable; see [171] (Example 4.23).*

Another way to find bounds on the rank of a form is by using the rank of its derivatives. A first easy bound on the symmetric-rank of a homogeneous polynomial $F \in \mathbb{k}[x_0, \ldots, x_n]$ (where $\mathbb{k}$ is a characteristic zero field) is directly given by the maximum between the symmetric-ranks of its derivatives; indeed, if $F = \sum_{i=1}^{r} L_i^d$, then, for any $j = 0, \ldots, n$,

$$\frac{\partial F}{\partial x_j} = \sum_{i=1}^{r} \frac{\partial L_i^d}{\partial x_j} = (d-1)\sum_{i=1}^{r} \frac{\partial L_i}{\partial x_j} L_i^{d-1}. \tag{26}$$

A more interesting bound is given in [172].

**Theorem 28** ([172] (Theorem 3.2)). *Let $1 \leq p \leq n$ be an integer, and let $F \in \Bbbk[x_0, \ldots, x_n]$ be a form, where $\Bbbk$ is a characteristic zero field. Set $F_k = \frac{\partial F}{\partial x_k}$, for $0 \leq k \leq n$. If:*

$$\mathrm{rk}_{\mathrm{sym}}\left(F_0 + \sum_{k=1}^{n} \lambda_k F_k\right) \geq m,$$

*for all $\lambda_k \in \Bbbk$, and if the forms $F_1, F_2, \ldots, F_p$ are linearly independent, then:*

$$\mathrm{rk}_{\mathrm{sym}}(F) \geq m + p.$$

The latter bound was lightly improved in [173] (Theorem 2.3).

Formula (26) can be generalized to higher order differentials. As a consequence, for any $G \in S^j V^*$, with $1 \leq j \leq d-1$, we have that $R_{\mathrm{sym}}(F) \geq R_{\mathrm{sym}}(G \circ F)$, and in particular, if $F \in \langle L_1^d, \ldots, L_r^d \rangle$, we have that $G \circ F \in \langle L_1^{d-j}, \ldots, L_r^{d-j} \rangle$. Since this holds for any $G \in S^j V^*$, we conclude that the image of the $(j, d-j)$-th catalecticant matrix is contained in $\langle L_1^{d-j}, \ldots, L_r^{d-j} \rangle$. Therefore,

$$R_{\mathrm{sym}}(F) \geq \dim_\Bbbk(\mathrm{Imm}\, Cat_{j,d-j}(F)) = \mathrm{rk}\, Cat_{j,d-j}(F). \tag{27}$$

The latter bound is very classical and goes back to Sylvester. By using the geometry of the hypersurface $V(F)$ in $\mathbb{P}^n$, it can be improved; see [76].

**Theorem 29** ([76] (Theorem 1.3)). *Let $F$ be a degree $d$ form with $n+1$ essential variables. Let $1 \leq j \leq d-1$. Use the convention that $\dim \emptyset = -1$. Then, the symmetric-rank of $F$ is such that:*

$$R_{\mathrm{sym}}(F) \geq \mathrm{rk}\, Cat_{j,d-j}(F) + \dim \Sigma_j(F) + 1,$$

*where $Cat_{j,d-j}(F)$ is the $(j, d-j)$-th catalecticant matrix of $F$ and:*

$$\Sigma_j(F) = \left\{ P \in V(F) \subset \mathbb{P}V : \frac{\partial^\alpha F}{\partial x^\alpha}(P) = 0, \quad \forall |\alpha| \leq j \right\}.$$

The latter result has been used to find lower bounds on the rank of the determinant and the permanent of the generic square matrix; see [76] (Corollary 1.4).

The bound (27) given by the ranks of catalecticant matrices is a particular case of a more general fact, which holds for general tensors.

Given a tensor $T \in V_1 \otimes \ldots \otimes V_d$, there are several ways to view it as a linear map. For example, we can "reshape" it as a linear map $V_i^* \to V_1 \otimes \ldots \otimes \widehat{V_i} \otimes \ldots \otimes V_d$, for any $i$, or as $V_i^* \otimes V_j^* \to V_1 \otimes \ldots \otimes \widehat{V_i} \otimes \ldots \otimes \widehat{V_j} \otimes \ldots \otimes V_d$, for any $i \neq j$, or more in general, as:

$$V_{i_1}^* \otimes V_{i_2}^* \otimes \ldots \otimes V_{i_s}^* \to V_1 \otimes \ldots \otimes \widehat{V_{i_1}} \otimes \ldots \otimes \widehat{V_{i_s}} \otimes \ldots \otimes V_d, \tag{28}$$

for any choice of $i_1, \ldots, i_s$. All these ways of reshaping the tensor are called flattenings. Now, if $T$ is a tensor of rank $r$, then all its flattenings have (as matrices) rank at most $r$. In this way, the ranks of the flattenings give lower bounds for the rank of a tensor, similarly as the ranks of catalecticant matrices gave lower bounds for the symmetric-rank of a homogeneous polynomial.

We also point out that other notions of flattening, i.e., other ways to construct linear maps starting from a given tensor, have been introduced in the literature, such as Young flattenings (see [85]) and Koszul flattenings (see [84]). These were used to find equations of certain secant varieties of Veronese and other varieties and to provide algebraic algorithms to compute decompositions.

We conclude this section with a very powerful method to compute lower bounds on ranks of tensors: the so-called substitution method. In order to ease the notation, we report the result in the case $T \in V_1 \otimes V_2 \otimes V_3$, with $\dim_\Bbbk V_i = n_i$. For a general result, see [174] (Appendix B).

**Theorem 30** (The substitution method ([174] Appendix B) or ([175] Section 5.3)). *Let $T \in V_1 \otimes V_2 \otimes V_3$. Write $T = \sum_{i=1}^{n_1} e_i \otimes T_i$, where the $e_i$'s form a basis of $V_1$ and the $T_i$'s are the corresponding "slices" of the tensor. Assume that $M_{n_1} \neq 0$. Then, there exist constants $\lambda_1, \ldots, \lambda_{n_1-1}$ such that the tensor:*

$$T' = \sum_{i=1}^{n_1-1} e_i \otimes (T_i - \lambda_i T_{n_1}) \in \Bbbk^{n_1-1} \otimes V_2 \otimes V_3,$$

*has rank at most $R(T) - 1$. If $T_{n_1}$ is a matrix of rank one, then equality holds.*

Roughly speaking, this method is applied in an iterative way, with each of the $V_i$'s playing the role of $V_1$ in the theorem, in order to reduce the tensor to a smaller one whose rank we are able to compute. Since, in the theorem above, $R(T) \geq R(T') + 1$, at each step, we get a plus one on the lower bound. For a complete description of this method and its uses, we refer to [175] (Section 5.3).

A remarkable use of this method is due to Shitov, who recently gave counterexamples to very interesting conjectures such as Comon's conjecture, on the equality between the rank and symmetric-rank of a symmetric tensor, and Strassen's conjecture, on the additivity of the tensor rank for sums of tensors defined over disjoint subvector spaces of the tensor space.

We will come back with more details on Strassen's conjecture, and its symmetric version, in the next section. We spend a few words more here on Comon's conjecture.

Given a symmetric tensor $F \in S^d V \subset V^{\otimes d}$, we may regard it as a tensor, forgetting the symmetries, and we could ask for its tensor rank, or we can take into account its symmetries and consider its symmetric-rank. Clearly,

$$R(F) \leq R_{\text{sym}}(F). \tag{29}$$

The question raised by Comon asks if whether such an inequality is actually an equality. Affirmative answers were given in several cases (see [176–180]). In [181], Shitov found an example (a cubic in 800 variables) where the inequality (29) is strict. As the author says, unfortunately, no symmetric analogs of this substitution method are known. However, a possible formulation of such analogs, which might lead to a smaller case where (29) is strict, was proposed.

**Conjecture 12** ([181] (Conjecture 7)). *Let $F, G \in S = \Bbbk[x_0, \ldots, x_n]$ of degree $d, d-1$, respectively. Let $L$ be a linear form. Then,*

$$R_{\text{sym}}(F + LG) \geq d + \min_{L' \in S_1} R_{\text{sym}}(F + L'G).$$

**Remark 21.** *A symmetric tensor $F \in S^d V$ can be viewed as a partially-symmetric tensor in $S^{d_1} V \otimes \ldots \otimes S^{d_m} V$, for any $\underline{d} = (d_1, \ldots, d_m) \in \mathbb{N}^m$ such that $d_1 + \ldots + d_m = d$. Moreover, if $\underline{d}' = (d'_1, \ldots, d'_{m'}) \in \mathbb{N}^{m'}$ is a refinement of $\underline{d}$, i.e., there is some grouping of the entries of $\underline{d}'$ to get $\underline{d}$, then we have:*

$$R_{\text{sym}}(F) \geq R_{\underline{d}}(F) \geq R_{\underline{d}'}(F), \tag{30}$$

which is a particular case of (29). In the recent paper [139], the authors investigated the partially-symmetric version of Comon's question, i.e., the question if, for a given $F$, (30) is an equality or not. Their approach consisted of bounding from below the right-hand side of (30) with the simultaneous rank of its partial derivatives of some given order and then studying the latter by using classical apolarity theory (see also Remark 18). If such a simultaneous rank coincides with the symmetric-rank of $F$, then also all intermediate ranks are the same. In particular, for each case in which Comon's conjecture is proven to be true, then also all partially-symmetric tensors coincide. For more details, we refer to [139], where particular families of homogeneous polynomials are considered.

## 5.3. Formulae for Symmetric Ranks

In order to find exact values of the symmetric-rank of a given polynomial, we can use one of the available algorithms for rank computations; see Section 2.3. However, as we already mentioned, the algorithms will give an answer only if some special conditions are satisfied, and the answer will be only valid for that specific form. Thus, having exact formulae working for a family of forms is of the utmost interest.

Formulae for the rank are usually obtained by finding an explicit (a posteriori sharp) upper bound and then by showing that the rank cannot be less than the previously-found lower bound.

An interesting case is the one of monomials. The lower bound of (27) is used to obtain a rank formula for the complex rank of any monomial, similarly as in Example 22; i.e., given a monomial $F = \mathbf{x}^\alpha$, whose exponents are increasingly ordered, we have that $F^\perp = (y_0^{\alpha_0+1}, \ldots, y_n^{\alpha_n+1})$, and then, one has to:

1. first, as in Example 20, exhibit the set of points apolar to $F$ given by the complete intersection $(y_1^{\alpha_1} - y_0^{\alpha_1}, \ldots, y_n^{\alpha_n} - y_0^{\alpha_0})$; this proves that the right-hand side of (31) is an upper bound for the rank;
2. second, as in Example 22, use Theorem 27 with $e = 1$ and $G = y_0$ to show that the right-hand side of (31) is a lower bound for the rank.

**Theorem 31** ([166] (Proposition 3.1)). *Let $1 \leq \alpha_0 \leq \alpha_1 \ldots \leq \alpha_n$. Then,*

$$R_{\text{sym}}(\mathbf{x}^\alpha) = \frac{1}{\alpha_0 + 1} \prod_{i=0}^{n} (\alpha_i + 1). \tag{31}$$

Another relevant type of forms for which we know the rank is the one of reduced cubic forms. The reducible cubics, which are not equivalent to a monomial (up to change of variables), can be classified into three canonical forms. The symmetric complex rank for each one was computed, as the following result summarizes: the first two were first presented in [76], while the last one is in [172]. In particular, for all three cases, we have that the lower bound given by Theorem 28 is sharp.

**Theorem 32** ([172] (Theorem 4.5)). *Let $F \in \mathbb{C}[x_0, \ldots, x_n]$ be a form essentially involving $n + 1$ variables, which is not equivalent to a monomial. If $F$ is a reducible cubic form, then one and only one of the following holds:*

1. *$F$ is equivalent to:*

$$x_0(x_0^2 + x_1^2 + \ldots + x_n^2),$$

*and $R_{\text{sym}}(F) = 2n$.*
2. *$F$ is equivalent to:*

$$x_0(x_1^2 + x_2^2 + \ldots + x_n^2),$$

*and $R_{\text{sym}}(F) = 2n$.*
3. *$F$ equivalent to:*

$$x_0(x_0x_1 + x_2x_3 + x_4^2 + \ldots + x_n^2),$$

*and $R_{\text{sym}}(F) = 2n + 1$.*

Another way to find formulae for symmetric-ranks relies on a symmetric version of Strassen's conjecture on tensors. In 1973, Strassen formulated a conjecture about the additivity of the tensor ranks [182], i.e., given tensors $T_i, \ldots, T_s$ in $V^{\otimes d}$ defined over disjoint subvector spaces, then,

$$R(\sum_{i=1}^{s} T_i) = \sum_{i=1}^{s} R(T_i).$$

After a series of positive results (see, e.g., [183–185]), Shitov gave a proof of the existence of a counter-example to the general conjecture in the case of tensors of order three [186]. Via a clever use of the substitution method we introduced in the previous section, the author described a way to construct a counter-example, but he did not give an explicit one.

However, as the author mentioned is his final remarks, no counter example is known for the symmetric version of the conjecture that goes as follows: given homogeneous polynomials $F_1, \ldots, F_s$ in different sets of variables, then:

$$R_{\text{sym}}(\sum_{i=1}^{s} F_i) = \sum_{i=1}^{s} R_{\text{sym}}(F_i).$$

In this case, Strassen's conjecture is known to be true in a variety of situations. The case of sums of coprime monomials was proven in [166] (Theorem 3.2) via apolarity theory by studying the Hilbert function of the apolar ideal of $F = \sum_{i=1}^{s} F_i$. Indeed, it is not difficult to prove that:

$$F^\perp = \bigcap_{i=1}^{s} F_i^\perp + (F_i + \lambda_{i,j} F_j : i \neq j), \tag{32}$$

where the $\lambda_{i,j}$'s are suitable coefficients.

In this way, since apolar ideals of monomials are easy to compute, it is possible to express explicitly also the apolar ideal of a sum of coprime monomials. Therefore, the authors applied an analogous strategy as the one used for Theorem 31 (by using more technical algebraic computations) to prove that Strassen's conjecture holds for sums of monomials.

In [187], the authors proved that Strassen's conjecture holds whenever the summands are in either one or two variables. In [171,173], the authors provided conditions on the summands to guarantee that additivity of the symmetric-ranks holds. For example, in [173], the author showed that whenever the catalecticant bound (27) (or the lower bound given by Theorem 29) is sharp for all the $F_i$'s, then Strassen's conjecture holds, and the corresponding bound for $\sum_{i=1}^{s} F_i$ is also sharp.

A nice list of cases in which Strassen's conjecture holds was presented in [171]. This was done again by studying the Hilbert function of the apolar ideal of $F = \sum_{i=1}^{s} F_i$, computed as described in (32), and employing the bound given by Theorem 27.

**Theorem 33** ([171] (Theorem 6.1)). *Let $F = F_1 + \ldots + F_m$, where the degree d forms $F_i$ are in different sets of variables. If, for $i = 1, \ldots, m$, each $F_i$ is of one of the following types:*

- *$F_i$ is a monomial;*
- *$F_i$ is a form in one or two variables;*
- *$F_i = x_0^a(x_1^b + \ldots + x_n^b)$ with $a + 1 \geq b$;*
- *$F_i = x_0^a(x_1^b + x_2^b)$;*
- *$F_i = x_0^a(x_0^b + \ldots + x_n^b)$ with $a + 1 \geq b$;*
- *$F_i = x_0^a(x_0^b + x_1^b + x_2^b)$;*
- *$F_i = x_0^a G(x_1, \ldots, x_n)$ where $G^\perp = (H_1, \ldots, H_m)$ is a complete intersection and $a \leq \deg H_i$;*
- *$F_i$ is a Vandermonde determinant;*

*then Strassen's conjecture holds for F.*

## 5.4. Identifiability of Tensors

For simplicity, in this section we work on the field $\mathbb{C}$ of the complex numbers. Let us consider tensors in $\mathbf{V} = \mathbb{C}^{n_1+1} \otimes \ldots \otimes \mathbb{C}^{n_d+1}$. A problem of particular interest when studying minimal decompositions of tensors is to count how many there are.

**Problem 8.** *Suppose a given tensor $T \in \mathbf{V}$ has rank $r$, i.e., it can be written as $T = \bigoplus_{i=1}^{r} v_i^1 \otimes \ldots \otimes v_i^d$. When is it that such a decomposition is unique (up to permutation of the summands and scaling of the vectors)?*

This problem has been studied quite a bit in the last two centuries (e.g., see [12,188–190]), and it is also of interest with respect to many applied problems (e.g., see [191–193]). Our main references for this brief exposition are [194,195].

Let us begin with a few definitions.

**Definition 26.** *A rank-r tensor $T \in \mathbf{V}$ is said to be identifiable over $\mathbb{C}$ if its presentation $T = \bigoplus_{i=1}^{r} v_i^1 \otimes \ldots \otimes v_i^d$ is unique (up to permutations of the summands and scaling of the vectors).*

It is interesting to study the identifiability of a generic tensor of given shape and rank.

**Definition 27.** *We say that tensors in $\mathbf{V}$ are r-generically identifiable over $\mathbb{C}$ if identifiability over $\mathbb{C}$ holds in a Zariski dense open subset of the space of tensors of rank $r$. Moreover, we say that the tensors in $\mathbf{V}$ are generically identifiable if they are $r_g$-generically identifiable, where $r_g$ denotes the generic rank in $\mathbf{V}$.*

Let us recall that the generic rank for tensors $\mathbf{V}$ is the minimum value for which there is a Zariski open non-empty set $U$ of $\mathbf{V}$ where each point represents a tensor with rank $\leq r_g$; see Section 3. Considering $\mathbf{n} = (n_1, \ldots, n_d)$, let $X_\mathbf{n} \subset \mathbb{P}\mathbf{V}$ be the Segre embedding of $\mathbb{P}^{n_1} \times \ldots \times \mathbb{P}^{n_d}$. As we already said previously, if $r_g$ is the generic rank for tensors in $\mathbf{V}$, then $\sigma_{r_g}(X_\mathbf{n})$ is the first secant variety of $X_\mathbf{n}$, which fills the ambient space. Therefore, to say that the tensors in $\mathbf{V}$ are generically identifiable over $\mathbb{C}$ amounts to saying the following: let $r_g$ be the generic rank with respect to the Segre embedding $X_\mathbf{n}$ in $\mathbb{P}\mathbf{V}$; then, for the generic point $[T] \in \mathbb{P}\mathbf{V}$, there exists a unique $\mathbb{P}^{r-1}$, which is $r_g$-secant to $X_\mathbf{n}$ in $r_g$ distinct points and passes through $[T]$. The $r_g$ points of $X_\mathbf{n}$ gives (up to scalar) the $r_g$ summands in the unique (up to permutation of summands) minimal decomposition of the tensor $T$.

When $\sigma_r(X_\mathbf{n}) \neq \mathbb{P}\mathbf{V}$, i.e., the rank $r$ is smaller than the generic one (we can say that $r$ is sub-generic), then we have that the set of tensors $T \in \mathbf{V}$ with rank $r$ is $r$-generically identifiable over $\mathbb{C}$ if there is an open set $U$ of $\sigma_r(X_\mathbf{n})$ such that for the points $[T]$ in $U$, there exists a unique $\mathbb{P}^{r-1}$, which is $r$-secant in $r$ distinct points to $X_\mathbf{n}$ and passes through $[T]$.

Obviously, the same problem is interesting also when treating symmetric or skew-symmetric tensors, i.e., when $n_1 = \ldots = n_d = n$ and $T \in S^d(\mathbb{C}^{n+1}) \subset \mathbf{V}$ or $T \in \bigwedge^d(\mathbb{C}^{n+1})$. From a geometric point of view, in these cases, we have to look at Veronese varieties or Grassmannians, respectively, and their secant varieties, as we have seen in the previous sections.

Generic identifiability is quite rare as a phenomenon, and it has been largely investigated; in particular, we refer to [12,188,196–201]. As an example of how generic identifiability seldom presents itself, we can consider the case of symmetric tensors.

It is classically known that there are three cases of generic identifiability, namely:

- binary forms of odd degree ($n = 1$ and $d = 2t + 1$), where the generic rank is $t + 1$ [189];
- ternary quintics ($n = 2$ and $d = 5$), where the generic rank is seven [190];
- quaternary cubics ($n = 3$ and $d = 3$), where the generic rank is five [189].

Recently, Galuppi and Mella proved that these are the only generically identifiable cases when considering symmetric-ranks of symmetric tensors; see [202].

When we come to partially-symmetric tensors (which are related to Segre–Veronese varieties, as we described in Section 4.2), a complete classification of generically-identifiable cases is not known, but it is known that it happens in the following cases; see [203].

- $S^{d_1}\mathbb{C}^2 \otimes \ldots \otimes S^{d_t}\mathbb{C}^2$, with $d_1 \leq \ldots d_t$ and $d_1 + 1 \geq r_g$, where $r_g$ is the generic partially-symmetric-rank, i.e., forms of multidegree $(d_1, \ldots, d_t)$ in $t$ sets of two variables; here, the generic partially-symmetric-rank is $t+1$ [204];
- $S^2\mathbb{C}^{n+1} \otimes S^2\mathbb{C}^{n+1}$, i.e., forms of multidegree $(2,2)$ in two sets of $n+1$ variables; here, the generic partially-symmetric-rank is $n+1$ [205];
- $S^2\mathbb{C}^3 \otimes S^2\mathbb{C}^3 \otimes S^2\mathbb{C}^3 \otimes S^2\mathbb{C}^3$, i.e., forms of multidegree $(2,2,2,2)$ in four sets of three variables; here, the generic partially-symmetric-rank is four (this is a classical result; see also [203]);
- $S^2\mathbb{C}^3 \otimes S^3\mathbb{C}^3$, i.e., forms of multidegree $(2,3)$ in two sets of three variables; here, the generic partially-symmetric-rank is four [206];
- $S^2\mathbb{C}^3 \otimes S^2\mathbb{C}^3 \otimes S^4\mathbb{C}^3$, i.e., forms of multidegree $(2,2,4)$ in three sets of three variables; here, the generic partially-symmetric-rank is seven [203].

When considering $r$-generically identifiable tensors for sub-generic rank, i.e., for $r < r_g$, things change completely, in as much as we do expect $r$-generically-identifiability in this case. Again, the symmetric case is the best known; in [201] (Theorem 1.1), it was proven that every case where $r$ is a sub-generic rank and $\sigma_r(X_{n,d})$ has the expected dimension for the Veronese variety $X_{n,d}$, $r$-generically identifiability holds with the only following exceptions:

- $S^6\mathbb{C}^3$, i.e., forms of degree six in three variables, having rank nine;
- $S^4\mathbb{C}^4$, i.e., forms of degree four in four variables, having rank eight;
- $S^3\mathbb{C}^6$, i.e., forms of degree three in six variables, having rank nine.

In all the latter cases, the generic forms have exactly two decompositions.

Regarding generic identifiability for skew-symmetric tensors, there are not many studies, and we refer to [207].

It is quite different when we are in the defective cases, namely, when we want to study $r$-generic identifiability and the $r$-th secant variety is defective. In this case, non-identifiability is expected; in particular, we will have that the number of decompositions for the generic tensor parametrized by a point of $\sigma_r(Y)$ is infinite.

## 5.5. Varieties of Sums of Powers

Identifiability deals with the case in which tensors have a unique (up to permutation of the summands) decomposition. When the decomposition is not unique, what can we say about all possible decompositions of the given tensor? In the case of symmetric tensors, that is homogeneous polynomial, an answer is given by studying varieties of sums of powers, the so-called VSP, defined by Ranestad and Schreyer in [208].

**Definition 28.** *Let F be a form in $n+1$ variables having Waring rank $r$, and let $\mathrm{Hilb}_r \mathbb{P}^n$ be the Hilbert scheme of $r$ points in $\mathbb{P}^n$; we define:*

$$\mathrm{VSP}(F,r) = \overline{\{\mathbb{X} = \{P_1, \ldots, P_r\} \in \mathrm{Hilb}_r \mathbb{P}^n : I_\mathbb{X} = \wp_1 \cap \ldots \cap \wp_r \subset F^\perp\}}.$$

For example, when identifiability holds, VSP is just one single point. It is interesting to note that, even for forms having generic rank, the corresponding VSP might be quite big, as in the case of binary forms of even degree.

Using Sylvester's algorithm we have a complete description of VSP for binary forms, and it turns out to be always a linear space.

**Example 24.** *Consider the binary form $F = x_0^2 x_1^2$. Since $F^\perp = (y_0^3, y_1^3)$, by Sylvester's algorithm, we have that the rank of F is three. Moreover, by the apolarity lemma, we have that $\mathrm{VSP}(F, 3)$ is the projectivization of the vector space $W = \langle y_0^3, y_1^3 \rangle$ because the generic form in W has three distinct roots.*

In general, the study of VSPsis quite difficult, but rewarding: VSPs play an important role in classification work by Mukai see [209–211]. For a review of the case of general plane curves of degree up to ten, that is for general ternary forms of degree up to ten, a complete description is given in [208], including results from Mukai and original results. We summarize them in the following.

**Theorem 34** ([208] (Theorem 1.7)). *Let $F \in S^d \mathbb{C}^3$ be a general ternary cubic with $d = 2t - 2, 2 \leq t \leq 5$, then:*

$$\mathrm{VSP}\left(F, \binom{t+1}{2}\right) \simeq G(t, V, \eta) = \{ E \in G(t, V) \mid \wedge^2 E \subset \eta \}$$

*where V is a $2t + 1$-dimensional vector space and $\eta$ is a net of alternating forms $\eta : \wedge^2 V \to \mathbb{C}^3$ on V. Moreover:*

- *if F is a smooth plane conic section, then $\mathrm{VSP}(F, 3)$ is a Fano three-fold of index two and degree five in $\mathbb{P}^6$.*
- *if F is a general plane quartic curve, then $\mathrm{VSP}(F, 6)$ is a smooth Fano three-fold of index one and genus 12 with anti-canonical embedding of degree 22;*
- *if F is a general plane sextic curve, then $\mathrm{VSP}(F, 10)$ is isomorphic to the polarized K3-surface of genus 20;*
- *if F is a general plane octic curve, then $\mathrm{VSP}(F, 15)$ is finite of degree 16, i.e., consists of 16 points.*

Very often, for a given specific form, we do not have such a complete description, but at least, we can get some relevant information, for example about the dimension of the VSP: this is the case for monomials.

**Theorem 35** ([212] (Theorem 2)). *Let $F \in \mathbb{C}[x_0, \ldots, x_n]$ be a monomial $F = \mathbf{x}^\alpha$ with exponents $0 < \alpha_0 \leq \ldots \leq \alpha_n$. Let $A = \mathbb{C}[y_0, \ldots, y_n]/(y_1^{\alpha_1+1}, \ldots, y_n^{\alpha_n+1})$. Then, $\mathrm{VSP}(F, \mathrm{R}_{\mathrm{sym}}(F))$ is irreducible and:*

$$\dim \mathrm{VSP}(F, \mathrm{R}_{\mathrm{sym}}(F)) = \sum_{i=1}^{n} \mathrm{HF}(A; d_i - d_0).$$

A complete knowledge of $\mathrm{VSP}(F, r)$ gives us a complete control on all sums of powers decompositions of F involving r summands. Such a complete knowledge comes at a price: a complete description of the variety of sums of powers might be very difficult to obtain. However, even less complete information might be useful to have and, possibly, easier to obtain. One viable option is given by Waring loci as defined in [213].

**Definition 29.** *The Waring locus of a degree d form $F \in S^d V$ is:*

$$\mathcal{W}_F = \{[L] \in \mathbb{P}V \; : \; F = L^d + L_2^d + \ldots + L_r^d, \; r = \mathrm{R}_{\mathrm{sym}}(F)\},$$

*i.e., the space of linear form which appears in some minimal sums of powers decomposition of F. The forbidden locus of F is defined as the complement of the Waring locus of F, and we denote it by $\mathcal{F}_F$.*

**Remark 22.** *In this definition, the notion of essential variables has a very important role; see Remark 12. In particular, it is possible to prove that if $F \in \mathbb{C}[x_0, \ldots, x_n]$ has less than $n + 1$ essential variables, say $x_0, \ldots, x_m$, then for any minimal decomposition $F = \sum_{i=1}^{r} L_i$, the $L_i$'s also involve only the variables $x_0, \ldots, x_m$. For this reason, if in general, we have $F \in S^d V$, which has less than $\dim_\mathbb{C} V$ essential variables, say that $W \subset V$ is the linear span of a set of essential variables, then $\mathcal{W}_F \subset \mathbb{P}W$.*

In [213], the forbidden locus, and thus the Waring locus, of several classes of polynomials was computed. For example, in the case of monomials, we have the following description.

**Theorem 36** ([213] (Theorem 3.3)). *If $F = \mathbf{x}^\alpha$, such that the exponents are increasingly ordered and $m = \min_i\{\alpha_i = \alpha_0\}$, then:*

$$\mathcal{F}_F = V(y_0 \cdots y_m).$$

The study of Waring and forbidden loci can have a two-fold application. One is to construct step-by-step a minimal decomposition of a given form $F$: if $[L]$ belongs to $\mathcal{W}_F$, then there exists some coefficient $\lambda \in \mathbb{C}$ such that $F' = F + \lambda L^d$ has rank smaller than $F$; then, we can iterate the process by considering $\mathcal{W}_{F'}$. In [214], this idea is used to present an algorithm to find minimal decompositions of forms in any number of variables and of any degree of rank $r \leq 5$ (the analysis runs over all possible configuration of $r$ points in the space whose number of possibilities grows quickly with the rank); a Macaulay2 package implementing this algorithm can be found in the ancillary files of the arXiv version of [214]. A second possible application relates to the search for forms of high rank: if $[L]$ belongs to $\mathcal{W}_F$, then the rank of $F + \lambda L^d$ cannot increase as $\lambda$ varies, but conversely, if the rank of $F + \lambda L^d$ increases, then $[L]$ belongs to the forbidden locus of $F$. Unfortunately, it is not always possible to use elements in the forbidden locus to increase the rank of a given form; however, this idea can give a place to look for forms of high rank. For example, since $\mathcal{F}_{x_0 x_1 x_2} = V(y_0 y_1 y_2)$, that is the forbidden locus of the monomial $F = x_0 x_1 x_2$ is the union of the three coordinate lines of $\mathbb{P}^2$, the only possible way to make the rank of $F$ increase is to consider $F + \lambda L^3$ where $L$ is a linear form not containing at least one of the variables. However, as some computations can show, the rank of $F + \lambda L^3$ does not increase for any value of $\lambda$ and for any choice of $L$ in the forbidden locus.

Another family of polynomial for which we have a complete description of Waring and forbidden loci are binary forms.

**Theorem 37** ([213] (Theorem 3.5)). *Let $F$ be a degree $d$ binary form, and let $G \in F^\perp$ be an element of minimal degree. Then,*

- *if $\operatorname{rk}(F) < \binom{d+1}{2}$, then $\mathcal{W}_F = V(G)$;*
- *if $\operatorname{rk}(F) > \binom{d+1}{2}$, then $\mathcal{F}_F = V(G)$;*
- *if $\operatorname{rk}(F) = \binom{d+1}{2}$ and $d$ is even, then $\mathcal{F}_F$ is finite and not empty;*
  *if $\operatorname{rk}(F) = \binom{d+1}{2}$ and $d$ is odd, then $\mathcal{W}_F = V(G)$.*

**Example 25.** The result about Waring and forbidden loci of binary forms can be nicely interpreted in terms of rational normal curves. If $F = x_0 x_1^2$, then:

$$\mathcal{F}_F = \{[y_1^3]\},$$

and this means that any plane containing $[F]$ and $[y_1^3]$ is not intersecting the twisted cubic curve in three distinct points; indeed, the line spanned by $[F]$ and $[y_1^3]$ is tangent to the twisted cubic curve.

Other families of homogeneous polynomials for which we have a description of Waring and forbidden loci are quadrics [213] (Corollary 3.2) (in which case, the forbidden locus is given by a quadric) and plane cubics [213] (Section 3.4). We conclude with two remarks coming from the treatment of the latter case:

- in all the previous cases, the Waring locus of $F$ is always either closed or open. However, this is not true in general. In fact, for the cusp $F = x_0^3 + x_1^2 x_2$, we have that the Waring locus is:

$$\mathcal{W}_F = \{[y_0^3]\} \cup \{[(ay_1 + by_2)^3] \ : \ a, b \in \mathbb{C} \text{ and } b \neq 0\},$$

that is the Waring locus is given by two disjoint components: a point and a line minus a point; therefore, $\mathcal{W}_F$ is neither open nor closed in $\mathbb{P}^2$;

- since the space of minimal decompositions of forms of high rank is high dimensional, it is expected that the Waring locus is very large, and conversely, the forbidden locus is reasonably small. For example, the forbidden locus of the maximal rank cubic $F = x_0(x_1^2 + x_0x_2)$ is just a point, i.e.,

$$\mathcal{F}_F = \{[y_0^3]\}.$$

There is an open conjecture stating that, for any form $F$, the forbidden locus $\mathcal{F}_F$ is not empty.

*5.6. Equations for the Secant Varieties*

A very crucial problem is to find equations for the secant varieties we have studied in the previous sections, mainly for Veronese, Segre and Grassmann varieties. Notice that having such equations (even equations defining only set-theoretically the secant varieties in question) would be crucial in having methods to find border ranks of tensors.

5.6.1. Segre Varieties

Let us consider first Segre varieties; see also [108]. In the case of two factors, i.e., $X_\mathbf{n}$ with $\mathbf{n} = (n_1, n_2)$, the Segre variety, which is the image of the embedding:

$$\mathbb{P}^{n_1} \times \mathbb{P}^{n_2} \to X_\mathbf{n} \subset \mathbb{P}^N, \quad N = (n_1+1)(n_2+1) - 1,$$

corresponds to the variety of rank one matrices, and $\sigma_s(X_\mathbf{n})$ corresponds to the variety of rank $s$ matrices, which is defined by the ideal generated by the $(s+1) \times (s+1)$ minors of the generic $(n_1+1) \times (n_2+1)$ matrix, whose entries are the homogenous coordinates of $\mathbb{P}^N$. In this case, the ideal is rather well understood; see, e.g., [215] and also the extensive bibliography given in the book of Weyman [216].

We will only refer to a small part of this vast subject, and we recall that the ideal $I_{\sigma_s(X_\mathbf{n})}$ is a perfect ideal of height $(n_1 + 1 - (s+1) - 1) \times (n_2 + 1 - (s+1) - 1) = (n_1 - s - 1) \times (n_2 - s - 1)$ in the polynomial ring with $N+1$ variables, with a very well-known resolution: the Eagon–Northcott complex. It follows from this description that all the secant varieties of the Segre embeddings of a product of two projective spaces are arithmetically Cohen–Macaulay varieties. Moreover, from the resolution, one can also deduce the degree, as well as other significant geometric invariants, of these varieties. A determinantal formula for the degree was first given by Giambelli. There is, however, a reformulation of this result, which we will use (see, e.g., [1] (p. 244) or [217] (Theorem 6.5)), where this lovely reformulation of Giambelli's Formula is attributed to Herzog and Trung:

$$\deg(\sigma_s(X_{(n_1,n_2)})) = \prod_{i=0}^{n_1-s} \frac{\binom{n_2+1+i}{s}}{\binom{s+i}{s}}.$$

Let us now pass to the case of the Segre varieties with more than two factors. Therefore, let $X_\mathbf{n} \subset \mathbb{P}^N$ with $\mathbf{n} = (n_1, \ldots, n_t)$, $N = \prod_{i=1}^{t}(n_i+1) - 1$ and $t \geq 3$, where we usually assume that $n_1 \geq \ldots \geq n_t$.

If we let $T$ be the generic $(n_1+1) \times \ldots \times (n_t+1)$ tensor whose entries are the homogeneous coordinates in $\mathbb{P}^N$, then it is well known that the ideal of $X_\mathbf{n}$ has still a determinantal representation, namely it is generated by all the $2 \times 2$ "minors" of $T$, that is the $2 \times 2$ minors of the flattenings of $T$. It is natural to ask if the flattenings can be used also to find equations of higher secant varieties of $X_\mathbf{n}$. If we split $1, \ldots, t$ into two subsets, for simplicity say $1, \ldots, \ell$ and $\ell+1, \ldots, t$, then we can form the composition:

$$\nu_{(n_1,\ldots,n_\ell)} \times \nu_{(n_{\ell+1},\ldots,n_t)} : (\mathbb{P}^{n_1} \times \ldots \times \mathbb{P}^{n_\ell}) \times (\mathbb{P}^{n_{\ell+1}} \times \ldots \times \mathbb{P}^{n_t}) \to \mathbb{P}^a \times \mathbb{P}^b,$$

where $a = \prod_{i=1}^{\ell}(n_i+1) - 1$, $b = \prod_{i=\ell+1}^{t}(n_i+1) - 1$, followed by:

$$\nu_{1,1} : \mathbb{P}^a \times \mathbb{P}^b \to \mathbb{P}^N, \quad N \text{ as above}.$$

Clearly $X_\mathbf{n} \subset \nu_{1,1}(\mathbb{P}^a \times \mathbb{P}^b)$, and hence, $\sigma_s(X_\mathbf{n}) \subset \sigma_s(\nu_{1,1}(\mathbb{P}^a \times \mathbb{P}^b))$.

Thus, the $(s+1) \times (s+1)$ minors of the matrix associated with the embedding $\nu_{1,1}$ will all vanish on $\sigma_s(X_\mathbf{n})$. That matrix, written in terms of the coordinates of the various $\mathbb{P}^{n_i}$, is what we have called a flattening of the tensor $T$.

As we have seen in (28), we can perform a flattening of $T$ for every partition of $1, \ldots, t$ into two subsets. The $(s+1) \times (s+1)$ minors of all of these flattenings will give us equations that vanish on $\sigma_s(X_\mathbf{n})$. In [90], it was conjectured that, at least for $s = 2$, these equations are precisely the generators for the ideal $I_{\sigma_2(X_\mathbf{n})}$ of $\sigma_2(X_\mathbf{n})$. The conjecture was proven in [218] for the special case of $t = 3$ (and set theoretically for all $t$'s). Then, Allman and Rhodes [16] proved the conjecture for up to five factors, while Landsberg and Weyman [96] found the generators for the defining ideals of secant varieties for the Segre varieties in the following cases: all secant varieties for $\mathbb{P}^1 \times \mathbb{P}^m \times \mathbb{P}^n$ for all $m, n$; the secant line varieties of the Segre varieties with four factors; the secant plane varieties for any Segre variety with three factors. The proofs use representation theoretic methods.

Note that for $s > 2$, one cannot expect, in general, that the ideals $I_{\sigma_s(X_\mathbf{n})}$ are generated by the $(s+1) \times (s+1)$ minors of flattenings of $T$. Indeed, in many cases, there are no such minors, e.g., it is easy to check that if we consider $\mathbf{n} = (1,1,1,1,1)$ and $X_\mathbf{n} \subset \mathbb{P}^{31}$, we get that the flattenings can give only ten $4 \times 8$ matrices and five $2 \times 16$ matrices. Therefore, we get quadrics, which generate $I_{X_\mathbf{n}}$, and quartic forms, which are zero on $\sigma_3(X_\mathbf{n})$, but no equations for $\sigma_4(X_\mathbf{n})$ or $\sigma_5(X_\mathbf{n})$, which, by a simple dimension count, do not fill all of $\mathbb{P}^{31}$.

There is a particular case when we know that the minors of a single flattening are enough to generate the ideal $I(\sigma_s(X_\mathbf{n}))$, namely the unbalanced case we already met in Theorem 16, for which we have the following result.

**Theorem 38** ([108]). *Let $X = X_\mathbf{n} \subset \mathbb{P}^M$ with $M = \prod_{i=1}^t (n_i + 1) - 1$; let $N = \prod_{i=1}^{t-1}(n_i + 1) - 1$; and let $Y_{(N,n_t)}$ be the Segre embedding of $\mathbb{P}^N \times \mathbb{P}^{n_t}$ into $\mathbb{P}^M$. Assume $n_t > N - \sum_{i=1}^{t-1} n_i + 1$. Then, for:*

$$N - \sum_{i=1}^{t-1} n_i + 1 \leq s \leq \min\{n_t, N\},$$

*we have that $\sigma_s(X) = \sigma_s(Y) \neq \mathbb{P}^M$, and its ideal is generated by the $(s+1) \times (s+1)$ minors of an $(n_t + 1) \times (N+1)$ matrix of indeterminates, i.e., the flattening of the generic tensor with respect to the splitting $\{1, \ldots, t-1\} \cup \{t\}$.*

We can notice that in the case above, $X$ is defective for $N - \sum_{i=1}^{t-1} n_i + 1 < s$; see Theorem 16, while when equality holds, $\sigma_s(X)$ has the expected dimension. Moreover, in the cases covered by the theorem, $\sigma_s(X)$ is arithmetically Cohen–Macaulay, and a minimal free resolution of its defining ideal is given by the Eagon–Northcott complex.

### 5.6.2. Veronese Varieties

Now, let us consider the case of Veronese varieties. One case for which we have a rather complete information about the ideals of their higher secant varieties is the family of rational normal curves, i.e., the Veronese embeddings of $\mathbb{P}^1$. In this case, the ideals in question are classically known; in particular, the ideal of $\sigma_s(X_{1,d})$ is generated by the $(s+1) \times (s+1)$ minors of catalecticant matrices associated with generic binary form of degree $d$, whose coefficients corresponds to the coordinates of the ambient coordinates. Moreover, we also know the entire minimal free resolution of these ideals, again given by the Eagon–Northcott complex.

Since the space of quadrics can be associated with the space of symmetric matrices, a similar analysis as the one done for the Segre varieties with two factors can be done in the case $d = 2$. In particular, the defining ideals for the higher secant varieties of the quadratic Veronese embeddings of $\mathbb{P}^n$, i.e., of $\sigma_s(X_{n,2})$, are defined by the $(s+1) \times (s+1)$ minors of the generic symmetric matrix of size $(n+1) \times (n+1)$.

For any $n,d$, the ideal of $\sigma_2(X_{n,d})$ is considered in [78], where it is proven that it is generated by the $3 \times 3$ minors of the first two catalecticant matrices of the generic polynomial of degree $d$ in $n+1$ variables; for the ideal of the $3 \times 3$ minors, see also [219].

In general for all $\sigma_s(X_{n,d})$'s, these kinds of equations, given by $(s+1) \times (s+1)$ minors of catalecticant matrices ([220]), are known, but in most of the cases, they are not enough to generate the whole ideal.

Notice that those catalecticant matrices can also be viewed this way (see [2]): consider a generic symmetric tensor (whose entries are indeterminates) $T$; perform a flattening of $T$ as we just did for generic tensors and Segre varieties; erase from the matrix that is thus obtained all the repeated rows or columns. What you get is a generic catalecticant matrix, and all of them are obtained in this way, i.e., those equations are the same as you get for generic tensors, symmetrized.

Only in a few cases, our knowledge about the equations of secant varieties of Veronese varieties is complete; see for example [25,78,85,108]. All recent approaches employ representation theory and the definition of Young flattenings. We borrow the following list of known results from [85].

| | | | |
|---|---|---|---|
| $\sigma_s(X_{n,2})$ | size $s+1$ minors generic symmetric matrix | ideal | classical |
| $\sigma_s(X_{1,d})$ | size $s+1$ minors of any generic catalecticant | ideal | classical |
| $\sigma_2(X_{n,d})$ | size 3 minors of generic $(1,d-1)$ and $(2,d-2)$-catalecticants | ideal | [78] |
| $\sigma_3(X_{n,3})$ | Aronhold equation + size 4 minors of generic $(1,2)$-catalecticant | ideal | Aronhold ($n=2$) [25] [85] |
| $\sigma_3(X_{n,d})$ ($d \geq 4$) | size 4 minors of generic $(1,3)$ and $(2,2)$-catalecticant | scheme | [221] ($n=2, d=4$) [85] |
| $\sigma_4(X_{2,d})$ | size 5 minors of generic $(\lfloor \frac{d}{2} \rfloor, \lceil \frac{d}{2} \rceil)$-catalecticant | scheme | [221] ($d=4$) [85] |
| $\sigma_5(X_{2,d})$ ($d \geq 6, d=4$) | size 6 minors of generic $(\lfloor \frac{d}{2} \rfloor, \lceil \frac{d}{2} \rceil)$-catalecticant | scheme | Clebsch ($d=4$) [25] [85] |
| $\sigma_s(X_{2,5})$ $s \leq 5$ | size $2s+2$ sub-Pfaffians of generic Young $((31),(31))$-flattening | irred.comp. | [85] |
| $\sigma_6(X_{2,5})$ $s \leq 5$ | size 14 sub-Pfaffians of generic Young $((31),(31))$-flattening | scheme | [85] |
| $\sigma_6(X_{2,d})$ | size 7 minors of generic $(\lfloor \frac{d}{2} \rfloor, \lceil \frac{d}{2} \rceil)$-catalecticant | scheme | [85] |
| $\sigma_7(X_{2,6})$ | symmetric flattenings + Young flattenings | irred. comp. | [85] |
| $\sigma_8(X_{2,6})$ | symmetric flattenings + Young flattenings | irred. comp. | [85] |
| $\sigma_9(X_{2,6})$ | determinant of generic $(3,3)$-catalecticant | ideal | classical |
| $\sigma_s(X_{2,7})$ ($s \leq 10$) | size $(2s+2)$ sub-Pfaffians of generic $((4,1),(4,1))$-Young flattening | irred. comp. | [85] |
| $\sigma_s(X_{2,2m})$ ($s \leq \binom{m+1}{2}$) | rank of $(a,d-a)$-catalecticant $\leq \min\left\{s, \binom{a+2}{2}\right\}$ for $1 \leq a \leq m$, open and closed | scheme | [25,85] |
| $\sigma_s(X_{2,2m+1})$ ($s \leq \binom{m+1}{2}+1$) | rank of $(a,d-a)$-catalecticant $\leq \min\left\{s, \binom{a+2}{2}\right\}$ for $1 \leq a \leq m$, open and closed | scheme | [25,85] |
| $\sigma_s(X_{n,2m})$ ($s \leq \binom{m+n-1}{n}$) | size $s+1$ minors of generic $(m,m)$-catalecticant | irred. component | [25,85] |
| $\sigma_s(X_{2,2m+1})$ ($s \leq \binom{m+n}{n}$) | size $(\lfloor n/2 \rfloor)j+1$ minors of a Young flattening | irred. comp. | [25,85] |

Note that the knowledge of equations that define the $\sigma_s(X_{n,d})$'s, also just set-theoretically, would give the possibility to compute the symmetric border rank for any tensor in $S^d V$.

For the sake of completeness, we mention that equations for secant varieties in other cases can be found in [85] (Grassmannian and other homogeneous varieties), in [108] (Segre–Veronese varieties and Del Pezzo surfaces) and in [83] (Veronese re-embeddings of varieties).

## 5.7. The Real World

For many applications, it is very interesting to study tensor decompositions over the real numbers.

The first thing to observe here is that since $\mathbb{R}$ is not algebraically closed, the geometric picture is much more different. In particular, a first difference is in the definition of secant varieties, where, instead of considering the closure in the Zariski topology, we need to consider the Euclidean topology. In this way, we have that open sets are no longer dense, and there is not a definition of "generic rank": if $X$ is variety in $\mathbb{P}^N_\mathbb{R}$, the set $U_r(X) = \{P \in \mathbb{P}^N_\mathbb{R} \mid R_X(P) = r\}$ might be non-empty interior for several values of $r$. Such values are called the typical ranks of $X$.

It is known that the minimal typical (real) rank of $X$ coincides with the generic (complex) rank of the complexification $X \otimes \mathbb{C}$; see [158] (Theorem 2).

The kind of techniques that are used to treat this problem are sometimes very different from what we have seen in the case of algebraically-closed fields.

Now, we want to overview the few known cases on real symmetric-ranks.

The typical ranks of binary forms are completely known. Comon and Ottaviani conjectured in [222] that typical ranks take all values between $\left\lfloor \frac{d+2}{2} \right\rfloor$ and $d$. The conjecture was proven by Blekherman in [223].

In [224], the authors showed that any value between the minimal and the maximal typical rank is also a typical rank. Regarding real symmetric-ranks, they proved that: the typical real rank of ternary cubics is four; the typical ranks of quaternary cubics are only five and six; and they gave bounds on typical ranks of ternary quartics and quintics.

Another family of symmetric tensors for which we have some results on real ranks are monomials. First of all, note that the apolarity lemma (Lemma 5) can still be employed, by making all algebraic computations over the complex number, but then looking for ideals of reduced points apolar to the given homogeneous polynomial and having only real coefficients. This was the method used in [225] to compute the real rank of binary monomials.

Indeed, if $M = x_0^{\alpha_1} x_1^{\alpha_1}$, then, as we have already seen, $M^\perp = (y_0^{\alpha_0+1}, y_1^{\alpha_1+1})$. Now, Waring decompositions of $M$ are in one-to-one relation with reduced sets of points (which are principal ideals since we are in $\mathbb{P}^1$), whose ideal is contained in $M^\perp$. Now, if we only look for sets of points that are also completely real, then we want to understand for which degree $d$ it is possible to find suitable polynomials $H_0$ and $H_1$ such that $G = y_0^{\alpha_0+1} H_0 + y_1^{\alpha_1+1} H_1$ is of degree $d$ and have only distinct real roots.

The authors observe the following two elementary facts, which hold for any univariate $g(y) = c_d y^d + c_{d-1} y^{d-1} + \ldots + c_1 y + c_0$, as a consequence of the classic Descartes' rule of signs:

- if $c_i = c_{i-1} = 0$, for some $i = 1, \ldots, d$, then $f$ does not have $d$ real distinct roots; see [225] (Lemma 4.1);
- for any $j = 1, \ldots, d-1$, there exists $c_i$'s such that $f$ has $d$ real distinct roots and $c_j = 0$; see [225] (Lemma 4.2).

As a consequence of this, we obtain that:

- if $d < \alpha_0 + \alpha_1$, then $G$ (or rather, its dehomogenization) has two consecutive null coefficients; hence, it cannot have $d$ real distinct roots;
- if $d = \alpha_0 + \alpha_1$, then only the coefficient corresponding to $y_0^{\alpha_0} y_1^{\alpha_1}$ of $G$ (or rather its dehomogenization) is equal to zero; hence, it is possible to find $H_0$ and $H_1$ such that $G$ has $d$ real distinct roots.

Therefore, we get the following result.

**Theorem 39** ([225] (Proposition 3.1)). *If $\alpha_0, \alpha_1$ are not negative integers, then $R^{\mathbb{R}}_{\text{sym}}(x_0^{\alpha_0} x_1^{\alpha_1}) = \alpha_0 + \alpha_1$.*

Note that, comparing the latter result with Theorem 31, we can see that for binary monomials, the real and the complex rank coincide if and only if the least exponent is one. However, this is true in full generality, as shown in [226].

**Theorem 40** ([226] (Theorem 3.5)). *Let $M = x_0^{\alpha_0} \cdots x_n^{\alpha_n}$ be a degree $d$ monomial with $\alpha_0 = \min_i\{\alpha_i\}$. Then,*

$$R^{\mathbb{R}}_{\text{sym}}(M) = R^{\mathbb{C}}_{\text{sym}}(M) \quad \text{if and only if} \quad \alpha_0 = 1.$$

Note that the real rank of monomials is not known in general, and as far as we know, the first unknown case is the monomial $x_0^2 x_1^2 x_2^2$, whose real rank is bounded by $11 \leq R^{\mathbb{R}}_{\text{sym}}(x_0^2 x_1^2 x_2^2) \leq 13$; here, the upper bound is given by [226] (Proposition 3.6 and Example 3.6), and the lower bound is given by [227] (Example 6.7).

We conclude with a result on real ranks of reducible real cubics, which gives a (partial) real counterpart to Theorem 32.

**Theorem 41** ([172] (Theorem 5.6)). *If $F \in \mathbb{R}[x_0, \ldots, x_n]$ is a reducible cubic form essentially involving $n + 1$ variables, then one and only one of the following holds:*

- *$F$ is equivalent to $x_0(\sum_{i=1}^{n} \epsilon_i x_i^2)$, where $\epsilon_i \in \{-1, +1\}$, for $1 \leq i \leq n$, and:*

$$2n \leq R^{\mathbb{R}}_{\text{sym}}(F) \leq 2n + 1.$$

    *Moreover, if $\sum_i \epsilon_i = 0$, then $R^{\mathbb{R}}_{\text{sym}}(F) = 2n$.*
- *$F$ is equivalent to $x_0(\sum_{i=0}^{n} \epsilon_i x_i^2)$, where $\epsilon_i \in \{-1, +1\}$, for $1 \leq i \leq n$, and:*

$$2n \leq R^{\mathbb{R}}_{\text{sym}}(F) \leq 2n + 1.$$

    *Moreover, if $\epsilon_0 = \ldots = \epsilon_n$, then $R^{\mathbb{R}}_{\text{sym}}(F) = 2n$. If $\epsilon_0 \neq \epsilon_1$ and $\epsilon_1 = \ldots = \epsilon_n$, then $R^{\mathbb{R}}_{\text{sym}}(F) = 2n + 1$.*
- *$F$ is equivalent to $(\alpha x_0 + x_p)(\sum_{i=0}^{n} \epsilon_i x_i^2)$, for $\alpha \neq 0$, where $\epsilon_0 = \ldots = \epsilon_{p-1} = 1$ and $\epsilon_p = \ldots = \epsilon_n = -1$ for $1 \leq p \leq n$, and:*

$$2n \leq R^{\mathbb{R}}_{\text{sym}}(F) \leq 2n + 3.$$

    *Moreover, if $\alpha = -1$ or $\alpha = 1$, then $2n + 1 \leq R^{\mathbb{R}}_{\text{sym}}(F) \leq 2n + 3$.*

**Author Contributions:** All the authors contributed equally to the writing and preparation of the paper.

**Funding:** A.B. acknowledges the Narodowe Centrum Nauki, Warsaw Center of Mathematics and Computer Science, Institute of Mathematics of university of Warsaw which financed the 36th Autumn School in Algebraic Geometry "Power sum decompositions and apolarity, a geometric approach" 1–7 September 2013 Łukęcin, Poland where a partial version of the present paper was conceived. M.V.C. and A.G. acknowledges financial support from MIUR (Italy). A.O. acknowledges financial support from the Spanish Ministry of Economy and Competitiveness, through the María de Maeztu Programme for Units of Excellence in R&D (MDM-2014-0445).

**Conflicts of Interest:** The authors declare no conflict of interest.

## References

1. Harris, J. *Algebraic Geometry, A First Course*; Graduate Texts in Math; Springer: New York, NY, USA, 1992.
2. Palatini, F. Sulle varietà algebriche per le quali sono di dimensione minore dell' ordinario, senza riempire lo spazio ambiente, una o alcuna delle varietà formate da spazi seganti. *Atti Accad. Torino Cl. Sci. Mat. Fis. Nat.* **1909**, *44*, 362–375.
3. Terracini, A. Sulle $v_k$ per cui la varietà degli $s_h$ $(h+1)$-seganti ha dimensione minore dell'ordinario. *Rend. Circ. Mat. Palermo* **1911**, *31*, 392–396. [CrossRef]

4. Terracini, A. Sulla rappresentazione delle forme quaternarie mediante somme di potenze di forme lineari. *Atti della Reale Accademia delle Scienze di Torino* **1916**, *51*, 107–117.
5. Scorza, G. Sopra la teoria delle figure polari delle curve piane del 4° ordine. *Annali di Matematica* **1899**, *2*, 155–202. [CrossRef]
6. Scorza, G. Sulla determinazione delle varietà a tre dimensioni di $s_r$ ($r \geq 7$) i cui $s_3$ tangenti si tagliano a due a due. *Rend. Circ. Mat. Palermo* **1908**, *25*, 193–204. [CrossRef]
7. Zak, F.L. *Tangents and Secants of Algebraic Varieties*; American Mathematical Society: Providence, RI, USA, 1993.
8. Alexander, J.; Hirschowitz, A. Polynomial interpolation in several variables. *J. Algebr. Geom.* **1995**, *4*, 201–222.
9. Waring, E. *Meditationes Algebricae*; American Mathematical Sociey: Providence, RI, USA, 1991.
10. de Lathauwer, L.; Castaing, J. Tensor-based techniques for the blind separation of DS–CDMA signals. *Signal Process.* **2007**, *87*, 322–336. [CrossRef]
11. Smilde, A.; Bro, R.; Geladi, P. *Multi-Way Analysis: Applications in the Chemical Sciences*; John Wiley & Sons: Hoboken, NJ, USA, 2005.
12. Strassen, V. Rank and optimal computation of generic tensors. *Linear Algebra Its Appl.* **1983**, *52*, 645–685. [CrossRef]
13. Valiant, L.G. Quantum computers that can be simulated classically in polynomial time. In Proceedings of the Thirty-Third Annual ACM Symposium On Theory of Computing, Crete, Greece, 6–8 July 2001; pp. 114–123.
14. Bernardi, A.; Carusotto, I. Algebraic geometry tools for the study of entanglement: an application to spin squeezed states. *J. Phys. A* **2012**, *45*, 105304. [CrossRef]
15. Eisert, J.; Gross, D. *Lectures on Quantum Information*; Wiley-VCH: Weinheim, Germany, 2007.
16. Allman, E.S.; Rhodes, J.A. Phylogenetic ideals and varieties for the general Markov model. *Adv. Appl. Math.* **2008**, *40*, 127–148. [CrossRef]
17. Comon, P. Independent component analysis, a new concept? *Signal Process.* **1994**, *36*, 287–314. [CrossRef]
18. Sidiropoulos, N.D.; Giannakis, G.B.; Bro, R. Blind parafac receivers for ds-cdma systems. *IEEE Trans. Signal Process.* **2000**, *48*, 810–823. [CrossRef]
19. Schultz, T.; Seidel, H. Estimating crossing fibers: A tensor decomposition approach. *IEEE Trans. Vis. Comput. Graph.* **2008**, *14*, 1635–1642. [CrossRef] [PubMed]
20. Comon, P.; Jutten, C. *Handbook of Blind Source Separation: Independent Component Analysis and Applications*; Academic Press: Cambridge, MA, USA, 2010.
21. Catalisano, M.V.; Geramita, A.V.; Gimigliano, A. On the secant varieties to the tangential varieties of a Veronesean. *Proc. Am. Math. Soc.* **2002**, *130*, 975–985. [CrossRef]
22. Catalisano, M.V.; Geramita, A.V.; Gimigliano, A. Ranks of tensors, secant varieties of Segre varieties and fat points. *Linear Algebra Appl.* **2002**, *355*, 263–285. [CrossRef]
23. Carlini, E.; Grieve, N.; Oeding, L. Four lectures on secant varieties. In *Connections between Algebra, Combinatorics, and Geometry*; Springer: Berlin, Germany, 2014; pp. 101–146.
24. Geramita, A.V. Inverse systems of fat points: Waring's problem, secant varieties of Veronese varieties and parameter spaces for Gorenstein ideals. In *The Curves Seminar at Queen's, Vol. X (Kingston, ON, 1995)*; Queen's Univ.: Kingston, ON, Canada, 1996; pp. 2–114.
25. Iarrobino, A.; Kanev, V. Power sums, Gorenstein algebras, and determinantal loci. In *vLecture Notes in Mathematics*; Springer: Berlin, Germany, 1999; Volume 1721.
26. Landsberg, J.M. *Tensors: Geometry and Applications*; American Mathematical Society: Providence, RI, USA, 2012; Volume 128.
27. Comas, G.; Seiguer, M. On the rank of a binary form. *Found. Comput. Math.* **2011**, *11*, 65–78. [CrossRef]
28. Bernardi, A.; Gimigliano, A.; Idà, M. Computing symmetric-rank for symmetric tensors. *J. Symb. Comput.* **2011**, *46*, 34–53. [CrossRef]
29. Brachat, J.; Comon, P.; Mourrain, B.; Tsigaridas, E. Symmetric tensor decomposition. *Linear Alg. Its Appl.* **2010**, *433*, 1851–1872. [CrossRef]
30. Bernardi, A.; Taufer, D. Waring, tangential and cactus decompositions. *arXiv* **2018**, arXiv:1812.02612.
31. Taufer, D. Symmetric Tensor Decomposition. Master's Thesis, Galileian School of Higher Education, University of Padova, Padua, Italy, 2017.
32. Davenport, H. On Waring's problem for fourth powers. *Ann. Math.* **1939**, *40*, 731–747. [CrossRef]
33. Wakeford, E.K. On canonical forms. *Proc. Lond. Math. Soc.* **1920**, *2*, 403–410. [CrossRef]

34. Pucci, M. The Veronese variety and catalecticant matrices. *J. Algebra* **1998**, *202*, 72–95. [CrossRef]
35. Parolin, A. Varietà secanti alle varietà di Segre e di Veronese e loro applicazioni. Ph.D. Thesis, Università di Bologna, Bologna, Italy, 2004.
36. Brambilla, M.C.; Ottaviani, G. On the Alexander-Hirschowitz theorem. *J. Pure Appl. Algebra* **2008**, *212*, 1229–1251. [CrossRef]
37. Clebsch, A. Ueber curven vierter ordnung. *J. Reine Angew. Math.* **1861**, *59*, 125–145. [CrossRef]
38. Richmond, H.W. On canonical forms. *Quart. J. Pure Appl. Math.* **1904**, *33*, 331–340.
39. Palatini, F. Sulla rappresentazione delle forme ed in particolare della cubica quinaria con la somma di potenze di forme lineari. *Atti Acad. Torino* **1903**, *38*, 43–50.
40. Palatini, F. Sulla rappresentazione delle forme ternarie mediante la somma di potenze di forme lineari. *Rend. Accad. Lincei* **1903**, *5*, 378–384.
41. Campbell, J.E. Note on the maximum number of arbitrary points which can be double points on a curve, or surface, of any degree. *Messenger Math.* **1891**, *21*, 158–164.
42. Terracini, A. Sulla rappresentazione delle coppie di forme ternarie mediante somme di potenze di forme lineari. *Annali Matematica Pura Applicata* **1915**, *24*, 1–10. [CrossRef]
43. Roé, J.; Zappalà, G.; Baggio, S. Linear systems of surfaces with double points: Terracini revisited. *Le Matematiche* **2003**, *56*, 269–280.
44. Bronowski, J. *The Sum of Powers as Canonical Expression*; Cambridge Univ Press: Cambridge, UK, 1933; Volume 29, pp. 69–82.
45. Hirschowitz, A. La methode d'Horace pour l'interpolation à plusieurs variables. *Manuscr. Math.* **1985**, *50*, 337–388. [CrossRef]
46. Alexander, J. Singularités imposables en position générale à une hypersurface projective. *Compos. Math.* **1988**, *68*, 305–354.
47. Alexander, J.; Hirschowitz, A. Un lemme d'Horace différentiel: Application aux singularités hyperquartiques de $\mathbb{P}^5$. *J. Algebr. Geom.* **1992**, *1*, 411–426.
48. Alexander, J.; Hirschowitz, A. Generic hypersurface singularities. *Proc. Math. Sci.* **1997**, *107*, 139–154. [CrossRef]
49. Chandler, K.A. A brief proof of a maximal rank theorem for generic double points in projective space. *Trans. Am. Math. Soc.* **2001**, *353*, 1907–1920. [CrossRef]
50. Ådlandsvik, B. Varieties with an extremal number of degenerate higher secant varieties. *J. Reine Angew. Math.* **1988**, *392*, 16–26.
51. Eisenbud, D.; Harris, J. *The Geometry of Schemes*; Springer Science & Business Media: Berlin, Germany, 2006; Volume 197.
52. Hartshorne, R. *Algebraic Geometry*; Springer Science & Business Media: Berlin, Germany, 2013; Volume 52.
53. Mumford, D. *The red Book of Varieties and Schemes: Includes the Michigan Lectures (1974) on Curves and Their Jacobians*; Springer Science & Business Media: Berlin, Germany, 1999; Volume 1358.
54. Geramita, A.V. Catalecticant varieties. In *Commutative Algebra and Algebraic Geometry (Ferrara)*; Dekker: New York, NY, USA, 1999; pp. 143–156.
55. Ardila, F.; Postnikov, A. Combinatorics and geometry of power ideals. *Trans. Am. Math. Soc.* **2010**, *362*, 4357–4384. [CrossRef]
56. Fröberg, R. An inequality for Hilbert series of graded algebras. *Math. Scand.* **1985**, *56*, 117–144. [CrossRef]
57. Iarrobino, A. Inverse system of a symbolic power III: thin algebras and fat points. *Compos. Math.* **1997**, *108*, 319–356. [CrossRef]
58. Chandler, K.A. The geometric interpretation of Fröberg–Iarrobino conjectures on infinitesimal neighborhoods of points in projective space. *J. Algebra* **2005**, *286*, 421–455. [CrossRef]
59. Segre, B. Alcune questioni su insiemi finiti di punti in geometria algebrica. In Proceedings of the Atti del Convegno Internazionale di Studi Accursiani, Bologna, Italy, 21–26 October 1963; pp. 15–33.
60. Harbourne, B. The geometry of rational surfaces and Hilbert functions of points in the plane. In *Proceedings of the 1984 Vancouver Conference in Algebraic Geometry*; American Mathematical Society: Providence, RI, USA, 1986; pp. 95–111.
61. Gimigliano, A. On Linear Systems of Plane Curves. Ph.D. Thesis, Queen's University, Kingston, ON, Canada, 1987.

62. Hirschowitz, A. Une conjecture pour la cohomologie des diviseurs sur les surfaces rationnelles génériques. *J. Reine Angew. Math* **1989**, *397*, 208–213.
63. Ciliberto, C.; Harbourne, B.; Miranda, R.; Roé, J. Variations on Nagata's conjecture. In Proceedings of the Celebration of Algebraic Geometry, Cambridge, MA, USA, 25–28 August 2011; pp. 185–203.
64. Ciliberto, C. Geometric aspects of polynomial interpolation in more variables and of Waring's problem. In *European Congress of Mathematics, Vol. I (Barcelona, 2000)*; Birkhäuser: Basel, Switzerland, 2001; pp. 289–316..
65. Ciliberto, C.; Miranda, R. Linear systems of plane curves with base points of equal multiplicity. *Trans. Am. Math. Soc.* **2000**, *352*, 4037–4050. [CrossRef]
66. Ciliberto, C.; Miranda, R. The Segre and Harbourne–Hirschowitz conjectures. In *Applications of Algebraic Geometry to Coding Theory, Physics and Computation*; Springer: Berlin, Germany, 2001; pp. 37–51.
67. Gimigliano, A. Our thin knowledge of fat points. In *Queen's Papers in Pure and Applied Mathematics*; Queen's University: Kingston, ON, Canada, 1989.
68. Beauville, A. *Surfaces Algébriques Complexes*; Société Mathématique de France: Marseille, France, 1978; Volume 54.
69. Nagata, M. On the 14-th problem of hilbert. *Am. J. Math.* **1959**, *81*, 766–772. [CrossRef]
70. Bernardi, A.; Catalisano, M.V.; Gimigliano, A.; Idà, M. Secant varieties to osculating varieties of Veronese embeddings of $\mathbb{P}^n$. *J. Algebra* **2009**, *321*, 982–1004. [CrossRef]
71. Alexander, J.; Hirschowitz, A. An asymptotic vanishing theorem for generic unions of multiple points. *Invent. Math.* **2000**, *140*, 303–325. [CrossRef]
72. Murnaghan, F.D. *The Theory of Group Representations*; Johns Hopkins Univ.: Baltimore, MD, USA, 1938.
73. Arrondo, E.; Bernardi, A. On the variety parameterizing completely decomposable polynomials. *J. Pure Appl. Algebra* **2011**, *215*, 201–220. [CrossRef]
74. Arrondo, E.; Paoletti, R. Characterization of Veronese varieties via projection in Grassmannians. In *Projective Varieties with Unexpected Properties*; Walter de Gruyter: Berlin, Germany, 2005; pp. 1–12.
75. de Silva, V.; Lim, L.H. Tensor rank and the ill-posedness of the best low-rank approximation problem. *SIAM J. Matrix Anal. Appl.* **2008**, *30*, 1084–1127. [CrossRef]
76. Landsberg, J.M.; Teitler, Z. On the ranks and border ranks of symmetric tensors. *Found. Comput. Math.* **2010**, *10*, 339–366. [CrossRef]
77. Carlini, E. Reducing the number of variables of a polynomial. In *Algebraic Geometry and Geometric Modeling*; Springer: Berlin, Germany, 2006; pp. 237–247.
78. Kanev, V. Chordal varieties of veronese varieties and catalecticant matrices. *J. Math. Sci.* **1999**, *94*, 1114–1125. [CrossRef]
79. Ballico, E.; Bernardi, A. Decomposition of homogeneous polynomials with low rank. *Math. Z.* **2012**, *271*, 1141–1149. [CrossRef]
80. Ballico, E.; Bernardi, A. Stratification of the fourth secant variety of Veronese varieties via the symmetric rank. *Adv. Pure Appl. Math.* **2013**, *4*, 215–250. [CrossRef]
81. Buczyńska, W.; Buczyński, J. On differences between the border rank and the smoothable rank of a polynomial. *Glasg. Math. J.* **2015**, *57*, 401–413. [CrossRef]
82. Bernardi, A.; Brachat, J.; Mourrain, B. A comparison of different notions of ranks of symmetric tensors. *Linear Algebra Appl.* **2014**, *460*, 205–230. [CrossRef]
83. Buczyski, J.; Ginensky, A.; Landsberg, J.M. Determinantal equations for secant varieties and the Eisenbud–Koh–Stillman conjecture. *J. Lond. Math. Soc.* **2013**, *88*, 1–24. [CrossRef]
84. Oeding, L.; Ottaviani, G. Eigenvectors of tensors and algorithms for Waring decomposition. *J. Symb. Comput.* **2013**, *54*, 9–35. [CrossRef]
85. Landsberg, J.M.; Ottaviani, G. Equations for secant varieties of veronese and other varieties. *Annali di Matematica Pura ed Applicata* **2013**, *192*, 569–606. [CrossRef]
86. Mourrain, B.; Pan, V.Y. Multivariate polynomials, duality, and structured matrices. *J. Complex.* **2000**, *16*, 110–180. [CrossRef]
87. Golub, G.H.; van Loan, C.F. *Matrix Computations*; JHU Press: Baltimore, MD, USA, 2012; Volume 3.
88. Bürgisser, P.; Clausen, M.; Shokrollahi, M.A. *Algebraic Complexity Theory*, Volume 315 of *Grundlehren der Mathematischen Wissenschaften [Fundamental Principles of Mathematical Sciences]*; With the collaboration of Thomas Lickteig; Springer: Berlin, Germany, 1997.

89. Geiger, D.; Heckerman, D.; King, H.; Meek, C. Stratified exponential families: graphical models and model selection. *Ann. Stat.* **2001**, *29*, 505–529.
90. Garcia, L.D.; Stillman, M.; Sturmfels, B. Algebraic geometry of Bayesian networks. *J. Symb. Comput.* **2005**, *39*, 331–355. [CrossRef]
91. Landsberg, J.M. Geometry and the complexity of matrix multiplication. *Bull. Am. Math. Soc.* **2008**, *45*, 247–284. [CrossRef]
92. Abo, H.; Ottaviani, G.; Peterson, C. Induction for secant varieties of Segre varieties. *Trans. Am. Math. Soc.* **2009**, *36*, 767–792. [CrossRef]
93. Chiantini, L.; Ciliberto, C. Weakly defective varieties. *Trans. Am. Math. Soc.* **2002**, *354*, 151–178. [CrossRef]
94. Catalisano, M.V.; Geramita, A.V.; Gimigliano, A. Publisher's erratum to: "Ranks of tensors, secant varieties of Segre varieties and fat points" [Linear Algebra Appl. 355 (2002), 263–285; MR1930149 (2003g:14070)]. *Linear Algebra Appl.* **2003**, *367*, 347–348. [CrossRef]
95. Landsberg, J.M.; Manivel, L. Generalizations of Strassen's equations for secant varieties of Segre varieties. *Commun. Algebra* **2008**, *36*, 405–422. [CrossRef]
96. Landsberg, J.M.; Weyman, J. On the ideals and singularities of secant varieties of Segre varieties. *Bull. Lond. Math. Soc.* **2007**, *39*, 685–697. [CrossRef]
97. Draisma, J. A tropical approach to secant dimensions. *J. Pure Appl. Algebra* **2008**, *212*, 349–363. [CrossRef]
98. Friedland, S. On the Generic Rank of 3-Tensors. Available online: http://front.math.ucdavis.edu/0805.1959 (accessed on 9 October 2018).
99. Catalisano, M.V.; Geramita, A.V.; Gimigliano, A. Higher secant varieties of Segre-Veronese varieties. In *Projective Varieties with Unexpected Properties*; Walter de Gruyter GmbH & Co. KG: Berlin, Germany, 2005; pp. 81–107.
100. Gałazka, M. Multigraded Apolarity. *arXiv* **2016**, arXiv:1601.06211.
101. Ballico, E.; Bernardi, A.; Gesmundo, F. A Note on the Cactus Rank for Segre–Veronese Varieties. *arXiv* **2017**, arXiv:1707.06389.
102. Gallet, M.; Ranestad, K.; Villamizar, N. Varieties of apolar subschemes of toric surfaces. *Ark. Mat.* **2018**, *56*, 73–99. [CrossRef]
103. Catalisano, M.V.; Geramita, A.V.; Gimigliano, A. Higher secant varieties of the Segre varieties $\mathbb{P}^1 \times \ldots \times \mathbb{P}^1$. *J. Pure Appl. Algebra* **2015**, *201*, 367–380. [CrossRef]
104. Catalisano, M.V.; Geramita, A.V.; Gimigliano, A. Secant varieties of $\mathbb{P}^1 \times \ldots \times \mathbb{P}^1$ ($n$-times) are not defective for $n > 5$. *J. Algebr. Geom.* **2011**, *20*, 295–327. [CrossRef]
105. Catalisano, M.V.; Ellia, P.; Gimigliano, A. Fat points on rational normal curves. *J. Algebra* **1999**, *216*, 600–619. [CrossRef]
106. Lickteig, T. Typical tensorial rank. *Linear Algebra Appl.* **1985**, *69*, 95–120. [CrossRef]
107. Gesmundo, F. An asymptotic bound for secant varieties of Segre varieties. *Ann. Univ. Ferrara* **2013**, *59*, 285–302. [CrossRef]
108. Catalisano, M.V.; Geramita, A.V.; Gimigliano, A. On the ideals of secant varieties to certain rational varieties. *J. Algebra* **2008**, *319*, 1913–1931. [CrossRef]
109. Ehrenborg, R.; Rota, G.-G. Apolarity and canonical forms for homogeneous polynomials. *Eur. J. Comb.* **1993**, *14*, 157–181. [CrossRef]
110. Catalisano, M.V.; Geramita, A.V.; Gimigliano, A. Secant varieties of Grassmann varieties. *Proc. Am. Math. Soc.* **2005**, *133*, 633–642. [CrossRef]
111. Abo, H.; Ottaviani, G.; Peterson, C. Non-defectivity of Grassmannians of planes. *J. Algebr. Geom.* **2012**, *21*, 1–20. [CrossRef]
112. Baur, K.; Draisma, J.; de Graaf, W.A. Secant dimensions of minimal orbits: Computations and conjectures. *Exp. Math.* **2007**, *16*, 239–250. [CrossRef]
113. McGillivray, B. A probabilistic algorithm for the secant defect of Grassmann varieties. *Linear Algebra Appl.* **2006**, *418*, 708–718. [CrossRef]
114. Boralevi, A. A note on secants of Grassmannians. *Rend. Istit. Mat. Univ. Trieste* **2013**, *45*, 67–72.
115. Rischter, R. Projective and Birational Geometry of Grassmannians and Other Special Varieties. Ph.D. Thesis, Universidade Federal de Itajubá (UNIFEI), Itajubá, Brazil, 2017.
116. Arrondo, E.; Bernardi, A.; Marques, P.; Mourrain, B. Skew-symmetric decomposition. *arXiv* **2018**, arXiv:1811.12725.

117. Westwick, R. Irreducible lengths of trivectors of rank seven and eight. *Pac. J. Math.* **1979**, *80*, 575–579. [CrossRef]
118. Segre, C. Sui complessi lineari di piani nello spazio a cinque dimensioni. *Annali di Matematica Pura ed Applicata* **1917**, *27*, 75–23. [CrossRef]
119. Schouten, J.A. Klassifizierung der alternierenden Grössen dritten Grades in 7 dimensionen. *Rend. Circ. Mat. Palermo* **1931**, *55*, 137–156. [CrossRef]
120. Lascoux, A. Degree of the dual of a Grassmann variety. *Commun. Algebra* **1981**, *9*, 1215–1225. [CrossRef]
121. Abo, H.; Brambilla, M.C. On the dimensions of secant varieties of Segre-Veronese varieties. *Ann. Mat. Pura Appl.* **2013**, *192*, 61–92. [CrossRef]
122. Bocci, C. Special effect varieties in higher dimension. *Collect. Math.* **2005**, *56*, 299–326.
123. London, F. Ueber die polarfiguren der ebenen curven dritter ordnung. *Math. Ann.* **1890**, *36*, 535–584. [CrossRef]
124. Abo, H. On non-defectivity of certain Segre-Veronese varieties. *J. Symb. Comput.* **2010**, *45*, 1254–1269. [CrossRef]
125. Abo, H.; Brambilla, M.C. Secant varieties of Segre-Veronese varieties $\mathbb{P}^m \times \mathbb{P}^n$ embedded by $\mathcal{O}(1,2)$. *Exp. Math.* **2009**, *18*, 369–384. [CrossRef]
126. Ballico, E. On the non-defectivity and non weak-defectivity of Segre-Veronese embeddings of products of projective spaces. *Port. Math.* **2006**, *63*, 101–111.
127. Ballico, E.; Bernardi, A.; Catalisano, M.V. Higher secant varieties of $\mathbb{P}^n \times \mathbb{P}^1$ embedded in bi-degree $(a,b)$. *Commun. Algebra* **2012**, *40*, 3822–3840. [CrossRef]
128. Carlini, E.; Chipalkatti, J. On Waring's problem for several algebraic forms. *Comment. Math. Helv.* **2003**, *78*, 494–517. [CrossRef]
129. Abrescia, S. About defectivity of certain Segre-Veronese varieties. *Can. J. Math.* **2008**, *60*, 961–974. [CrossRef]
130. Catalisano, M.V.; Geramita, A.V.; Gimigliano, A. Segre-Veronese embeddings of $\mathbb{P}^1 \times \mathbb{P}^1 \times \mathbb{P}^1$ and their secant varieties. *Collect. Math.* **2007**, *58*, 1–24.
131. Dionisi, C.; Fontanari, C. Grassman defectivity à la Terracini. *Le Matematiche* **2001**, *56*, 245–255.
132. Bernardi, A.; Carlini, E.; Catalisano, M.V. Higher secant varieties of $\mathbb{P}^n \times \mathbb{P}^m$ embedded in bi-degree $(1,d)$. *J. Pure Appl. Algebra* **2011**, *215*, 2853–2858. [CrossRef]
133. Abo, H.; Brambilla, M.C. New examples of defective secant varieties of Segre-Veronese varieties. *Collect. Math.* **2012**, *63*, 287–297. [CrossRef]
134. Ottaviani, G. Symplectic bundles on the plane, secant varieties and Lüroth quartics revisited. *Quaderni di Quaderni di Matematica* **2008**, *21*, arXiv:math/0702151v1.
135. Laface, A.; Postinghel, E. Secant varieties of Segre-Veronese embeddings of $(\mathbb{P}^1)^r$. *Math. Ann.* **2013**, *356*, 1455–1470. [CrossRef]
136. Carlini, E.; Catalisano, M.V. On rational normal curves in projective space. *J. Lond. Math. Soc.* **2009**, *80*, 1–17. [CrossRef]
137. Araujo, C.; Massarenti, A.; Rischter, R. On non-secant defectivity of Segre-Veronese varieties. *Trans. Am. Math. Soc.* **2018**. . [CrossRef]
138. Teitler, Z. Geometric lower bounds for generalized ranks. *arXiv* **2014**, arXiv:1406.5145.
139. Gesmundo, F.; Oneto, A.; Ventura, E. Partially symmetric variants of Comon's problem via simultaneous rank. *arXiv* **2018**, arXiv:1810.07679.
140. Ballico, E. On the secant varieties to the tangent developable of a Veronese variety. *J. Algebra* **2005**, *288*, 279–286. [CrossRef]
141. Abo, H.; Vannieuwenhoven, N. Most secant varieties of tangential varieties to Veronese varieties are nondefective. *Trans. Am. Math. Soc.* **2018**, *370*, 393–420. [CrossRef]
142. Ballico, E.; Fontanari, C. On the secant varieties to the osculating variety of a Veronese surface. *Cent. Eur. J. Math.* **2003**, *1*, 315–326. [CrossRef]
143. Bernardi, A.; Catalisano, M.V.; Gimigliano, A.; Idà, M. Osculating varieties of Veronese varieties and their higher secant varieties. *Can. J. Math.* **2007**, *59*, 488–502. [CrossRef]
144. Bernardi, A.; Catalisano, M.V. Some defective secant varieties to osculating varieties of Veronese surfaces. *Collect. Math.* **2006**, *57*, 43–68.
145. Shin, Y.S. Secants to the variety of completely reducible forms and the Hilbert function of the union of star-configurations. *J. Algebra Its Appl.* **2012**, *11*, 1250109. [CrossRef]

146. Abo, H. Varieties of completely decomposable forms and their secants. *J. Algebra* **2014**, *403*, 135–153. [CrossRef]
147. Catalisano, M.V.; Geramita, A.V.; Gimigliano, A.; Harbourne, B.; Migliore, J.; Nagel, U.; Shin, Y.S. Secant varieties of the varieties of reducible hypersurfaces in $\mathbb{P}^n$. *arXiv* **2015**, arXiv:1502.00167.
148. Torrance, D.A. Generic forms of low chow rank. *J. Algebra Its Appl.* **2017**, *16*, 1750047. [CrossRef]
149. Catalisano, M.V.; Chiantini, L.; Geramita, A.V.; Oneto, A. Waring-like decompositions of polynomials, 1. *Linear Algebra Its Appl.* **2017**, *533*, 311–325. [CrossRef]
150. Geramita, A.V.; Schenck, H.K. Fat points, inverse systems, and piecewise polynomial functions. *J. Algebra* **1998**, *204*, 116–128. [CrossRef]
151. Mammana, C. Sulla varietà delle curve algebriche piane spezzate in un dato modo. *Ann. Scuola Norm. Super. Pisa (3)* **1954**, *8*, 53–75.
152. Ananyan, T.; Hochster, M. Small subalgebras of polynomial rings and Stillman's conjecture. *arXiv* **2016**, arXiv:1610.09268.
153. Bik, A.; Draisma, J.; Eggermont, R.H. Polynomials and tensors of bounded strength. *arXiv* **2018**, arXiv:1805.01816.
154. Fröberg, R.; Ottaviani, G.; Shapiro, B. On the Waring problem for polynomial rings. *Proc. Natl. Acad. Sci. USA* **2012**, *109*, 5600–5602. [CrossRef]
155. Lundqvist, S.; Oneto, A.; Reznick, B.; Shapiro, B. On generic and maximal k-ranks of binary forms. *J. Pure Appl. Algebra* **2018**. [CrossRef]
156. Nicklasson, L. On the Hilbert series of ideals generated by generic forms. *Commun. Algebra* **2017**, *45*, 3390–3395. [CrossRef]
157. Jelisiejew, J. An upper bound for the Waring rank of a form. *Archiv der Mathematik* **2014**, *102*, 329–336. [CrossRef]
158. Blekherman, G.; Teitler, Z. On maximum, typical and generic ranks. *Math. Ann.* **2015**, *362*, 1021–1031. [CrossRef]
159. Reznick, B. On the length of binary forms. In *Quadratic and Higher Degree Forms*; Springer: Berlin, Germany, 2013; pp. 207–232.
160. Segre, B. *The Non-Singular Cubic Surfaces*; Oxford University Press: Oxford, UK, 1942.
161. De Paris, A. A proof that the maximum rank for ternary quartics is seven. *Le Matematiche* **2015**, *70*, 3–18.
162. Kleppe, J. Representing a Homogenous Polynomial as a Sum of Powers of Linear Forms. Ph.D. Thesis, University of Oslo, Oslo, Norway, 1999.
163. De Paris, A. Every ternary quintic is a sum of ten fifth powers. *Internat. J. Algebra Comput.* **2015**, *25*, 607–631. [CrossRef]
164. Buczyński, J.; Teitler, Z. Some examples of forms of high rank. *Collect. Math.* **2016**, *67*, 431–441. [CrossRef]
165. De Paris, A. The asymptotic leading term for maximum rank of ternary forms of a given degree. *Linear Algebra Appl.* **2016**, *500*, 15–29. [CrossRef]
166. Carlini, E.; Catalisano, M.V.; Geramita, A.V. The solution to the Waring problem for monomials and the sum of coprime monomials. *J. Algebra* **2012**, *370*, 5–14. [CrossRef]
167. Buczyński, J.; Han, K.; Mella, M.; Teitler, Z. On the locus of points of high rank. *Eur. J. Math.* **2018**, *4*, 113–136. [CrossRef]
168. Ballico, E.; Bernardi, A. Curvilinear schemes and maximum rank of forms. *Matematiche (Catania)* **2017**, *72*, 137–144.
169. Buczyńska, W.; Buczyński, J. Secant varieties to high degree Veronese reembeddings, catalecticant matrices and smoothable Gorenstein schemes. *J. Algebr. Geom.* **2014**, *23*, 63–90. [CrossRef]
170. Ranestad, K.; Schreyer, F.-O. On the rank of a symmetric form. *J. Algebra* **2011**, *346*, 340–342. [CrossRef]
171. Carlini, E.; Catalisano, M.V.; Chiantini, L.; Geramita, A.V.; Woo, Y. Symmetric tensors: Rank, Strassen's conjecture and e-computability. *Ann. Scuola Norm. Sup. Pisa* **2018**, *18*, 1–28.
172. Carlini, E.; Guo, C.; Ventura, E. Real and complex Waring rank of reducible cubic forms. *J. Pure Appl. Algebra* **2016**, *220*, 3692–3701. [CrossRef]
173. Teitler, Z. Sufficient conditions for Strassen's additivity conjecture. *Ill. J. Math.* **2015**, *59*, 1071–1085.
174. Alexeev, B.; Forbes, M.A.; Tsimerman, J. Tensor rank: some lower and upper bounds. In Proceedings of the 2011 IEEE 26th Annual Conference on Computational Complexity (CCC), San Jose, CA, USA, 8–11 June 2011; pp. 283–291.

175. Landsberg, J.M. *Geometry and Complexity Theory*; Cambridge University Press: Cambridge, UK, 2017; Volume 169.
176. Ballico, E.; Bernardi, A. Tensor ranks on tangent developable of Segre varieties. *Linear Multilinear Algebra* **2013**, *61*, 881–894. [CrossRef]
177. Seigal, A. Ranks and symmetric-ranks of cubic surfaces. *arXiv* **2018**, arXiv:1801.05377.
178. Friedland, S. Remarks on the symmetric-rank of symmetric tensors. *SIAM J. Matrix Anal. Appl.* **2016**, *37*, 320–337. [CrossRef]
179. Comon, P.; Golub, G.; Lim, L.H.; Mourrain, B. Symmetric tensors and symmetric tensor rank. *SIAM J. Matrix Anal. Appl.* **2008**, *30*, 1254–1279. [CrossRef]
180. Zhang, X.; Huang, Z.H.; Qi, L. Comon's conjecture, rank decomposition, and symmetric-rank decomposition of symmetric tensors. *SIAM J. Matrix Anal. Appl.* **2016**, *37*, 1719–1728. [CrossRef]
181. Shitov, Y. A counterexample to Comon's conjecture. *SIAM J. Appl. Algebra Geometry* **2018**, *2*, 428–443. [CrossRef]
182. Strassen, V. Vermeidung von divisionen. *J. Reine Angew. Math.* **1973**, *264*, 184–202.
183. Landsberg, J.M.; Michałek, M. Abelian tensors. *J. Math. Pures Appl.* **2017**, *108*, 333–371. [CrossRef]
184. Ja'Ja', J.; Takche, J. On the validity of the direct sum conjecture. *SIAM J. Comput.* **1986**, *15*, 1004–1020. [CrossRef]
185. Feig, E.; Winograd, S. On the direct sum conjecture. *Linear Algebra Appl.* **1984**, *63*, 193–219. [CrossRef]
186. Shitov, Y. A counterexample to Strassen's direct sum conjecture. *arXiv* **2017**, arXiv:1712.08660.
187. Carlini, E.; Catalisano, M.V.; Chiantini, L. Progress on the symmetric Strassen conjecture. *J. Pure Appl. Algebra* **2015**, *219*, 3149–3157. [CrossRef]
188. Kruskal, J.B. Three-way arrays: rank and uniqueness of trilinear decompositions, with application to arithmetic complexity and statistics. *Linear Algebra Appl.* **1977**, *18*, 95–138. [CrossRef]
189. Sylvester, J.J. *The Collected Mathematical Papers, Vol. I*; Cambridge University Press: Cambridge, UK, 1904.
190. Hilbert, D. Letter adresseé à M. Hermite. *Gesam. Abh.* **1888**, *2*, 148–153, .
191. Anandkumar, A.; Ge, R.; Hsu, D.; Kalade, S.M.; Telgarsky, M. Tensor decompositions for learning latent variable models. *J. Mach. Learn. Res.* **2014**, *15*, 2773–2832.
192. Appellof, C.J.; Davidson, E.R. Strategies for analyzing data from video fluorometric monitoring of liquid chromatographic effluents. *Anal. Chem.* **1981**, *53*, 2053–2056. [CrossRef]
193. Allman, E.S.; Matias, C.; Rhodes, J.A. Identifiability of parameters in latent structure models with many observed variables. *Ann. Stat.* **2009**, *37*, 3099–3132. [CrossRef]
194. Angelini, E.; Bocci, C.; Chiantini, L. Real identifiability vs complex identifiability. *Linear Multilinear Algebra* **2018**, *66*, 1257–1267. [CrossRef]
195. Angelini, E. On complex and real identifiability of tensors. *Riv. Mat. Univ. Parma* **2017**, *8*, 367–377.
196. Chiantini, L.; Ottaviani, G. On generic identifiability of 3-tensors of small rank. *SIAM J. Matrix Anal. Appl.* **2012**, *33*, 1018–1037. [CrossRef]
197. Bocci, C.; Chiantini, L. On the identifiability of binary Segre products. *J. Algebr. Geom.* **2013**, *22*, 1–11. [CrossRef]
198. Bhaskara, A.; Charikar, M.; Vijayaraghavan, A. Uniqueness of tensor decompositions with applications to polynomial identifiability. In Proceedings of the Conference on Learning Theory, Barcelona, Spain, 13–15 June 2014; pp. 742–778.
199. Bocci, C.; Chiantini, L.; Ottaviani, G. Refined methods for the identifiability of tensors. *Ann. Mat. Pura Appl.* **2014**, *193*, 1691–1702. [CrossRef]
200. Chiantini, L.; Ottaviani, G.; Vannieuwenhoven, N. An algorithm for generic and low-rank specific identifiability of complex tensors. *SIAM J. Matrix Anal. Appl.* **2014**, *35*, 1265–1287. [CrossRef]
201. Chiantini, L.; Ottaviani, G.; Vannieuwenhoven, N. On generic identifiability of symmetric tensors of subgeneric rank. *Trans. Am. Math. Soc.* **2017**, *369*, 4021–4042. [CrossRef]
202. Galuppi, F.; Mella, M. Identifiability of homogeneous polynomials and Cremona Transformations. *J. Reine Angew. Math.* **2018**. [CrossRef]
203. Angelini, E.; Galluppi, F.; Mella, M.; Ottaviani, G. On the number of Waring decompositions for a generic polynomial vector. *J. Pure Appl. Algebra* **2018**, *222*, 950–965. [CrossRef]
204. Ciliberto, C.; Russo, F. Varieties with minimal secant degree and linear systems of maximal dimension on surfaces. *Adv. Math.* **2006**, *200*, 1–50. [CrossRef]

205. Weierstrass, K. *Zur Theorie Der Bilinearen and Quadratischen Formen*; Monatsh. Akad. Wisa.: Berlin, Germany, 1867; pp. 310–338.
206. Roberts, R.A. Note on the plane cubic and a conic. *Proc. Lond. Math. Soc.* **1889**, *21*, 62–69.
207. Bernardi, A.; Vanzo, D. A new class of non-identifiable skew-symmetric tensors. *Annali Matematica Pura Applicata* **2018**, *197*, 1499–1510. [CrossRef]
208. Ranestad, K.; Schreyer, F.-O. Varieties of sums of powers. *J. Reine Angew. Math.* **2000**, *525*, 147–181. [CrossRef]
209. Mukai, S. Polarized K3 surfaces of genus 18 and 20. In *Complex Projective Geometry*; Cambridge Univ. Press: Cambridge, UK, 1992; pp. 264–276.
210. Mukai, S. Curves and symmetric spaces, 1. *Am. J. Math* **1995**, *117*, 1627–1644. [CrossRef]
211. Mukai, S. Fano 3-folds, complex projective geometry. *Lond. Math. Soc. Lecture Note* **1992**, *179*, 255–263.
212. Buczyńska, W.; Buczyński, J.; Teitler, Z. Waring decompositions of monomials. *J. Algebra* **2013**, *378*, 45–57. [CrossRef]
213. Carlini, E.; Catalisano, M.V.; Oneto, A. Waring loci and the Strassen conjecture. *Adv. Math.* **2017**, *314*, 630–662. [CrossRef]
214. Mourrain, B.; Oneto, A. On minimal decompositions of low rank symmetric tensors. *arXiv* **2018**, arXiv:1805.11940.
215. Lascoux, A. Syzygies des varietes determinatales. *Adv. Math.* **1978**, *30*, 202–237. [CrossRef]
216. Weyman, J. *Cohomology of Vector Bundles and Syzygies*; Cambridge University Press: Cambridge, UK, 2003.
217. Bruns, W.; Conca, A. Groebner bases and determinantal ideals. In *Singularities and Computer Algebra*; Kluwer Academic Publishers: Dordrecht, The Netherlands, 2003; pp. 9–66.
218. Landsberg, J.M.; Manivel, L. On the ideals of secant varieties of Segre varieties. *Found. Comput. Math.* **2004**, *4*, 397–422. [CrossRef]
219. Raicu, C. $3 \times 3$ Minors of Catalecticants. *Math. Res. Lett.* **2013**, *20*, 745–756. [CrossRef]
220. Catalano-Johnson, M.L. The possible dimensions of the higher secant varieties. *Am. J. Math.* **1996**, *118*, 355–361. [CrossRef]
221. Schreyer, F.-O. Geometry and algebra of prime Fano 3-folds of genus 12. *Compos. Math.* **2001**, *127*, 297–319. [CrossRef]
222. Comon, P.; Ottaviani, G. On the typical rank of real binary forms. *Linear Multilinear Algebra* **2012**, *60*, 657–667. [CrossRef]
223. Blekherman, G. Typical real ranks of binary forms. *Found. Comput. Math.* **2015**, *15*, 793–798. [CrossRef]
224. Bernardi, A.; Blekherman, G.; Ottaviani, G. On real typical ranks. *Bollettino dell'Unione Matematica Italiana* **2018**, *11*, 293–307. [CrossRef]
225. Boij, M.; Carlini, E.; Geramita, A. Monomials as sums of powers: the real binary case. *Proc. Am. Math. Soc.* **2011**, *139*, 3039–3043. [CrossRef]
226. Carlini, E.; Kummer, M.; Oneto, A.; Ventura, E. On the real rank of monomials. *Math. Z.* **2017**, *286*, 571–577. [CrossRef]
227. Michałek, M.; Moon, H.; Sturmfels, B.; Ventura, E. Real rank geometry of ternary forms. *Annali di Matematica Pura ed Applicata* **2017**, *196*, 1025–1054. [CrossRef]

© 2018 by the authors. Licensee MDPI, Basel, Switzerland. This article is an open access article distributed under the terms and conditions of the Creative Commons Attribution (CC BY) license (http://creativecommons.org/licenses/by/4.0/).

MDPI  
St. Alban-Anlage 66  
4052 Basel  
Switzerland  
Tel. +41 61 683 77 34  
Fax +41 61 302 89 18  
www.mdpi.com  

*Mathematics* Editorial Office  
E-mail: mathematics@mdpi.com  
www.mdpi.com/journal/mathematics  

www.ingramcontent.com/pod-product-compliance
Lightning Source LLC
LaVergne TN
LVHW071954080526
838202LV00064B/6743